Lecture Notes in Artificial

Edited by J. G. Carbonell and J. Siek

Subseries of Lecture Notes in Computer Science

Sašo Džeroski   Ljupčo Todorovski (Eds.)

Computational Discovery of Scientific Knowledge

Introduction, Techniques, and Applications in Environmental and Life Sciences

Series Editors

Jaime G. Carbonell, Carnegie Mellon University, Pittsburgh, PA, USA
Jörg Siekmann, University of Saarland, Saarbrücken, Germany

Volume Editors

Sašo Džeroski
Department of Knowledge Technologies
Jožef Stefan Institute
Ljubljana, Slovenia
E-mail: Saso.Dzeroski@ijs.si

Ljupčo Todorovski
Department of Knowledge technologies
Jožef Stefan Institute
Jjubliana, Slovenia
E-mail: Ljupco.Todorovski@ijs.si

Library of Congress Control Number: 2007931366

CR Subject Classification (1998): I.2, H.2.8, H.2-3, J.2-3, F.1

LNCS Sublibrary: SL 7 – Artificial Intelligence

ISSN 0302-9743
ISBN-10 3-540-73919-X Springer Berlin Heidelberg New York
ISBN-13 978-3-540-73919-7 Springer Berlin Heidelberg New York

Springer is a part of Springer Science+Business Media

springer.com

Typesetting: Camera-ready by author, data conversion by Scientific Publishing Services, Chennai, India
Printed on acid-free paper SPIN: 12099696 06/3180 5 4 3 2 1 0

Preface

Advances in technology have enabled the collection of data from scientific observations, simulations, and experiments at an ever-increasing pace. For the scientist and engineer to benefit from these enhanced data-collecting capabilities, it is becoming clear that semi-automated data analysis techniques must be applied to find the useful information in the data. Techniques from both data mining and computational discovery can be used to that end.

Computational scientific discovery focuses on applying computational methods to automate scientific activities, such as finding laws from observational data. It has emerged from the view that science is a problem-solving activity and that problem solving can be cast as search through a space of possible solutions. Early research on computational discovery within the fields of artificial intelligence and cognitive science focused on reconstructing episodes from the history of science. This typically included identifying data and knowledge available at the time and implementing a computer program that models the scientific activities and processes that led to the scientist's insight.

Recent efforts in this area have focused on individual scientific activities (such as formulating quantitative laws) and have produced a number of new scientific discoveries, many of them leading to publications in the relevant scientific literatures. The discoveries made using computational tools include qualitative laws of metallic behavior, quantitative conjectures in graph theory, and temporal laws of ecological behavior. Work in this paradigm has emphasized formalisms used to communicate among scientists, including numeric equations, structural models, and reaction pathways.

However, in recent years, research on data mining and knowledge discovery has produced another paradigm. Data mining is concerned with finding patterns (regularities) in data. Even when applied to scientific domains, such as astronomy, biology, and chemistry, this framework employs formalisms developed by artificial intelligence researchers themselves, such as decision trees, rule sets, and Bayesian networks. Although such methods can produce predictive models that are highly accurate, their outputs are not cast in terms familiar to scientists, and thus typically are not very communicable.

Mining scientific data focuses on building highly predictive models, rather than producing knowledge in any standard scientific notation. In contrast, much of the work in computational scientific discovery has put a strong emphasis on formalisms used by scientists to communicate scientific knowledge, such as numeric equations, structural models, and reaction pathways. In this sense, computational scientific discovery is complementary to mining scientific data.

The book provides an introduction to computational approaches to the discovery of communicable scientific knowledge and gives an overview of recent advances in this area. The primary focus is on discovery in scientific and engineering disciplines, where communication of knowledge is often a central concern.

This volume has its origins in the symposium "Computational Discovery of Communicable Knowledge," organized by Pat Langley, held March 24-25, 2001 at Stanford University. A detailed report on the symposium can be found in the Proceedings of the DS-2001 Conference (S. Džeroski, and P. Langley. Computational discovery of communicable knowledge: Symposium report. In *Proceedings of the Fourth International Conference on Discovery Science*, pages 45-49. Springer, Berlin, 2001). Many of the presentations from that symposium have a corresponding chapter in the book. To achieve a more representative coverage of recent research in computational discovery, we have invited a number of additional contributions as well.

The book is organized as follows. The first chapter introduces the field of computational scientific discovery and provides a brief overview thereof. It also provides a more detailed overview of the contents of the book. The majority of the contributed chapters fall within two broad categories, which correspond to Parts I and II of the book, respectively. The first describes a number of computational discovery methods for system identification and automated modelling, while the second discusses a number of methods for computational discovery developed for biomedical and bioinformatics applications.

In the first part of the book, the focus is on establishing models of dynamic systems, i.e., systems that change their state over time. The models are mostly based on equations, in particular ordinary differential equations that represent a standard formalism for modelling dynamic systems in many engineering and scientific areas. This is in contrast to the bulk of previous research on equation discovery, which focuses on algebraic equations. Topics covered in this part include a reasoning tool for nonlinear system identification, the use of different forms of domain knowledge when inducing models of dynamic systems (including the use of existing models for theory revision, partial knowledge of the target model, knowledge on basic processes, and knowledge on measurement scales of the system variables), and applications to Earth sciences.

While the first part of the book focuses on a class of methods and covers a variety of scientific fields and areas, the second focuses on the field of biomedicine. The first three chapters are in line with the first part of the book and continue with the theme of model formation, but use representation formalisms specific to the biomedical field, such as chemical reaction networks and genetic pathways. The last two chapters present approaches to forming scientific hypotheses by connecting disconnected scientific literatures on the same topic. This part also includes a chapter on using learning in logic for predicting gene function.

We would like to conclude with some words of thanks. Pat Langley organized the symposium that motivated this volume and encouraged us to edit it. More importantly, he has pioneered research on computational scientific discovery and provided unrelenting support to our research in this area. We would also like to thank the participants of the symposium. Finally, we would like to thank all the contributors to this volume for their excellent contributions and their patience with the editors.

May 2007

Sašo Džeroski
Ljupčo Todorovski

Table of Contents

Computational Discovery of Scientific Knowledge

Sašo Džeroski[1], Pat Langley[2], and Ljupčo Todorovski[1]

[1] Department of Knowledge Technologies, Jozef Stefan Institute
Jamova 39, SI-1000 Ljubljana, Slovenia
Saso.Dzeroski@ijs.si, Ljupco.Todorovski@ijs.si
[2] Computational Learning Laboratory
Center for the Study of Language and Information
Stanford University, Stanford, CA 94305 USA
langley@isle.org

Abstract. This chapter introduces the field of computational scientific discovery and provides a brief overview thereof. We first try to be more specific about what scientific discovery is and also place it in the broader context of the scientific enterprise. We discuss the components of scientific behavior, that is, the knowledge structures that arise in science and the processes that manipulate them. We give a brief historical review of research in computational scientific discovery and discuss the lessons learned, especially in relation to work in data mining that has recently received substantial attention. Finally, we discuss the contents of the book and how it fits in the overall framework of computational scientific discovery.

1 Introduction

This book deals with computational approaches to scientific discovery. Research on computational scientific discovery aims to develop computer systems which produce results that, if a human scientist did the same, we would refer to as discoveries. Of course, if we hope to develop computational methods for scientific discovery, we must be more specific about the nature of such discoveries and how they relate to the broader context of the scientific enterprise.

The term science refers both to scientific knowledge and the process of acquiring such knowledge. It includes any systematic field of study that relates to observed phenomena (as opposed to mathematics) and that involves claims which can be tested empirically (as opposed to philosophy). We will attempt to characterize science more fully later in the chapter, but one thing is clear: Science is about knowledge.

Science is perhaps the most complex human intellectual activity, which makes it difficult to describe. Shrager and Langley (1990) analyze it in terms of the knowledge structures that scientists consider and the processes or activities they use to transform them. Basic knowledge structures that arise in science include observations, laws, and theories, and related activities include data collection, law formation, and theory construction.

S. Džeroski and L. Todorovski (Eds.): Computational Discovery, LNAI 4660, pp. 1–14, 2007.

There are two primary reasons why we might want to study scientific discovery from a computational perspective:

- to understand how humans perform this intriguing activity, which belongs to the realm of cognitive science; and
- to automate or assist in facets of the scientific process, which belongs to the realm of artificial intelligence.

Science is a highly complex intellectual endeavor, and discovery is arguably the most creative part of the scientific process. Thus, efforts to automate it completely would rightfully be judged as audacious, but, as Simon (1966) noted, one can view many kinds of scientific discovery as examples of problem solving through heuristic search. Most research in automating scientific discovery has focused on small, well-defined tasks that are amenable to such treatment and that allow measurable progress.

Traditional accounts of science (Klemke et al., 1998) focus on the individual, who supposedly observes nature, hypothesizes laws or theories, and tests them against new observations. Most computational models of scientific discovery share this concern with individual behavior. However, science is almost always a collective activity that is conducted by interacting members of a scientific community. The most fundamental demonstration of this fact is the emphasize placed on communicating one's findings to other researchers in journal articles and conference presentations.

This emphasis on exchanging results makes it essential that scientific knowledge be *communicable*. We will not attempt to define this term, but it seems clear that contributions are more communicable if they are cast in established formalisms and if they make contact with concepts that are familiar to most researchers in the respective field of study. The research reported in this book focuses on computational discovery of such communicable knowledge.

In the remainder of this chapter, we first examine more closely the scientific method and its relation to scientific discovery. After this, we discuss the components of scientific behavior, that is, the knowledge structures that arise in science and the processes that manipulate them. We then give a brief historical review of research in computational scientific discovery and discuss the lessons learned, especially in relation to work in data mining that has recently received substantial attention. Finally, we discuss the contents of the book and how it fits in the overall framework of computational scientific discovery.

2 The Scientific Method and Scientific Discovery

The Merriam-Webster Dictionary (2003) defines science as: "a) knowledge or a system of knowledge covering general truths or the operation of general laws, especially as obtained and tested through the scientific method, and b) such knowledge or such a system of knowledge concerned with the physical world and its phenomena". The scientific method, in turn, is defined as the "principles and procedures for the systematic pursuit of knowledge involving the recognition and formulation of a problem, the collection of data through observation and experiment, and the formulation and testing of hypotheses".

While there is consensus that science revolves around knowledge, there are different views in the philosophy of science (Klemke et al., 1998; Achinstein, 2004) about the nature of its content. The 'causal realism' position is that scientific knowledge is ontological, in that it identifies entities in the world, their causal powers, and the mechanisms through which they exert influence. In contrast, the 'constructive empiricism' tradition states that, scientific theories are objective, testable, and predictive. We believe that both frameworks are correct, in that they describe different facets of the truth.

The *scientific method* (Gower, 1996), dedicated to the systematic pursuit of reliable knowledge, incorporates a number of steps. First we must ask some meaningful question or identify a significant problem. We must next gather information relevant to the question, which might include existing scientific knowledge or new observations. We then formulate a hypothesis that could plausibly answer the question.

Next we must test this proposal by making observations and determining whether they are consistent with the hypothesis' predictions. When observations are consistent with the hypothesis, they lend it support and we may consider publishing it. If other scientists can reproduce our results, then the community comes to consider it as reliable knowledge. In contrast, if the observations are inconsistent, we should reject the hypothesis and either abandon it or, more typically, modify it, at which point the testing process continues. Hypotheses can take many different forms, including taxonomies, empirical laws, and explanatory theories, but all of them can be evaluated by comparing their implications to observed phenomena.

Most analyses of the scientific method come from philosophers of science, who have focused mainly on the evaluation of hypotheses and largely ignored their generation and revision. Unfortunately, what we refer to as discovery resides in just these activities. Thus, although there is a large literature on normative methods for making predictions from hypotheses, checking their consistency, and determining whether they are valid, there are remarkably few treatments of their production. Some (e.g., Popper (1959)) have even suggested that rational accounts of the discovery process are impossible. A few philosophers (e.g., Darden (2006); Hanson (1958); Lakatos (1976)) have gone against this trend and made important contributions to the topic, but most efforts have come from artificial intelligence and cognitive science.

Briefly, scientific discovery is the process by which a scientist creates or finds some hitherto unknown knowledge, such as a class of objects, an empirical law, or an explanatory theory. The knowledge in question may also be referred to as a scientific discovery. An important aspect of many knowledge structures, such as laws and theories, is their generality, in that they apply to many specific situations or many specific observations. We maintain that generality is an essential feature of a meaningful discovery, as will become apparent in the next section when we discuss types of scientific knowledge.

A defining aspect of discovery is that the knowledge should be new and previously unknown. Naturally, one might ask 'new to whom?'. We take the position that the knowledge should be unknown to the scientist in question with respect to the observations and background knowledge available to him when he made the discovery. This means that two or more scientists can make the same discovery independently, sometimes years apart, which has indeed happened in practice many times throughout the history of science. In this view, scientific discovery concerns a change in an individual's knowledge, which means that developing computer systems that reproduce events from the history of science can still provide important insights into the nature of discovery processes.

3 The Elements of Scientific Behavior

To describe scientific behavior, we follow Shrager and Langley (1990) and use as basic components knowledge structures and the activities that transform them. The former represent the raw materials and products of science, while the latter concern the process of producing scientific knowledge. The account below mostly follows the earlier treatise, but the definitions of several knowledge structures and activities have changed, reflecting improvements in our understanding over the past 15 years.

3.1 Scientific Knowledge Structures

Science is largely about understanding the world in which we live. To this end, we gather information about the world. Observation is the primary means of collecting this information, and observations are the primary input to the process of scientific discovery.

Observations (or data) represent recordings of the environment made by sensors or measuring instruments. Typically, the state of the environment varies over time or under different conditions, and one makes recordings for these different states, where what constitutes a state depends on the object of scientific study. We will refer to each of these recordings as an observation.

We can identify three important types of scientific knowledge – taxonomies, laws, and theories – that constitute the major products of the scientific enterprise. The creation of new taxonomies, laws, and theories, as well as revising and improving existing ones, make up the bulk of scientific discovery, making them some of the key activities in science.

- *Taxonomies* define or describe concepts for a domain, along with specialization relations among them. A prototypical example is the taxonomy for biological organisms, which are grouped into species, genera, families, and so forth, but similar structures play important roles in particle physics, chemistry, astronomy, and many other sciences. Taxonomies specify the concepts and terms used to state laws and theories.
- *Laws* summarize relations among observed variables, objects, or events. For example, Black's heat law states that mixing two substances produces a

temperature increases in one substance and a decrease in the other until they reach equilibrium. The law also describes a precise numeric relationship between the initial and final temperatures. The first statement is qualitative in form, whereas the latter is quantitative. Some laws may be quite general, whereas others may be very specific.
- *Theories* are statements about the structures or processes that arise in the environment. A theory is stated using terms from the domain's taxonomy and interconnects a set of laws into a unified theoretical account. For example, Boyle's law describes the inverse relation between the pressure and volume of a gas, whereas Charles' law states the direct relation between its temperature and pressure. The kinetic theory of gases provides a unifying account for both, explaining them in terms of Newtonian interactions among unobserved molecules.

Note that all three kinds of knowledge are important and present in the body of scientific knowledge. Different types of knowledge are generated at different stages in the development of a scientific discipline. Taxonomies are generated early in a field's history, providing the basic concepts for the discipline. After this, scientists formulate empirical laws based on their observations. Eventually, these laws give rise to theories that provide a deeper understanding of the structures and processes studied in the discipline.

A knowledge structure that a scientist has proposed, but that has not yet been tested with respect to observations, is termed an hypothesis. Note that taxonomies, laws, and theories can all have this status. As mentioned earlier, hypotheses must be evaluated to determine whether they are consistent with observations (and background knowledge). If it is consistent, we say that a hypothesis has been corroborated and it comes to be viewed as scientific knowledge. If an hypothesis is inconsistent with the evidence, then we either reject or modify it, giving rise to a new hypothesis that is further tested and evaluated.

Background knowledge is knowledge about the environment separate from that specifically under study. It typically includes previously generated scientific knowledge in the domain of study. Such knowledge differs from theories or laws at the hypothesis stage, in that the scientist regards it with relative certainty rather than as the subject of active evaluation. Scientific knowledge begins its life cycle as a hypothesis which (if corroborated) becomes background knowledge.

Besides the basic data and knowledge types considered above, several other types of structures play important roles in science. These include models, predictions, and explanations. These occupy an intermediate position, as they are derived from laws and theories and, as such, they are not primary products of the scientific process.

- *Models* are special cases of laws and theories that apply to particular situations in the environment and only hold under certain environmental conditions. These conditions specify the particular experimental or observational setting, with the model indicating how the law or theory applies in the setting. By applying laws and theories to a particular setting, models make it possible to use these for making predictions.

- *Predictions* represent expectations about the behavior of the environment under specific conditions. In science, a model is typically used to make a prediction, and then an actual observation is made of the behavior in the environment. Postdictions are analogous to predictions, except that the scientist generates them after making the observations he or she intends to explain. A prediction/postdiction that is consistent with the respective observation is successful and lends support to the model (and the respective law/theory) that produced it.
- *Explanations* are narratives that connect a theory to a law (or a model to a prediction) by a chain of inferences appropriate to the field. In such cases, we say that the theory explains the law. In some disciplines, inference chains must be deductive or mathematical. If a law cannot be explained by a theory (or a prediction by a model), we have an anomaly that brings either the theory or the observation into question.

3.2 Scientific Activities

Scientific processes and activities are concerned with generating and manipulating scientific data and knowledge structures. Here we consider the processes and activities in the same order as we discussed the structures that they generate in the previous subsection.

The process of observation involves inspecting the environmental setting by focusing an instrument, sometimes simply the agent's senses, on that setting. The result is a concrete description of the setting, expressed in terms from the agent's taxonomy and guided by the model of the setting. Since one can observe many things in any given situation, the observer must select some aspects to record and some to ignore.

As we have noted, scientific discovery is concerned with generating scientific knowledge in the form of taxonomies, laws and theories. These can be generated directly from observations (and possibly background knowledge), but, quite often, scientists modify an existing taxonomy, law, or theory to take into account anomalous observations that it cannot handle.

- *Taxonomy formation (and revision)* involves the organization of observations into classes and subclasses, along with the definition of those classes. This process may operate on, or take into account, an existing taxonomy or background knowledge. For instance, early chemists organized certain chemicals into the classes of acids, alkalis, and salts to summarize regularities in their taste and behavior. As time went on, they refined this taxonomy and modified the definitions of each class.
- *Inductive law formation (and revision)* involves the generation of empirical laws that cover observed data. The laws are stated using terms from the agent's taxonomy, and they are constrained by a model of the setting and possibly by the scientist's background knowledge. In some cases, the scientist may generate an entirely new law; in others, he may modify or extend an existing law.

- *Theory formation (and revision)* stands in the same relation to empirical laws as does law formation to data. Given one or more laws, this activity generates a theory from which one can derive the laws for a given model by explanation. Thus, a theory interconnects a set of laws into a unified account. Theory revision responds to anomalous phenomena or laws that cannot be explained by an existing theory, producing a revised theory that explains the anomaly while maintaining the ability to cover existing laws.

While some scientific activities revolve around inductive reasoning, others instead rely on deduction. Scientists typically derive predictions from laws or models, and sometimes they even deduce laws from theoretical principles.

- In contrast to inductive law discovery from observations, *deductive law formation* starts with a theory and uses an explanatory framework to deduce both a law and an explanation of how that law follows from the theory.
- The *prediction* process takes a law, along with a particular setting, and produces a prediction about what one will observe in the setting. Typically, a scientist derives a model from the law, taking into account the setting's particularities, and derives a prediction from the model. The analogous process of *postdiction* takes place in cases where the scientist must account for existing observations. Prediction and postdiction stand in the same relation to each other as deductive law formation and explanation.
- The process of *explanation* connects a theory to a law (or a law to a prediction) by specifying the deductive reasoning that derives the law from the theory. In the context of evaluation, a successful explanation lends support to the theory or law. If explanation fails, then an anomaly results that may trigger a revision of the theory or law. Explanation and deductive law formation are closely related, although explanation aims to account for a law that is already known. Also, in some fields explanation relies on abductive reasoning that leads the scientist to posit unobserved structures or processes, rather than deduction from given premises.

To assess the validity of theories or laws, scientists compare their predictions or postdictions with observations. This produces either consistent results or anomalies, which may serve to stimulate further theory or law formation or revision. This process is called *evaluation* and generally follows experimentation and observation.

Experimentation involves experimental design and manipulation. *Experimental design* specifies settings in which the scientist will collect measurements. Typically, he varies selected aspects of the environment (the independent variables) to determine their effect on other aspects (the dependent variables). He then constructs a physical setting (this is called *manipulation*) that corresponds to the desired environmental conditions and carries out the experiment.

Observation will typically follow or will be interleaved with systematic experimentation, in which case we call it active observation. However, there are fields and phenomena where experimental control is difficult, and sometimes impossible. In such cases the scientist can still collect data to test his hypotheses through passive observation.

4 History of Research on Computational Discovery of Scientific Knowledge

4.1 A Brief Historical Account of Computational Scientific Discovery

Now that we have considered the goals of research on computational discovery and the elements it involves, we can provide some historical context for the work reported in this volume. The idea that one might automate the discovery of scientific knowledge has a long history, going back at least to the writings of Francis Bacon (1620) and John Stuart Mill (1900). However, the modern treatment of this task came from Herbert Simon, who proposed viewing scientific discovery as an instance of heuristic problem solving. In this paradigm, one uses mental operators to transform one knowledge state into another, invoking rules of thumb to select from applicable operators, choose among candidate states, and decide when one has found an acceptable solution. Newell et al. (1958) proposed this framework as both a theory of human problem solving and an approach to building computer programs with similar abilities.

Simon (1966) suggested that, despite the mystery normally attached to scientific discovery, one might explain it in similar terms. He noted that scientific theories can be viewed as knowledge states, and that mental operations can transform them in response to observations. He even outlined an approach to explaining creative phenomena such as scientific insight using these and other established psychological mechanisms. Simon's early papers on this topic only outlined an approach to modeling discovery as problem-space search, but they set a clear research agenda that is still being explored today.

The late 1970s saw two research efforts that transformed Simon's early proposals into running computer programs. The AM system (Lenat, 1978) rediscovered a variety of concepts and conjectures in number theory, starting from basic concepts and heuristics for combining them. The Bacon system (Langley, 1979; Langley et al., 1983) rediscovered a number of numeric laws from the history of physics and chemistry, starting from experimental data and heuristics for detecting regularities in them. Despite many differences, both systems utilized data-driven induction of descriptive laws and were demonstrated on historical examples. Together, they provided the first compelling evidence that computational scientific discovery was actually possible. There is no question that these early systems had many limitations, but they took the crucial first steps toward understanding the discovery process.

The following decade saw a number of research teams build on and extend the ideas developed in AM and Bacon. A volume edited by Shrager and Langley (1990) includes representative work from this period that had previously been scattered throughout the literature in different fields. This collection reported work on discovery of descriptive laws, but it also included chapters on new topics, including the formation of explanatory models, hypothesis-driven experimental design, and model revision. On reading this book, one gets the general impression of an active research community exploring a variety of ideas that address different facets of the complex endeavor we know as science.

The early work on computational discovery focused on reconstructions from the history of science that were consistent with widely accepted theories of human cognition. This was an appropriate strategy, in that these examples let researchers test their methods on relatively simple problems for which answers were known, yet that were relevant because they had once been challenging to human scientists. Such evaluations were legitimate because it was quite possible to develop methods that failed on historical examples, and many approaches were ruled out in this manner. However, critics often argued that the evidence for computational discovery methods would be more compelling when they had uncovered new scientific knowledge rather than rediscovered existing results.

The period from 1990 to 2000 produced a number of novel results along these lines, a number of which have been reviewed by Valdez-Perez (1996) and Langley (2000). These successes have involved a variety of scientific disciplines, including astronomy, biology, chemistry, metallurgy, ecology, linguistics, and even mathematics, and they run the gamut of discovery tasks, including the formation of taxonomies, qualitative laws, numeric equations, structural models, and process explanations. What they hold in common is that each led to the discovery of new knowledge that was deemed significant enough to appear in the literature of the relevant field, which is the usual measure of scientific success. The same techniques have also proved successful in engineering disciplines, in which analogous modeling tasks also arise. These results provide clear evidence that our computational methods are capable of making new discoveries, and thus respond directly to early criticisms.

Another development during this period was the emergence of the data mining movement, which held its first major conference in 1995. This paradigm has emphasized the efficient induction of accurate predictive models from very large data sets. Typical applications involved records of commercial transactions, but some data-mining work has instead dealt with scientific domains. Although research in this area is sometimes referred to as "knowledge discovery" (Fayyad et al., 1996), the resulting models are generally encoded as decision trees, logical rules, Bayesian networks, or other formalisms invented by computer scientists. Thus, it contrasts with the smaller but older movement of computational scientific discovery, which focuses on knowledge cast in formalisms used by practicing scientists and which is less concerned with large data sets than with making the best use of available observations.

4.2 Lessons Learned for the Computational Discovery of Scientific Knowledge

Developments in both data mining and computational scientific discovery make it clear that technologies for knowledge discovery are mature enough for application, but this does not mean there remains no need for additional research. In another paper, Langley (2002) recounts some lessons that have emerged from work in scientific domains, which we review here.

1. The output of a discovery system should be communicated easily to domain scientists. This issue deserves mention because traditional notations developed by machine learning researchers, such as decision trees or Bayesian networks, differ substantially from formalisms typical to the natural sciences, such as numeric equations and reaction pathways. Most work on computational scientific discovery attempts to generate knowledge in an established notation, but communicability is a significant enough issue that it merits special attention.
2. Discovery systems should take advantage of background knowledge to constrain their search. Most research in computational scientific discovery and data mining emphasizes the construction of knowledge from scratch, whereas human scientists often utilize their prior knowledge to make tasks tractable. For instance, science is an incremental process that involves the gradual improvement and extension of previous knowledge, which suggests the need for more work on methods for revising scientific laws, models, and theories. In addition, scientists often use theoretical constraints to guide their construction of models, so more work on this topic is needed as well.
3. Computational methods for scientific discovery should be able to infer knowledge from small data sets. Despite the rhetoric common in papers on data mining, scientific data are often rare and difficult to obtain. This suggests an increased focus on ways to reduce the variance of discovered models and mitigate the tendency to overfit the data, as opposed to developing methods for processing large data sets efficiently.
4. Discovery systems should produce models that move beyond description to provide explanations of data. Early work focused on discovery of descriptive regularities that summarized data, and most work on data mining retains this focus. However, mature sciences are generally concerned with explanatory accounts that incorporate theoretical variables, entities, or processes, and we increased work on methods that support such deeper scientific reasoning.
5. Computational discovery systems should support interaction with domain scientists. Most discovery research has focused on automated systems, yet few scientists want computers to replace them. Rather, they want computational tools that can assist them in constructing and revising their models. To this end, we need more work on interactive systems that let users play at least an equal role in the discovery process.

The chapters in this book respond directly to the first four of these issues, which suggests that they are now receiving the attention they deserve from researchers in the area. However, the fifth topic is not represented, and we hope it will become a more active topic in the future.

5 Overview of the Book

The chapters of the book present state-of-the-art approaches to computational scientific discovery, representing recent progress in the area. These approaches correspond to various scientific activities and deal with different scientific knowledge structures. Note, however, that the main focus of this edited volume is on inductive model formation from observed data. This is in contrast with a previous related book (Shrager & Langley, 1990) where most of the research presented concerned the formation and revision of scientific theories and laws.

In the first part of the book, titled "Equation Discovery & Dynamic Systems Identification", the focus is on establishing models of dynamic systems, i.e., systems that change their state over time. The models are mostly based on equations, in particular ordinary differential equations that represent a standard formalism for modeling dynamic systems in many engineering and scientific areas. This is in contrast to the bulk of previous research on equation discovery, which focuses on algebraic equations. The first two chapters by Stole, Easley, and Bradley present the PRET reasoning tool for nonlinear system identification, i.e., for solving the task of establishing equation-based models of dynamic systems. PRET integrates qualitative reasoning, numerical simulation, geometric reasoning, constraint reasoning, backward chaining, reasoning with abstraction levels, declarative meta-control, and truth maintenance to identify a proper model structure and its parameters for the modeling task at hand. Background knowledge for building models guides the reasoning engine. While the first chapter focuses mainly on general modeling knowledge that is valid in different scientific and engineering domains, the focus of the second chapter is on representing and use of knowledge specific to the domain of interest. The second chapter also presents PRET's heuristics for performing active observation of the modeled dynamic system.

The following chapter by Todorovski and Džeroski provides an overview of equation discovery approaches to inducing models of dynamic systems. Equation discovery deals with the task of automated discovery of quantitative laws, expressed in the form of equations, in collections of measured data. It has advanced greatly from the early stage, when the focus was on reconstructing well-known laws from scientific textbooks, and state-of-the-art approaches deal with establishing new laws and models from observed data. Among the most important recent research directions in this area has been the use of domain knowledge in addition to measured data in the equation discovery process. The chapter shows how modeling knowledge specific to the domain at hand can be integrated in the process of equation discovery for establishing and revising comprehensible models of real-world dynamic systems.

The chapter by Washio and Motoda also presents an approach to formulating equation-based models and laws from observed data. They use results from measurement theory (in particular the Buckingam theorem) about how to properly combine variables measured using different measurement units and scales. These rules are used to constrain the space of candidate models and laws for the observed phenomena. The second part of the chapter discusses the conditions that equations have to satisfy in order to be considered communicable knowledge.

The next two chapters deal with establishing models from Earth science data. The first chapter by Saito and Langley presents an approach to revising existing scientific models cast as sets of equations. The revision is guided by the goal of reducing the model error on newly acquired data and allows for revising parameter values, intrinsic properties, and functional forms used in the model equations. The second chapter by Schwabacher et al. shows how standard machine learning methods can be used to induce models that are represented in formalisms specific to the scientific fields of artificial intelligence and machine learning and yet understandable and communicable to Earth scientists.

In the next chapter, Colton reviews research on computational discovery in pure mathematics, where the focus is on theory and law formation. The author puts special emphasis on his own work in the area of taxonomy formation in mathematics, especially with respect to identifying important classes of numbers.

The last chapter in the first part of the book by Zhao et al. presents a spatial aggregation method for identifying spatio-temporal objects in observations. The method recursively aggregates data into objects and artifacts at higher levels of abstraction. Although the presented method does not correspond directly to any of the scientific activities presented in this introduction, it can be a very useful tool for aiding the processes of taxonomy, law, and model formation.

While the first part of the book focuses on a class of methods and covers a variety of scientific fields and areas, the focus of the second part is on computational scientific discovery in biomedicine and bioinformatics. The first three chapters are in line with the first part of the book and continue with the theme of model formation. However, the model representation formalisms change from equations to formalisms specific to biomedicine, such as chemical reaction networks and genetic pathways.

The chapter by Koza et al. deals with the problem of inducing chemical reaction networks from observations of compounds concentration through time. The authors show that chemical reaction networks can be transformed to (systems of) ordinary differential equations. They present and evaluate a genetic programming approach to inducing a restricted class of equations that correspond to chemical reaction networks.

The chapter by Zupan et al. presents a reasoning system for inferring genetic networks, i.e., networks of gene influences on one another and on biological outcomes of interest. The system uses abduction and qualitative simulation to transform observations into constraints that have to satisfied by a network that would describe observed experimental data best. The following chapter by Garrett et al. also represents genetic networks as qualitative models and uses qualitative simulation to match them against observed data. The authors present and evaluate a method for inducing qualitative models from observational data that is based on inductive logic programming.

The chapter by King et al. deals with the application of inductive logic programming methods to the task of analyzing a complex bioinformatic database in the domain of functional genomics. The authors discuss the importance of integrating background knowledge in the process of scientific data analysis and show that inductive logic programming tools provide an appropriate environment for the integration of knowledge and data in the process of scientific discovery. The work presented in the chapter is the initial step that later lead to the development of a robot scientist, capable of automatically performing a variety of scientific activities. The robot scientist project is one of the most exciting recent developments in the field of computational scientific discovery (King et al., 2004).

Finally, the last two chapters present approaches to forming hypotheses by connecting disconnected scientific literatures on the same topic. Weber presents a general model that, based on connections already published in the scientific literature between a symptom and a disease on one hand and connections between an active substance (chemical compound) and a symptom on the other hand, establishes a hypothesis that the chemical compound can be used for treatment of the disease. The hypothesis is of interest, if the relation between the disease and the compound has not been established before while evidences for the other two relations are well presented in scientific literature. In the final chapter, Hristovski et al. present an interactive system for literature discovery and apply it to the task of identifying gene markers for a particular disease. The system uses association rule mining to find relations between medical concepts from a bibliographic database and uses them to discover new relations that have not been reported in the medical literature yet.

References

Achinstein, P. (ed.): Science rules: A historical introduction to scientific methods. The Johns Hopkins University Press, Baltimore (2004)

Bacon, F.: The new organon and related writings. Liberal Arts Press, New York (1620/1960)

Darden, L.: Reasoning in biological discoveries. Cambridge University Press, Cambridge (2006)

Fayyad, U., Piatetsky-Shapiro, G., Smyth, P.: From data mining to knowledge discovery in databases. AI Magazine 17, 37–54 (1996)

Gower, B.: Scientific method. Routledge, Florence, KY (1996)

Hanson, N.R.: Patterns of discovery. Cambridge University Press, Cambridge (1958)

King, R.D., Whelan, K.E., Jones, F.M., Reiser, P.G.K., Bryant, C.H., Muggleton, S., Kell, D.B., Oliver, S.G.: Functional genomic hypothesis generation and experimentation by a robot scientist. Nature 427, 247–252 (2004)

Klemke, E.D., Hollinger, R., Rudge, D.W., Kline, A.D. (eds.): Introductory readings in the philosophy of science. 3rd edn. Prometheus Books, Amherst, NY (1998)

Lakatos, I.: Proofs and refutations: The logic of mathematical discovery. Cambridge University Press, Cambridge (1976)

Langley, P.: Rediscovering physics with Bacon.3. In: Proceedings of the Sixth International Joint Conference on Artificial Intelligence, pp. 505–507. Morgan Kaufmann, Tokyo (1979)

Langley, P.: The computational support of scientific discovery. International Journal of Human-Computer Studies 53, 393–410 (2000)

Langley, P.: Lessons for the computational discovery of scientific knowledge. In: Proceedings of First International Workshop on Data Mining Lessons Learned, pp. 9–12. University of New South Wales, Sydney (2002)

Langley, P., Bradshaw, G.L., Simon, H.A.: Rediscovering chemistry with the bacon system. In: Michalski, R.S., Carbonell, J.G., Mitchell, T.M. (eds.) Machine learning: An artificial intelligence approach, Morgan Kaufmann, San Mateo (1983)

Lenat, D.B.: The ubiquity of discovery. Artificial Intelligence 9, 257–285 (1978)

Merriam-Webster.: Merriam-webster's collegiate dictionary. 11th edn. Merriam-Webster, Springfield, MA (2003)

Mill, J.S.: A system of logic ratiocinative and inductive being a connected view of the principles of evidence and the methods of scientific investigation. 8th edn. Longmans, Green, & Co., London (1900)

Newell, A., Shaw, J.C., Simon, H.A.: Chess-playing programs and the problem of complexity. IBM Journal of Research and Development 2, 320–325 (1958)

Popper, K.R.: The logic of scientific discovery. Hutchinson, London (1959)

Shrager, J., Langley, P. (eds.): Computational models of scientific discovery and theory formation. Morgan Kaufmann, San Mateo (1990)

Simon, H.A.: Scientific discovery and the psychology of human problem solving. In: Colodny, R.G. (ed.) Mind and cosmos: Essays in contemporary science and philosophy, University of Pittsburgh Press, Pittsburgh (1966)

Valdez-Perez, R.E.: Computer science research on scientific discovery. Knowledge Engineering Review 11, 51–66 (1996)

Part I
Equation Discovery and Dynamic Systems Identification

Communicable Knowledge in Automated System Identification

Reinhard Stolle[1] and Elizabeth Bradley[2,*]

[1] Palo Alto Research Center (PARC),
Palo Alto, California, USA
stolle@parc.com
[2] Department of Computer Science,
University of Colorado, Boulder, Colorado, USA
lizb@cs.colorado.edu

Abstract. We describe the program PRET, an engineering tool for *nonlinear system identification*, which is the task of inferring a (possibly nonlinear) ordinary differential equation model from external observations of a target system's behavior. PRET has several characteristics in common with programs from the fields of machine learning and computational scientific discovery. However, since PRET is intended to be an engineer's tool, it makes different choices with regard to the tradeoff between model accuracy and parsimony. The choice of a good model depends on the engineering task at hand, and PRET is designed to let the user communicate the task-specific modeling constraints to the program. PRET's inputs, its outputs, and its internal knowledge base are instances of communicable knowledge—knowledge that is represented in a form that is meaningful to the domain experts that are the intended users of the program.

1 Introduction

Models of dynamic systems are essential tools in a variety of disciplines ranging from science and engineering to economics and the social sciences (Morrison, 1991). A good model facilitates various types of reasoning about the modeled system, such as prediction of future behavior, explanation of observed behavior, understanding of correlations and influences between variables, and hypothetical reasoning about alternative scenarios.

Building good models is a routine, but difficult, task. The modeler must derive an intensional (and finite) description of the system from extensional (and possibly infinite) observations of its behavior. Traditional examples of such finite descriptions are structural models, reaction pathways, and numeric equations.

Strictly speaking, every formalization of the properties of a dynamic system constitutes a model thereof. The spectrum ranges from models that use a

* Supported by NSF NYI #CCR-9357740, ONR #N00014-96-1-0720, and a Packard Fellowship in Science and Engineering from the David and Lucile Packard Foundation.

S. Džeroski and L. Todorovski (Eds.): Computational Discovery, LNAI 4660, pp. 17–43, 2007.

language that is very close to the domain of the system to models that use a language that is well-suited to describe the system mathematically. An example of the domain-centered end of this spectrum might be formal instructions on how to build an electrical circuit (e.g., a wiring diagram). These instructions would use terms like resistor, capacitor, inductor, connection, and switch. The other extreme might be differential equations; for example, the ordinary differential equation (ODE)

$$1.23\ddot{x} + 3\dot{x} + 46x = 0 \tag{1}$$

models an electrical circuit consisting of a resistor (R), an inductor (L), and a capacitor (C), but the form of the equation gives no hint of that correspondence.

A practitioner uses his or her domain knowledge to establish the correspondence between the mathematical formulation of the model and its domain-centered interpretation: the interpretation of the variable x—as current or voltage, respectively—in the above equation governs whether the ODE models a series or parallel circuit. A model for a parallel RLC circuit, for example, is the equation

$$LC\ddot{v} + \frac{L}{R}\dot{v} + v = 0 \tag{2}$$

where v is the voltage variable. Domain-centered models are useful for building physical systems or recognizing the function of existing physical systems, among other things. Mathematical models are useful for the precise simulation, prediction and understanding of dynamic systems. Trained experts routinely use their domain knowledge and expertise to move back and forth between these different model types during different phases of the reasoning process.

In this chapter, we describe the modeling program PRET, which automatically constructs ODE models for given dynamic systems. The next section relates the task of *modeling in an engineering setting* to other modeling and discovery approaches that make use of related techniques. Sections 3, 4 and 5 describe the program and how it automates the modeling process. PRET uses a generate-and-test paradigm, which is described in Section 3. The "generate" phase of the generate-and-test cycle is described in detail in (Easley & Bradley, this volume). The emphasis of the remainder of this current chapter is then on the "test" phase. In Section 4 we show that PRET's inputs and internal knowledge base are instances of communicable knowledge. Finally, Section 5 explains how PRET orchestrates its reasoning process, fluidly shifting back and forth between various reasoning modes.

An introductory example of a PRET run is offered in Section 3. The sole purpose of this simple example is to illustrate the basic functionality of PRET and the main ideas behind its design. PRET has been successfully applied on a variety of systems, ranging from textbook problems to difficult real-world applications like vehicle suspensions, water resource systems, and various robotics applications (forced pendulum, radio-controlled car, etc.). Such examples, which show the power of the program and indicate its intended application space are more complicated; they are better discussed after both phases of the generate-and-test cycle have been described. We refer the reader to (Easley & Bradley, this volume) and to (Bradley et al., 2001).

2 Communicable Models for Engineering Tasks

2.1 Explicit and Implicit Models

Typically, a modeler builds models out of simple components, assuming that the overall behavior follows from the behavior of the components and their interaction (Falkenhainer & Forbus, 1991). The basic building blocks of models—called model fragments—usually correspond to well-understood concepts in the modeling domain. For example, in the context of the example illustrated by Eqn. (2), the term $\frac{v}{R}$ corresponds to the concept "current through a resistor." Similarly, the composition of models from model fragments corresponds to well-understood principles in the modeling domain. For example, the model of a series circuit that consists of a single loop of components may be composed out of the model fragments for those components according to Kirchhoff's voltage law: the sum of all voltages in a loop is zero.

We call the type of models that is explained in the previous paragraph *explicit* models. We use this term in order to emphasize that a model and its fragments explicitly represent entities and concepts that are well-understood in the target domain and that can be reasoned about explicitly using an established body of domain knowledge.

Research on connectionist computing, Bayesian networks, data mining, and knowledge discovery has produced different kinds of intensional description of dynamic systems. These new kinds of model use data structures that prevail in the field of Artificial Intelligence (AI), such as decision trees, Bayesian networks, rule sets, or neural networks. We call these descriptions *implicit* models to emphasize that they are not necessarily compositional and their ingredients do not immediately correspond to concepts and entities with which practitioners of the modeling domain are familiar.[1] The dynamic systems community has also developed a variety of ways to model and predict the dynamics of a low-dimensional system using implicit models (Farmer & Sidorowich, 1987; Casdagli & Eubank, 1992); these methods match up well in practice against traditional statistical and neural-net based techniques (Weigend & Gershenfeld, 1993).

Both implicit and explicit models are very useful, but for different reasons and for different purposes. Implicit models can be extremely powerful tools because they can simulate and predict the behavior of dynamic systems with high accuracy. Furthermore, in many modeling tasks the desired model does not need to resemble the target system structurally. Instead, the modeler merely wants to

[1] The distinction between explicit and implicit models concerns only the *result* of the modeling process, namely the intensional description of the target system. It is independent of the *search method* used to find the model. It is possible to construct explicit models using AI search methods like genetic programming (Koza et al., this volume) or backpropagation (Saito & Langley, this volume).

capture—or replicate—the input/output behavior of the target system. In such cases, implicit models are a very practical choice because they can be learned (or "trained") from numerical sample data, avoiding a combinatorial search through the space of explicit compositional models. However, since implicit models do not make use of the formalisms of the target domain, they are often less useful for tasks that involve explanations and understanding with respect to the body of knowledge that is familiar to the domain practitioner.

Explicit models, on the other hand, are *communicable* to domain practitioners; they communicate knowledge about the target system in a form that makes sense in the context of a general body of knowledge about the domain. An explicit model facilitates various types of reasoning about the target system, such as, for example, hypotheses about alternative scenarios. An engineer may recognize a particular term in an ODE model of a mechanical system as a "friction term." This direct correspondence between a model fragment and real-world knowledge about the phenomenon "friction" allows the engineer to anticipate the effects of changing the friction term by, say, adding a drop of oil to the target system. As another example, consider an ODE model of a robot arm that makes explicit reference to the gravitational constant g. Again, the direct correspondence to the real-world phenomenon "gravity" facilitates reasoning about the deployment of the robot in a different gravitational environment—on Mars, for example.

2.2 Scientific Theories and Engineering Models

In the AI literature, work on automatically finding a model for a given dynamic system falls under the rubrics of "reasoning about physical systems," "automated modeling," "machine learning," and "scientific discovery."[2]

The purpose of this chapter is to describe the automated modeling program PRET and to place it in the landscape of related modeling and discovery systems that produce communicable output. PRET—like several other systems that discover communicable models—not only produces a communicable model as its output, but also uses communicable knowledge during the process of computing that output. The advantages and implications of this approach are explained in more detail in the next sections. What makes PRET different and unique is its focus on automated system identification, which is the process of modeling in the context of a particular engineering task. In this section, we examine how this engineering focus distinguishes PRET from programs that discover scientific theories. In particular, we argue that this focus imposes a task-specific tradeoff between parsimony and accuracy on the modeling process. Furthermore, we describe how PRET's combination of traditional system identification techniques[3]

[2] For a survey of approaches to computational scientific discovery, see (Langley, 2000).

[3] Perhaps the most important of these techniques—and one that is unique in the AI/modeling literature—is *input-output modeling*, in which PRET interacts directly and autonomously with its target systems, using sensors and actuators to perform experiments whose results are useful to the model-building process. See (Easley & Bradley, 1999b).

and AI techniques—especially its qualitative, "abstract-level first" techniques—allows it to find the right balance point with respect to this tradeoff.

In the research areas of Qualitative Reasoning (QR) and Qualitative Physics (QP) (Weld & de Kleer, 1990), a model of a physical system is mainly used as a representation that allows an automated system to *reason* about the physical system (Forbus, 1984; Kuipers, 1993). QR/QP reasoners are usually concerned with the physical system's structure, function, or behavior. For example, qualitative simulation (Kuipers, 1986) builds a tree of qualitative descriptions of possible future evolutions of the system. Typically, the system's structural and functional properties are known, and the task of *modeling* (Nayak, 1995) consists of finding a formal representation of these properties that is most suitable to the intended reasoning process. Such models frequently highlight qualitative and abstract properties of the system so as to facilitate efficient qualitative inferences. Modeling of systems with known functional and structural properties is generally called *clear-box modeling.*

The goal of Scientific Discovery (e.g., (Langley et al., 1987)) and System Identification (Ljung, 1987) is to investigate physical systems whose structural, functional properties are not—or are only partially—known. Modeling a target system, then, is the process of inferring an intensional (and finite) description—a *model*—of the system from extensional (and possibly infinite) observations of its behavior. For example, a typical system identification task is to observe a driven pendulum's behavior over time and infer from these time series measurements an ordinary differential equation system that accounts for the observed behavior. This process is usually referred to as *black-box modeling.* It amounts to inverting simulation, which is the process of predicting a system's behavior over time, given the equations that govern the system's dynamics.

Whereas the desired model in a system identification task usually takes the form of a set of differential equations, the field of scientific discovery comprises a wider range of tasks with a broader variety of possible models. According to Langley (Langley, 2000), a scientific discovery program typically tries to discover regularities in a space of entities and concepts that has been designed by a human. Such regularities may take the form of qualitative laws, quantitative laws, process models, or structural models (which may even postulate unobserved entities). The discovery of process models amounts to explaining phenomena that involve change over time; it is the kind of scientific discovery that comes closest to system identification. Most of the scientific discovery literature, e.g., (Huang & Żytkow, 1997; Langley et al., 1987; Todorovski & Džeroski, 1997; Washio et al., 1999; Żytkow, 1999), revolves around the discovery of natural laws. Predator-prey systems or planetary motion are prominent examples. System identification, on the other hand, is typically performed in an engineering context—building a controller for a robot arm, for example.

For the purposes of this paper, we distinguish between *theories* and *models.* Żytkow's terminology (Żytkow, 1999) views theories as analytical and models as synthetic products. We prefer to draw the distinction along the generality/specificity axis. A theory and a model are similar in the sense that both are derived from observations of target systems. However, theories aim at a more-general and more-comprehensive description of a wider range of observations. Constructing a theory includes the definition (or postulation) of relevant entities and quantities; the laws of the theory, then, express relationships that hold between these entities and quantities. The developer of a theory tries to achieve a tight correspondence between the postulated structural setup of the entities and the laws that describe the behavior of the entities.

One of the major goals of theory development is an increased *understanding* of the observed phenomena: the quality of a theory depends not only on how accurately the theory accounts for the observations, but also on how well the theory connects to other theories, on whether it generalizes or concretizes previous theories, and on how widely it is applicable. Therefore, research in scientific discovery must address the question about whether the discovered theory accurately models the target system (e.g., nature), or whether it just happens to match the observations that were presented to the discovery program. Likewise, machine learning systems routinely use validation techniques (such as cross-validation) in order to ensure the "accuracy" of the learned model.

Engineering modeling is much less general and much more task-specific. A given *domain theory* sets up the space of possible quantities of interest.[4] A model, then, is a mathematical account of the behavioral relationships between these quantities. Some or all model fragments may or may not correspond to a structural fragment of the modeled system. For example, one may recognize a particular term as a "friction term" or a "gravity term." Whether such correspondences exist, however, is of secondary concern. The primary concern is to accurately describe the *behavior* of the system within a fairly limited context and with a specific task (e.g., controller design) in mind.

2.3 Parsimony in Engineering Modeling

In the previous paragraphs, we described the distinction between theory development and engineering modeling. As one may expect, the dividing line between these two kinds of observation interpretation is somewhat fuzzy. Even though many engineering models are very specific compared to the scope of a scientific theory, engineers also broaden their exploration beyond a single system, in order, for example, to build a cruise control that works for most cars, not just a particular Audi on a warm day. Furthermore, one might argue that the distinction between "accounting for" and "explaining" an observation is arbitrary. There may not be a big difference between saying "the resistor explains the dissipation of energy" and saying "this term (which corresponds to the resistor) accounts for that behavior (which corresponds to the dissipation of energy)."

Nevertheless, it is important to note that explanation and understanding are stated *goals* of scientific discovery; in system identification, they are often merely

[4] By choosing a very specific domain theory and by setting up a specific space of possible model fragments, a human may actually convey substantial information about the target system to the automatic modeling program. As mentioned in Section 3.4, we call this compromise between clear- and black-box modeling *grey-box modeling*.

byproducts of the modeling process. Furthermore, scientific discovery approaches often introduce higher-level concepts (e.g., the label "linear friction force" for the mathematical term $c\dot{x}$) to achieve structural coherence and consistency with background knowledge and/or other theories. Such higher-level concepts are useful in automated system identification as well, but PRET's approach is to restrict the search space in an effective, efficient grey-box modeling approach. Finally, in system identification, the task at hand provides an objective measure as to when a model is "good enough"—as opposed to a scientific theory, which is always only a step toward a further theory: something that is more general, more widely applicable, more accurate, and/or expressed in more fundamental terms.

The long-term vision in scientific discovery is even more ambitious than the previous paragraph suggests. Rather than "just" manipulating existing concepts, quantities, and entities in order to construct theories, scientific discovery programs may even invent or construct new concepts or entities on the fly. Furthermore, Chapter 10 of (Langley et al., 1987) speculates about the automated "discovery of research problems, the invention of scientific instruments, and the discovery and application of good problem representations." Such tasks are clearly outside the scope of modeling from an engineering perspective.

This difference in long-term vision between theory developers and model constructors has important consequences concerning the *parsimony* of the developed theory or constructed model. Both system identification and scientific discovery strive for a simple representation of the target system. However, from a scientific discovery viewpoint, parsimony must be achieved within the constraint that the theory be behaviorally accurate and structurally coherent with background knowledge and other theories, as described in the previous paragraphs. In system identification, however, parsimony is critically important. Modelers work hard to build abstract, minimal models that account for the observations. Typically, the desired model is the one that is just concrete enough to capture the behavior that is relevant for the task at hand.

The remainder of this chapter describes how PRET performs system identification in order to find an ODE model of a given target system. It does so *in an engineering context*—with all the implications for parsimony and structural accuracy that have been described in this section.

3 Automated Modeling with PRET

3.1 ODEs: The Formalism of Choice for Engineering Modeling

PRET is an engineering tool for *nonlinear system identification*, which is the task of inferring a (possibly nonlinear) ODE model from external observations of a target system's behavior. For several reasons, ODEs are the modeling formalism of choice for many engineering tasks. First, ODEs are explicit and communicable, i.e., they can be interpreted within the context of domain knowledge, which establishes a structural correspondence between the model fragments and domain-specific phenomena and/or entities. Model builders and users have developed a body of knowledge that uses explicit ODE models as the core language to represent and reason about systems. This body of knowledge contains theorems, techniques and procedures about systems and their corresponding models.

Second, ODEs often mark just the right trade-off between complexity and precision. Partial differential equations, for example, can describe dynamic systems more accurately, but they are also vastly more difficult to deal with. Qualitative differential equations (Kuipers, 1986) are an example of the other side of this trade-off point; they can be constructed more easily than ODEs can, but their usefulness for engineering tasks (such as controller design, for example) is also more limited.

Third, there is a large body of ODE theory and associated knowledge that is independent of any particular scientific application domain. Therefore, ODEs and the associated techniques are widely applicable across multiple domains. For example, Eqn. (1) can model both a series and a parallel RLC circuit. Furthermore, the same equation can also model a series or parallel *mechanical* system consisting of a mass, a spring and a damper. See (Easley & Bradley, this volume) for a more-detailed explanation.

PRET's inputs are a set of observations of the outputs of the target system, some optional hypotheses about the physics involved, and a set of tolerances within which a successful model must match the observations; its output is an ordinary differential equation model of the internal dynamics of that system. See Fig. 1 for a block diagram.

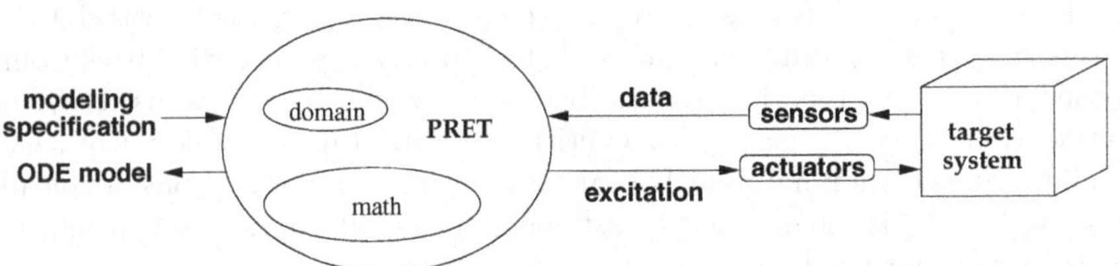

Fig. 1. PRET combines AI and formal engineering techniques to build ODE models of nonlinear dynamic systems. It builds models using domain-specific knowledge, tests them using an encoded ODE theory, and interacts directly and autonomously with target systems using sensors and actuators.

PRET uses a small, powerful domain theory to build models and a larger, more-general mathematical theory to test them. It is designed to work in any domain that admits ODE models; adding a new domain is simply a matter of coding one or two simple domain rules. Its architecture wraps a layer of AI techniques around a set of traditional formal engineering methods. Models are represented using a component-based modeling framework (Easley & Bradley, 1999a) that accommodates different domains, adapts smoothly to varying amounts of domain knowledge, and allows expert users to create model-building frameworks for new application domains easily (Easley & Bradley, 2000). An input-output modeling subsystem (Easley & Bradley, 1999b) allows PRET to observe target systems actively, manipulating actuators and reading sensors to perform experiments whose results augment its knowledge in a manner that is useful to the modeling problem that it is trying to solve.

The program's entire reasoning process is orchestrated by a special first-order logic inference system, which automatically chooses, invokes, and interprets the results of the techniques that are appropriate for each point in the model-building procedure. This combination of techniques lets PRET shift fluidly back and forth between domain-specific reasoning, general mathematics, and actual physical experiments in order to navigate efficiently through an exponential search space of possible models.

3.2 System Identification Phases

PRET's combination of symbolic and numeric techniques reflects the two phases that are interleaved in the general system identification process: first, *structural identification*, in which the form of the differential equation is determined, and then *parameter estimation*, in which values for the coefficients are obtained. If structural identification produces an incorrect ODE model, no coefficient values can make its solutions match the sensor data. In this event, the structural identification process must be repeated—often using information about why the previous attempt failed—until the process converges to a solution, as shown diagrammatically in Fig. 2.

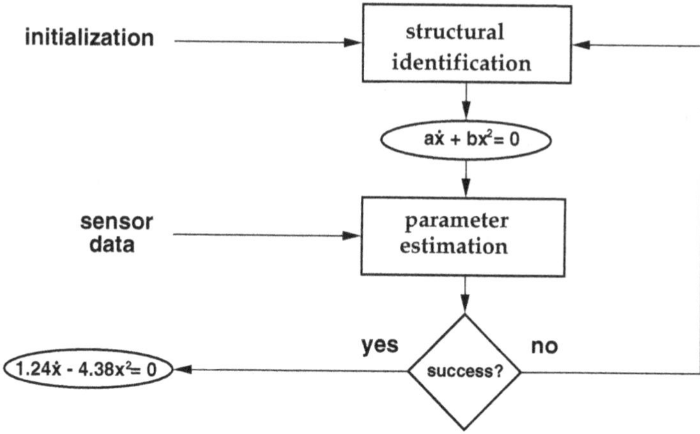

Fig. 2. The system identification (SID) process. Structural identification yields the general form of the model; in parameter estimation, values for the unknown coefficients in that model are determined. PRET automates both phases of this process.

In linear physical systems, structural identification and parameter estimation are fairly well understood. The difficulties—and the subtleties employed by practitioners—arise where noisy or incomplete data are involved, or where efficiency is an issue. See (Juang, 1994; Ljung, 1987) for some examples. In *nonlinear* systems, however, both procedures are vastly more difficult—the type of material that is covered only in the last few pages of standard textbooks. Unlike system identification software used in the control theory community, PRET is not just an automated parameter estimator; rather, it uses sophisticated reasoning techniques to automate the structural phase of model building as well.

3.3 The Generate-and-Test Paradigm

PRET's basic paradigm is "generate and test". It first uses its encoded domain theory—the upper ellipse in Fig. 1—to assemble combinations of user-specified and automatically generated ODE fragments into a candidate model. In a mechanics problem, for instance, the generate phase uses Newton's laws to combine force terms; in electronics, it uses Kirchhoff's laws to sum voltages in a loop or currents in a cutset.

The challenge in the design of the algorithms for PRET's generate phase was to aviod a combinatorial explosion of the search space. We achieve this objective by allowing the user to provide knowledge about the domain and the target system that may help to limit the space of possible models. The manner in which the generate phase makes use of this kind of communicable knowledge is the topic of (Easley & Bradley, this volume).

In order to test a candidate model, PRET performs a series of inferences about the model and the observations that the model is to match. This process is guided by two important assumptions: that abstract reasoning should be chosen over lower-level techniques, and that any model that cannot be proved wrong is right. PRET's inference engine uses an encoded mathematical theory (the lower ellipse in Fig. 1) to search for contradictions in the sets of facts inferred from the model and from the observations. An ODE that is linear, for instance, cannot account for chaotic behavior; such a model should fail the test if the target system has been observed to be chaotic. Furthermore, establishing whether an ODE is linear is a matter of simple symbolic algebra, so PRET's inference engine should not resort to a numerical integration to establish this contradiction. Like the domain theory, PRET's ODE theory is designed to be easily extended by an expert user.

The test phase of the generate-and-test cycle is the topic of this chapter. It uses the user's observations about the target system and PRET's internal ODE theory in order to rule out bad candidate models quickly. Both the observations and the ODE theory are expressed as communicable knowledge; they make direct use of standard engineering concepts and vocabulary.

3.4 A Simple, Introductory Example

To make these ideas more concrete, this section works through a simple example of a PRET run. This example illustrates how a user specifies the inputs to PRET, how this information is used in the generate-and-test cycle, and what PRET's main strategies are to find good models quickly. The purpose of this section is to give the reader an overview of how PRET works and what the main challenges were in the design of PRET's architecture and set of techniques and tactics. A more precise discussion of how candidate models are generated is presented in (Easley & Bradley, this volume). A more precise discussion of the reasoning that is used to test candidate models against the observations of the target system is presented in Sections 4 and 5 of this current chapter.

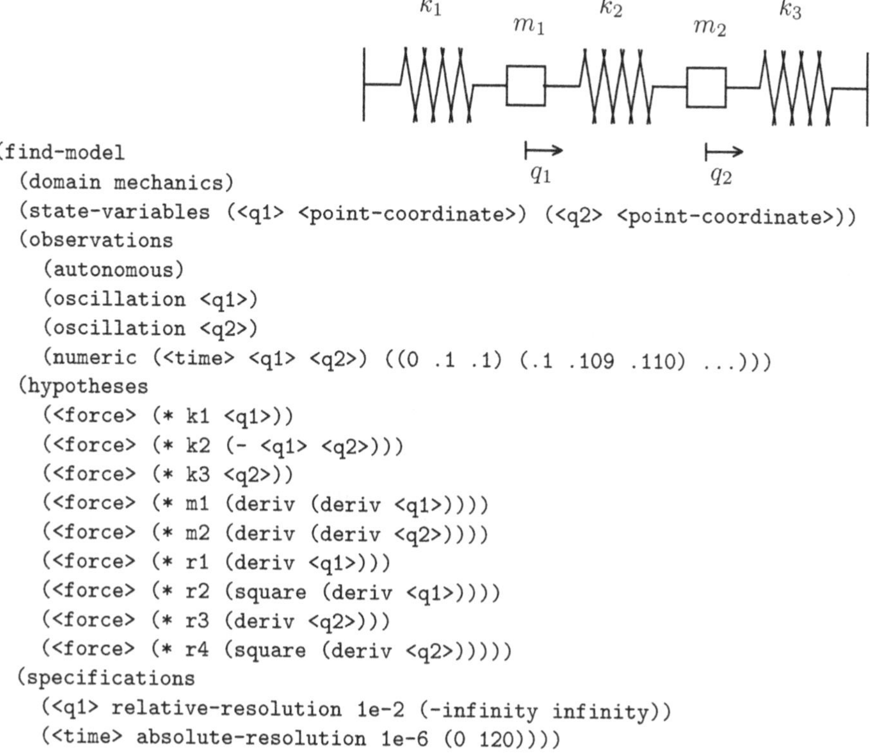


```
(find-model
  (domain mechanics)
  (state-variables (<q1> <point-coordinate>) (<q2> <point-coordinate>))
  (observations
    (autonomous)
    (oscillation <q1>)
    (oscillation <q2>)
    (numeric (<time> <q1> <q2>) ((0 .1 .1) (.1 .109 .110) ...)))
  (hypotheses
    (<force> (* k1 <q1>))
    (<force> (* k2 (- <q1> <q2>)))
    (<force> (* k3 <q2>))
    (<force> (* m1 (deriv (deriv <q1>))))
    (<force> (* m2 (deriv (deriv <q2>))))
    (<force> (* r1 (deriv <q1>)))
    (<force> (* r2 (square (deriv <q1>))))
    (<force> (* r3 (deriv <q2>)))
    (<force> (* r4 (square (deriv <q2>)))))
  (specifications
    (<q1> relative-resolution 1e-2 (-infinity infinity))
    (<time> absolute-resolution 1e-6 (0 120))))
```

Fig. 3. Modeling a simple spring/mass system using communicable formalisms. The vocabulary and concepts in which the user specifies the modeling problem are drawn from the engineering application domain. In this example call to PRET, the user first sets up the problem, then makes five observations about the position coordinates q_1 and q_2, hypothesizes nine different force terms, and finally specifies resolution and range criteria that a successful model must satisfy. Angle brackets (e.g., `<time>`) identify state variables and other special keywords that play roles in PRET's use of its domain theory. The `teletype` font identifies terms that play roles in a user's interaction with PRET.

Consider the spring/mass system shown at the top right of Fig. 3. The coefficients m_1 and m_2 represent the two mass elements in the system; the coefficients k_1, k_2, and k_3 represent the three spring elements. The state variables q_1 and q_2 measure the positions of the mass elements.

To instruct PRET to build a model of this system, a user would enter the `find-model` call at the left of the figure. This call contains four types of information: the domain, state variables, observations, hypotheses, and specifications.

The `domain` statement instantiates the relevant domain theory; the next two lines inform PRET that the system has two point-coordinate state variables.[5] `Observations` are measured automatically by sensors and/or interpreted by the user; they may be symbolic or numeric and can take on a variety of formats and degrees of precision. For example, the first observation in Fig. 3 informs PRET that the system to be modeled is autonomous.[6] The second observation states that the state variable q_1 oscillates.[7] `Numeric` observations are physical measurements made directly on the system.

An optional list of `hypotheses` about the physics involved—e.g., a set of ODE terms[8] ("model fragments") that describe different kinds of friction—may be supplied as part of the `find-model` call; these may conflict and need not be mutually exclusive, whereas observations are always held to be true.

Finally, `specifications` indicate the quantities of interest and their resolutions. The ones at the end of Fig. 3, for instance, require any successful model to match q_1 to within 1% 120 seconds of the system's evolution. Note that PRET uses tolerances (maximal error) as its accuracy criterion, which may seem unorthodox from a, say, machine learning perspective. Again, this choice is rooted in PRET's design rationale as an engineering tool, and it is further explained in Section 6.

It should be noted that this spring/mass example is representative neither of PRET's power nor of its intended applications. Linear systems of this type are very easy to model (Ljung, 1987); no engineer would use a software tool to do generate-and-test and guided search on such an easy problem. We chose this simple system to make this presentation brief and clear.

To construct a model from the information in this `find-model` call, PRET uses the `mechanics` domain rule `(point-sum <force> 0)` from its knowledge base to combine hypotheses into an ODE. In the absence of any domain knowledge—omitted here, again, to keep this example short and clear—PRET simply selects the first hypothesis, producing the ODE $k_1 q_1 = 0$. The model tester, implemented as a custom first-order logic inference engine (Stolle, 1998), uses a set of general rules about ODE properties to draw inferences from the model and from the observations. In this case, PRET uses its ODE theory to establish a

[5] As described in (Bradley et al., 2001), PRET uses a variety of techniques to infer this kind of information from the target system itself; to keep this example simple, we bypass those facilities by giving it the information up front.

[6] That is, it does not explicitly depend on time.

[7] Oscillation means that the corresponding phase-space trajectory contains a limit cycle (or spiral, in the case of damped oscillation). Again, PRET can infer this sort of qualitative observation from numeric observations of the target system itself; see (Easley & Bradley, this volume).

[8] The functor `deriv` stands for "derivative." Furthermore, expressions are in prefix notation. For example, the expression `(* r2 (square (deriv <q1>)))` represents the term $r_2 \dot{q_1}^2$.

contradiction between the model's order (the model's highest derivative) and its oscillatory behavior. The way PRET handles this first candidate model demonstrates the power of its abstract-reasoning-first approach: only a few steps of inexpensive qualitative reasoning suffice to let it quickly discard the model.

PRET tries all combinations of `<force>` hypotheses at single point coordinates, but all these models are ruled out for qualitative reasons. It then proceeds with ODE systems that consist of *two* force balances—one for each point coordinate. One example is

$$\begin{aligned} k_1 q_1 + m_1 \ddot{q_1} &= 0 \\ m_2 \ddot{q_2} &= 0 \end{aligned}$$

PRET cannot discard this model by purely qualitative means, so it invokes its nonlinear parameter estimation reasoner (NPER), which uses knowledge derived in the structural identification phase to guide the parameter estimation process (e.g., choosing good approximate initial values and thereby avoiding local minima in regression landscapes) (Bradley et al., 1998). The NPER finds no appropriate values for the coefficients k_1, m_1, and m_2, so this candidate model is also ruled out. This, however, is a far more expensive proposition than the simple symbolic contradiction proof for the one-term model above—roughly five minutes of CPU time, as compared to a fraction of a second—which is exactly why PRET's inference guidance system is set up to use the NPER only as a last resort, after all of the more-abstract reasoning tools in its arsenal have failed to establish a contradiction.

After having discarded a variety of unsuccessful candidate models via similar procedures, PRET eventually tries the model

$$\begin{aligned} k_1 q_1 + k_2(q_1 - q_2) + m_1 \ddot{q_1} &= 0 \\ k_3 q_2 + k_2(q_1 - q_2) + m_2 \ddot{q_2} &= 0 \end{aligned}$$

Again, it calls the NPER, this time successfully. It then substitutes the returned parameter values for the coefficients and integrates the resulting ODE system with fourth-order Runge-Kutta, comparing the result to the numeric time-series observation. The difference between the numerical solution and the observation stays within the specified resolution, so this candidate model is returned as the answer. If the list of user-supplied hypotheses is exhausted before a successful model is found, PRET generates hypotheses automatically using Taylor-series expansions on the state variables—the standard engineering fallback in this kind of situation. This simple solution actually has a far deeper and more important advantage as well: it confers black-box modeling capabilities on PRET.

The technical challenge of this model-building process is efficiency; the search space is huge—particularly if one resorts to Taylor expansions—and so PRET must choose promising model components, combine them intelligently into candidate models, and identify contradictions as quickly and simply as possible. In particular, PRET's *generate* phase must exploit all available domain-specific knowledge insofar as possible. A modeling domain that is too small may omit a key model; an overly general domain has a prohibitively large search space.

By specifying the modeling domain, the user helps PRET identify what the possible or typical "ingredients" of the target system's ODE are likely to be, thereby narrowing down the search space of candidate models. This "grey-box" modeling approach differs from traditional black-box modeling, where the model must be inferred only from external observations of the target system's behavior. It is also more realistic, as described in more depth in (Easley & Bradley, this volume): the engineers who are PRET's target audience do not operate in a complete vacuum, and its ability to leverage the kinds of domain knowledge that such users typically bring to a modeling problem lets PRET tailor the search space to the problem at hand.

Within the space of models that are generated by the grey-box modeling approach described in the previous paragraph, PRET must identify a model that is as parsimonious as possible. At the same time, the chosen model must also meet the accuracy requirements specified by the user. This tradeoff between parsimony and accuracy is driven by the user's engineering task at hand; it distinguishes engineering modeling from automated scientific discovery, as explained in the previous section.

In our approach, the key to quickly identifying the right model is to classify model and system behavior at the highest possible abstraction level. PRET incorporates several different reasoning techniques that are appropriate in different situations and that operate at different abstraction levels and in different domains. Examples of such techniques are symbolic algebra, qualitative reasoning about symbolic model properties, and phase-portrait analysis. These methods are drawn from the standard repertory of system identification techniques. In fact, PRET's internal representation—just like its inputs and outputs—uses standard engineering vocabulary, abstracted into first-order predicate logic: the predicates have names like `linear-system`, `damped-oscillation`, and `divergence`, and their meanings are therefore familiar and easily communicable to domain experts. This framework and its associated reasoning modes are discussed in Section 4. Coordinating the invocation and interaction of PRET's various reasoning modes is a difficult problem. To effectively build and test models of nonlinear systems, PRET must determine which methods are appropriate to a given situation, invoke and coordinate them, and interpret their results. The *reasoning control* mechanism that lets PRET orchestrate this subtle and complex reasoning process is described in Section 5.

4 Communicable Reasoning About Dynamic Systems

PRET's inputs and outputs are designed to be in the form of communicable knowledge: the inputs are hypotheses (that is, potential model fragments), observations at various abstraction levels, and specifications concerning the ranges and resolutions of interest. All of these inputs are presented to PRET in a form that directly mirrors the form in which domain experts typically express this kind of information. PRET's grey-box modeling approach and its component-based representations allow users to tailor the search space of models by providing knowledge about system components and structures that are typical of the modeling domain. PRET's outputs are ODEs, which is the model of choice for a wide variety of scientific and engineering tasks. In addition to interacting with the user in a communicable domain-centered language, PRET's internal reasoning machinery also employs communicable knowledge that makes use of the vocabulary and the concepts of its domain, namely system identification. The communicability of PRET's internal reasoning is the topic of this section.

As is described in (Easley & Bradley, this volume), PRET uses component-based representations, user hypotheses, and domain knowledge to generate candidate models of the given target system. Using the reasoning framework described in this section, PRET tests such a model against observations of the target system.

Like a human expert, PRET makes use of a variety of reasoning techniques at various abstraction levels during the course of this process, ranging from detailed numerical simulation to high-level symbolic reasoning. These modes and their interactions are described in the following subsections. The advantages of representing PRET's ODE theory and the associated techniques in the form of communicable knowledge are summarized in Section 4.6.

The challenge in designing PRET's model tester was to work out a formalism that met two requirements: first, it had to facilitate easy formulation of the various reasoning techniques; second, it had to allow PRET to reason about which techniques are appropriate in which situations. In particular, reasoning about both physical systems and candidate models should take place at an abstract level first and resort to more-detailed reasoning later and only if necessary. To accomplish this, PRET judges models according to the opportunistic paradigm "valid, if not proven invalid": if a model is bad, there must be a reason for it. Or, conversely, if there is no reason to discard a model, it is a valid model. PRET's central task, then, is to quickly find inconsistencies between a candidate model and the target system. Section 5 briefly describes the reasoning control techniques that allow it to do so.

PRET's test phase uses six different classes of techniques in order to test a candidate model against a set of observations of a target system:

- qualitative reasoning,
- qualitative simulation,
- constraint reasoning,
- geometric reasoning,
- parameter estimation, and
- numerical simulation.

In our experience, this set of techniques provides PRET with the right tools to quickly test models against the given observations.[9] Parameter estimation and numerical simulation are low-level, computationally expensive methods that ensure that no incorrect model passes the test. Intelligent use of the other, more-abstract techniques in the list above allows PRET to avoid these costly low-level techniques insofar as possible; most candidate models can be discarded

[9] See, e.g., the example section of (Bradley et al., 2001).

by purely qualitative techniques or by semi-numerical techniques in conjunction with constraint reasoning.

4.1 Qualitative Reasoning

Reasoning about abstract features of a physical system or a candidate model is typically faster than reasoning about their detailed properties. Because of this, PRET uses a "high-level first" strategy: it tries to rule out models by purely *qualitative* techniques (de Kleer & Williams, 1991; Faltings & Struss, 1992; Forbus, 1996; Weld & de Kleer, 1990) before advancing to more-expensive semi-numerical or numerical techniques. Often, only a few steps of inexpensive qualitative reasoning suffice to quickly discard a model.

Some of PRET's qualitative rules, in turn, make use of other tools, e.g., symbolic algebra facilities from the commercial package MAPLE (Char et al., 1991). For example, PRET's encoded ODE theory includes the qualitative rule that nonlinearity is a necessary condition for chaotic behavior:

```
(<- (falsum)
    ((linear-system)       ;; ode is linear
     (chaotic)))           ;; target system is chaotic
```

This lets any linear model be discarded without performing more-complex operations[10] such as, for example, a numerical integration of the ODE. Table 1 gives some examples of observations and the facts that the logic system infers from them.

Table 1. Some observations and the corresponding inferences drawn by the logic system

Observ. about state var. x_i	**Implications for model $f(\boldsymbol{x}, t) = 0$**
autonomous	cannot explicitly contain t (i.e., $f(\boldsymbol{x}) = 0$)
chaotic	cannot be linear
chaotic and autonomous	order > 2
oscillation and autonomous	imaginary part of one pair of roots > 0
linear	should satisfy $\ddot{x}_i = 0$
constant	should satisfy $\dot{x}_i = 0$
conservative	$\nabla \cdot f = 0$
damped oscill. and autonomous	$\nabla \cdot f < 0$

These examples highlight an important feature of PRET's knowledge base: not only PRET's inputs and outputs but also its internal reasoning rules are communicable to domain experts.

[10] Determining whether or not an ODE is linear involves calculation of the Jacobian, which is a simple symbolic operation that PRET accomplishes via a single call to MAPLE.

PRET's QR facilities are not only important for accelerating the search for inconsistencies between the physical system and the model; they also allow the user to express incomplete information (Kuipers, 1992). For example, the user might not know the exact value of a friction coefficient, but he or she might know that it is constant and positive. This is useful not only in isolation, but in conjunction with the constraint reasoning mode, as described later in this section.

4.2 Qualitative Simulation

After using its qualitative reasoning facilities to the fullest possible extent and before resorting to the numerical level, PRET attempts to establish contradictions by reasoning about the states of the physical system (Kuipers, 1992). It does not do full qualitative simulation (Kuipers, 1986); rather, it envisions the state space of all possible combinations of qualitative values of state variables and parameters. Specifically, PRET's qualitative envisioning module constrains the possible ranges of parameters in the candidate model. If the constraints become inconsistent—i.e., the range of a parameter becomes the empty set—the model is ruled out.

Currently, the qualitative states contain only sign information $(-, 0, +)$. For example, for the model $ax + by = 0$, the state $(x, y) = (+, +)$ constrains (a, b) to the possibilities $(+, -)$ or $(0, 0)$ or $(-, +)$. This strategy is faster than full qualitative simulation, but it is also less accurate; it may let invalid models pass the test, but these models will later be ruled out by the numeric simulator. However, for the models that do fail the qualitative envisioning test, this test is much cheaper than a numeric simulation and point-by-point comparison would be.

4.3 Constraint Reasoning

Often, information *between* the purely qualitative and the purely numeric levels is also available. If a linear system oscillates, for example, the imaginary parts of at least one pair of the roots of its model's characteristic polynomial must be nonzero. If the oscillation is damped, the real parts of those roots must also be negative. Thus, if the model $a\ddot{x}+b\dot{x}+cx = 0$ is to match an `damped-oscillation` observation, the coefficients must satisfy the inequalities $4ac > b^2$ and $b/a > 0$.

PRET uses expression inference (Sussman & Steele, 1980) to merge and simplify such constraints (Jaffar & Maher, 1994). However, this approach works only for linear and quadratic expressions and some special cases of higher order, but the expressions that arise in model testing can be far more complex. For example, if the candidate model $\ddot{x} + a\dot{x}^4 + b\dot{x}^2 = 0$ is to match an observation that the system is conservative, the coefficients a and b must take on values such that the divergence $-4a\dot{x}^3 - 2b\dot{x}$ is zero, below a certain resolution threshold, for the specified range of interest of x. We are investigating techniques (e.g., (Faltings & Gelle, 1997)) for reasoning about more-general expressions like this.

4.4 Geometric Reasoning

Other qualitative forms of information that are useful in reasoning about models are the geometry and topology of a system's behavior, as plotted in the time or frequency domain, state space, etc. A bend of a certain angle in the frequency response, for instance, indicates that the ODE has a root at that frequency, which implies algebraic inequalities on coefficients, much like the facts inferred from the `damped-oscillation` above; asymptotes in the time domain have well-known implications for system stability, and state-space trajectories that cross imply that an axis is missing from that space.

In order to incorporate this type of reasoning, PRET processes the `numeric` observations—curve fitting, recognition of linear regions and asymptotes, and so on—using MAPLE functions (Char et al., 1991) and simple phase-portrait analysis techniques (Bradley, 1995), producing the type of abstract information that its inference engine can leverage to avoid expensive numerical checks. These methods, which are used primarily in the analysis of sensor data, are described in more detail in (Bradley & Easley, 1998). PRET does not currently reason about topology, but we are investigating how best to do so (Robins et al., 1998; Robins et al., 2000).

4.5 Parameter Estimation and Numerical Simulation

PRET's final check of any model requires a point-by-point comparison of a numerical integration of that ODE against all numerical observations of the target system. In order to integrate the ODE, however, PRET must first estimate values for any unknown coefficients.

Parameter estimation, the lower box in Figure 2, is a complex nonlinear global optimization problem. PRET's nonlinear parameter estimation reasoner (NPER) solves this problem using a new, highly effective global optimization method that combines qualitative reasoning and local optimization techniques. Space limitations preclude a thorough discussion of this approach here; see (Bradley et al., 1998) for more details.

4.6 Benefits of the Communicable Reasoning Framework

Representing PRET's ODE theory and the associated reasoning techniques in a communicable format that resembles a domain expert's vocabulary and conceptual framework, as described in this section, has several advantages:

1. Formulating the ODE theory is a reasonably straightforward undertaking: the rules in the knowledge base—e.g., the ones shown in Table 1—resemble the knowledge presented in typical dynamic systems textbooks.
2. Given a trace of PRET's reasoning, it is easy to understand why a particular candidate model was ruled out.[11]

[11] We are currently working on ways in which such knowledge—essentially a contradiction proof—can be fed back to the generation phase automatically in order to help guide the choice of the next candidate model. Such approaches are often referred to as *discrepancy-driven refinement* (Addanki et al., 1991).

3. Similarly, it is easy to examine why a particular candidate model (the result of a PRET run) did *not* get ruled out. This point is important if PRET finds an ODE model that cannot be ruled out based on the union of the observations and PRET's ODE theory, but the model does not match the user's intuitions about the target system. This means that the model that passes PRET's validity test does not pass the user's mental validity test. This situation may arise for two reasons:
   (a) Incomplete observations: The user is aware of some aspects of the target system's behavior that do not match PRET's model, but the user did not provide the corresponding observations to PRET. See (Bradley et al., 1998) for an example.
   (b) Incomplete ODE theory: The user's knowledge about ODE theory lets him or her rule out PRET's model, but PRET's ODE theory does not include that knowledge.

   In the first case, the user simply starts another modeling run, this time supplying the additional observations that help refute the model that PRET has found in the previous run. In the second case, one has to "teach" PRET some more ODE theory, extending it so as to enable it to prove the contradiction (between the observations and the model) that the user sees and that PRET does not see.

   One of the most important advantages of PRET's communicable reasoning framework is its modularity and extensibility. It was intentionally designed so that working with it does not require knowledge of any of the inner workings of the program, which allows mathematics experts to easily modify and extend PRET's ODE theory. Implementing additional rules—similar to the ones shown in Table 1—is only a matter of a few lines of SCHEME code and/or a call to MAPLE.
4. Similarly to extending the ODE theory, the user may also want to extend PRET's arsenal of reasoning modes. Such extensions may result in a more-accurate assessment of candidate models and/or increased performance—because they may facilite high-level, abstract shortcuts for contradiction proofs. Adding a reasoning mode to PRET's repertoire amounts to writing two or three Horn clauses[12] that interpret the results of the reasoning mode by specifying the conditions under which those results contradict observations about the target system.

Please see (Stolle, 1998; Stolle & Bradley, 1998) for a complete discussion of PRET's reasoning modes.

5 Reasoning Control in PRET

PRET's challenge in properly orchestrating the reasoning modes described in the previous section was to test models against observations using the cheapest possible reasoning mode and, at the same time, to avoid duplication of effort. In order to accomplish this, the inference engine uses the following techniques.

[12] This is explained in the next section.

5.1 Resolution Theorem Proving

The observations and the ODE theory are expressed in the language of generalized Horn clause intuitionistic logic (McCarty, 1988). PRET's inference engine is a resolution-based theorem prover. For every candidate model, this prover combines basic facts about the target system, basic facts about the candidate model, and basic facts and rules from the ODE theory into one set of clauses, and then tries to derive `falsum`—which represents inconsistency—from that set.

The special formula `falsum` may only appear as the head of a clause. Such clauses are often called *integrity constraints*: they express fundamental reasons for inconsistencies, e.g., that a system cannot be oscillating *and* non-oscillating at the same time. For a detailed discussion of PRET's logic system see (Stolle, 1998; Stolle & Bradley, 1998; Hogan et al., 1998).

5.2 Declarative Meta Level Control

PRET uses declarative techniques not only for the representation of knowledge about dynamic systems and their models, but also for the representation of strategies that specify under which conditions the inference engine should focus its attention on particular pieces or types of knowledge. PRET provides meta-level language constructs that allow the implementer of the ODE theory to specify the *control strategy* that is to be used.

The intuition behind PRET's declarative control constructs is, again, that the search should be guided toward a cheap and quick proof of a contradiction. For example, PRET's meta control theory prioritizes stability reasoning about the target system depending on whether the system is known to be linear.[13] For a discussion of PRET's meta control constructs, see (Beckstein et al., 1996; Hogan et al., 1998).

5.3 Reasoning at Different Abstraction Levels

To every rule, the ODE theory implementer assigns a natural number, indicating its level of abstraction. The inference engine uses less-abstract ODE rules only if the more-abstract rules are insufficient to prove a contradiction.

This static abstraction level hierarchy facilitates strategies that cannot be expressed by the dynamic meta-level predicates alone: whereas the dynamic control rules impose an *order* on the subgoals and clauses of *one* particular (but complete) proof, the abstraction levels allow PRET to *omit* less-abstract parts of the ODE theory altogether. Since abstract reasoning usually involves less detail, this approach leads to short and quick proofs of the `falsum` whenever possible.

5.4 Storing and Reusing Intermediate Results

In order to avoid duplication of effort, PRET stores formulae that have been expensive to derive and that are likely to be useful again later in the reasoning

[13] If a system is known to be linear, its *overall* stability is easy to establish, whereas evaluating the stability of a nonlinear system is far more complicated and expensive.

process. Engineering a framework that lets PRET store just the right type and amount of knowledge is a surprisingly tricky endeavor. On the one hand, remembering every formula that has ever been derived is too expensive. On the other hand, many intermediate results are very expensive to derive and would have to be rederived multiple times if they were not stored for reuse.

PRET reuses previously derived knowledge in three ways. First, it remembers what it has found out about the physical system across all test phases of individual candidate models. The fact that a time series measured from the physical system contains a limit cycle, for example, can be reused across all candidate models. Second, every time PRET's reasoning proceeds to a less-abstract level, it needs all information that has already been derived at the more-abstract level, so it stores this information rather than rederiving it.[14] Finally, many of the reasoning modes described in Section 4 use knowledge that has been generated by previous inferences, which may in turn have triggered other reasoning modes. For instance, the NPER relies heavily on qualitative knowledge derived during the structural identification phase in order to avoid local extrema in regression landscapes. To facilitate this, PRET gives these modules access to the set of formulae that have been derived so far.

In summary, PRET's control knowledge is expressed as a declarative meta theory, which makes the formulation of control knowledge convenient, understandable, and extensible. None of the reasoning techniques described in Section 4 is new; expert engineers routinely use them when modeling dynamic systems, and versions of most have been used in at least one automated modeling tool. The set of techniques used by PRET's inference engine, the multimodal reasoning framework that integrates them, and the system architecture that lets PRET decide which one is appropriate in which situation, make the approach taken here novel and powerful.

6 Related Modeling Approaches

Section 2 described how engineering modeling fits into the more-general landscape of scientific discovery, modeling of physical systems, and machine learning. Sections 3, 4 and 5 have provided an overview of how PRET was designed as a unique tool to meet the particular challenges of modeling in an engineering setting. Given this background, we are now prepared to briefly review some more-closely related work in more detail.

Some of PRET's roots as an engineer's tool can be found in "the dynamicist's workbench" (Abelson et al., 1989; Abelson & Sussman, 1989). Its representational scheme and its reasoning about candidate models build on a large body of work in automated model building and reasoning about physical systems (see, for example, (Falkenhainer & Forbus, 1991; Forbus, 1984; Nayak, 1995; Addanki et al., 1991)). In particular, our emphasis on qualitative reasoning and qualitative

[14] This requires the developer to declare a number of predicates as *relevant* (Beckstein & Tobermann, 1992), which causes all succeeding subgoals with this predicate to be stored for later reuse. See (Hogan et al., 1998) for more discussion of this.

representations and their integration with numerical information and techniques falls largely into the category of qualitative physics. The project in this branch of the literature that is most closely related to PRET is the QR-based viscoelastic system modeling tool developed by Capelo et al. (1998), which also builds ODE models from time-series data. PRET is more general; it handles linear *and nonlinear* systems in a variety of domains using a richer set of model fragments that is designed to be adaptable.[15]

PRET takes a strict engineering approach to the questions of accuracy and parsimony. Its goal is to find an ODE system that serves as a useful model of the target system *in the context of engineering tasks*, such as controller design. PRET's notion of "accuracy" is relative only to the given observations: it finds an ODE system that matches the observations to within the user-specified precision, and does not try to second-guess these specifications or the user's choice of observations. It is the user's power and responsibility to ensure that the set of observations and specifications presented to PRET reflect the task at hand.

PRET's goal, then, is to construct the simplest model that matches the observed behavior to within the predefined `specifications`. Because evaluation criteria are always domain-specific, we believe that modeling tools should let their domain-expert users dictate them, and not simply build in an arbitrary set of thresholds and percentages. The notion of a *minimal* model that is tightly (some might say myopically) guided by its user's specifications represents a very different philosophy from traditional AI work in this area. Unlike some scientific discovery systems, PRET makes no attempt to exceed the range and resolution specifications that are prescribed by its user: a loose specification for a particular state variable, for instance, is taken as an explicit statement that an exact fit of that state variable is not important to the user, so PRET will not add terms to the ODE in order to model small fluctuations in that variable. Conversely, a single out-of-range data point will cause a candidate model to fail PRET's test.

These are not unwelcome side effects of the finite resolution; they are intentional and useful by-products of the abstraction level of the modeling process. A single outlying data point may appear benign if one reasons only about variances and means, but engineers care deeply about such single-point failures (such as the temperature dependence of O-ring behavior in space shuttle boosters), and a tool designed to support such reasoning must reflect those constraints.

It is, of course, possible to use PRET as a scientific discovery tool by supplying several sets of observations to it in separate runs and then unifying the results by hand. PRET can also be used to solve the kinds of cross-validation problems that arise in the machine learning literature: one would simply use it to perform several individual validation runs and then interpret the results.

Like the computational discovery work of (Schabacher et al., this volume) and (Saito & Langley, this volume), PRET makes direct contact with the applicable domain theory, and leverages that information in the model-building process. The theory and methods are of course different; PRET's domain is the

[15] Indeed, one of PRET's implemented modeling domains, `viscoelastics`, allows it to model the same problems as in (Capelo et al., 1998).

general mathematics of ODEs rather than the specifics of biological processes. Many of the research issues are similar, though: how best to combine concrete data and abstract models, how to communicate the results effectively to domain experts, etc.

Koza et al. (this volume) use genetic programming to automatically build directed graphs to model a variety of systems. While the goal is similar to PRET's—automatic construction of a mathematical model from observations—the models and techniques for deriving them are very different. PRET's target systems are nonlinear and dynamic, and ODEs are the best way to capture that behavior. Experts have used these kinds of models for many decades, so the associated domain-specific reasoning is fairly well-developed and can be exploited in an automated modeler. PRET takes this approach, rather than relying on general techniques like genetic programming, neural nets, regression rules, etc.

Like (Garret et al., this volume), PRET relies on mathematical logic to capture domain knowledge in a declarative form. Like (Washio & Motoda, this volume), PRET clearly separates domain-specific facts and general knowledge, making the priorities and connections explicit, and expressing each in a manner that is appropriate for their use.

Other automated analysis tools target nonlinear dynamic systems. The spatial aggregation framework of (Zhao et al., this volume) and Yip's KAM tool (Yip, 1991), among others, reason about the state-space geometry of their solutions. PRET's sensor data analysis facilities—see (Easley & Bradley, this volume) and (Bradley & Easley, 1998)—do essentially the same thing, but PRET then goes on to leverage that information to deduce what internal system dynamics *produced* that state-space geometry. Its ability to solve this kind of inverse problem—deducing a general, nonlinear ODE from partial information about its solutions—is one of PRET's unique strengths.

The branch of scientific discovery/machine learning research that is most closely related to PRET is the work of Todorovski (this volume) and Džeroski (this volume). This line of work began with LAGRANGE (Džeroski & Todorovski, 1995), which builds ODE and/or algebraic models of dynamic systems by applying regression techniques to time-series data. PRET and LAGRANGE can model problems of similar complexity; they differ in that PRET can handle *incomplete* data and systems that depend in a *nonlinear* manner on their parameters, whereas LAGRANGE cannot.

LAGRAMGE (Todorovski & Džeroski, 1997), the successor to LAGRANGE, improved upon its predecessor by incorporating the same kinds of optimization algorithms (e.g., Levenberg-Marquart) on which PRET's nonlinear parameter estimator is based. This broadened LAGRAMGE's generality (and its search space) to include models that are nonlinear in the state variables and the parameters.

The main difference between PRET and LAGRAMGE lies in how the initial conditions for the optimization are chosen. Simplex-based nonlinear optimization methods are essentially a sophisticated form of hill-climbing, and so initial-condition choice is a key element in their success or failure. PRET core design principle is to leverage all available information about the system and the model insofar as possible, and this plays a particularly important role in parameter estimation. In particular, PRET uses the arsenal of qualitative and quantitative reasoning techniques that have been described in previous sections in order to intelligently choose initial conditions for its nonlinear optimization runs. This not only broadens the class of ODEs for which it attains a successful fit, but also speeds up the fitting process for individual runs.

Because the reasoning involved in PRET's choice of initial conditions is both qualitative and quantitative, and because both the landscapes and methods of the optimization process are nonlinear, it is only possible to prove that the set of models that is accessible to LAGRAMGE is a *proper* subset of those that are accessible to PRET. (Indeed, any stronger statement would amount to a general solution of the global nonlinear optimization problem.) Very few optimization landscape geometries are forgiving of bad initial-condition choices, however, and so we believe that the difference between the two sets of models—PRET-accessible and LAGRAMGE-accessible—is large. Apart from this difference, PRET and LAGRAMGE are quite similar, though the design choices and implementation details (e.g., knowledge representations, reasoning modes, etc.) are of course different.

7 Conclusion

PRET is designed to produce the type of formal engineering models that a human expert would create—quickly and automatically. Unlike existing system identification tools, PRET is not just a fancy parameter estimator; rather, it uses sophisticated knowledge representation and reasoning techniques to automate the structural identification phase of model building as well.

PRET's inputs, its outputs, and the knowledge used by its internal reasoning machinery are all expressed in a form that makes this knowledge easily communicable to domain experts. The declarative knowledge representation framework described in this chapter allows knowledge about dynamic systems and their models to be represented in a highly effective manner. Since PRET keeps its operational semantics equivalent to its declarative semantics and uses a simple and clear modeling paradigm, it is extremely easy for domain experts to understand and use it. This allows scientists and engineers to use PRET as an engineering tool in the context of engineering tasks, communicating with the program using the application domain's vocabulary and conceptual framework.

PRET has been able to successfully construct models of a dozen or so textbook problems (Rössler, Lorenz, simple pendulum, pendulum on a spring, etc.; see (Bradley et al., 1998; Bradley & Stolle, 1996)), as well as several interesting and difficult real-world examples, such as a well, a shock absorber, and a driven pendulum, which are described in (Bradley et al., 2001), and a commercial radio-controlled car, which is covered in (Bradley et al., 1998). These examples are representative of wide classes of dynamic systems, both linear and nonlinear. The research effort on this project has now turned to the application of this useful problem-solving tool, rather than improvement of its algorithms. Our current task, for instance, is to use PRET to deduce information about paleoclimate dynamics from radioisotope dating data.

Acknowledgements: Apollo Hogan, Brian LaMacchia, Abbie O'Gallagher, Janet Rogers, Ray Spiteri, Tom Wrensch, and particularly Matt Easley contributed code and/or ideas to PRET. The authors would like to thank Pat Langley for enlightening discussions.

Parts of this paper are short versions of material previously published in the journal *Artificial Intelligence* (Bradley et al., 2001) and are republished here with kind permission of the publisher, Reed-Elsevier.

References

Abelson, H., Eisenberg, M., Halfant, M., Katzenelson, J., Sussman, G.J., Yip, K.: Intelligence in scientific computing. Communications of the ACM 32, 546–562 (1989)

Abelson, H., Sussman, G.J.: The Dynamicist's Workbench I: Automatic preparation of numerical experiments. In: Symbolic computation: Applications to scientific computing. Frontiers in Applied Mathematics. vol. 5, Society for Industrial and Applied Mathematics, Philadelphia, PA (1989)

Addanki, S., Cremonini, R., Penberthy, J.S.: Graphs of models. Artificial Intelligence 51, 145–177 (1991)

Beckstein, C., Stolle, R., Tobermann, G.: Meta-programming for generalized Horn clause logic. In: Proceedings of the Fifth International Workshop on Metaprogramming and Metareasoning in Logic, pp. 27–42. Bonn, Germany (1996)

Beckstein, C., Tobermann, G.: Evolutionary logic programming with RISC. In: Proceedings of the Fourth International Workshop on Logic Programming Environments, pp. 16–21. Washington, D.C. (1992)

Bradley, E.: Autonomous exploration and control of chaotic systems. Cybernetics and Systems 26, 299–319 (1995)

Bradley, E., Easley, M.: Reasoning about sensor data for automated system identification. Intelligent Data Analysis 2, 123–138 (1998)

Bradley, E., Easley, M., Stolle, R.: Reasoning about nonlinear system identification. Artificial Intelligence 133, 139–188 (2001)

Bradley, E., O'Gallagher, A., Rogers, J.: Global solutions for nonlinear systems using qualitative reasoning. Annals of Mathematics and Artificial Intelligence 23, 211–228 (1998)

Bradley, E., Stolle, R.: Automatic construction of accurate models of physical systems. Annals of Mathematics and Artificial Intelligence 17, 1–28 (1996)

Capelo, A., Ironi, L., Tentoni, S.: Automated mathematical modeling from experimental data: An application to material science. IEEE Transactions on Systems, Man and Cybernetics – C 28, 356–370 (1998)

Casdagli, M., Eubank, S. (eds.): Nonlinear modeling and forecasting. Addison Wesley, Reading, MA (1992)

Char, B.W., Geddes, K.O., Gonnet, G.H., Leong, B.L., Monagan, M.B., Watt, S.M.: Maple V language reference manual. Springer, Heidelberg (1991)

de Kleer, J., Williams, B.C. (eds.): Artificial intelligence. Special Volume on Qualitative Reasoning About Physical Systems II, vol. 51. Elsevier Science, Amsterdam (1991)

Džeroski, S., Todorovski, L.: Discovering dynamics: From inductive logic programming to machine discovery. Journal of Intelligent Information Systems 4, 89–108 (1995)

Easley, M., Bradley, E.: Generalized physical networks for automated model building. In: Proceedings of the Sixteenth International Joint Conference on Artificial Intelligence, pp. 1047–1053. Stockholm, Sweden (1999a)

Easley, M., Bradley, E.: Reasoning about input-output modeling of dynamical systems. In: Proceedings of the Third International Symposium on Intelligent Data Analysis, pp. 343–355. Amsterdam, The Netherlands (1999b)

Easley, M., Bradley, E.: Meta-domains for automated system identification. In: Proceedings of the Eleventh International Conference on Smart Engineering System Design, pp. 165–170. St. Louis, MI (2000)

Falkenhainer, B., Forbus, K.D.: Compositional modeling: Finding the right model for the job. Artificial Intelligence 51, 95–143 (1991)

Faltings, B., Gelle, E.: Local consistency for ternary numeric constraints. In: Proceedings of the Fifteenth International Joint Conference on Artificial Intelligence, pp. 392–397. Nagoya, Japan (1997)

Faltings, B., Struss, P. (eds.): Recent advances in qualitative physics. MIT Press, Cambridge, MA (1992)

Farmer, J., Sidorowich, J.: Predicting chaotic time series. Physical Review Letters 59, 845–848 (1987)

Forbus, K.D.: Qualitative process theory. Artificial Intelligence 24, 85–168 (1984)

Forbus, K.D.: Qualitative reasoning. In: Tucker Jr., A.B. (ed.) CRC computer science and engineering handbook, ch. 32, pp. 715–733. CRC Press, Boca Raton, FL (1996)

Hogan, A., Stolle, R., Bradley, E.: Putting declarative meta control to work (Technical Report CU-CS-856-98). University of Colorado, Boulder (1998)

Huang, K.-M., Żytkow, J.M.: Discovering empirical equations from robot-collected data. In: Foundations of Intelligent Systems (Proceedings of the Tenth International Symposium on Methodologies for Intelligent systems), pp. 287–297, Charlotte, NC (1997)

Jaffar, J., Maher, M.J.: Constraint logic programming: A survey. Journal of Logic Programming 20, 503–581 (1994)

Juang, J.-N.: Applied system identification. Prentice Hall, Englewood Cliffs, N.J. (1994)

Kuipers, B.J.: Qualitative simulation. Artificial Intelligence 29, 289–338 (1986)

Kuipers, B.J.: Qualitative reasoning: Modeling and simulation with incomplete knowledge. Addison-Wesley, Reading, MA (1992)

Kuipers, B.J.: Reasoning with qualitative models. Artificial Intelligence 59, 125–132 (1993)

Langley, P.: The computational support of scientific discovery. International Journal of Human-Computer Studies 53, 393–410 (2000)

Langley, P., Simon, H.A., Bradshaw, G.L., Żytkow, J.M. (eds.): Scientific discovery: Computational explorations of the creative processes. MIT Press, Cambridge, MA (1987)

Ljung, L. (ed.): System identification; theory for the user. Prentice-Hall, Englewood Cliffs, N.J. (1987)

McCarty, L.T.: Clausal intuitionistic logic I. Fixed-point semantics. The Journal of Logic Programming 5, 1–31 (1988)

Morrison, F.: The art of modeling dynamic systems. John Wiley & Sons, New York (1991)

Nayak, P.P.: Automated modeling of physical systems (Revised version of Ph.D. thesis, Stanford University). LNCS, vol. 1003, Springer, Heidelberg (1995)

Robins, V., Meiss, J., Bradley, E.: Computing connectedness: An exercise in computational topology. Nonlinearity 11, 913–922 (1998)

Robins, V., Meiss, J., Bradley, E.: Computing connectedness: Disconnectedness and discreteness. Physica D 139, 276–300 (2000)

Stolle, R.: Integrated multimodal reasoning for modeling of physical systems. In: Doctoral dissertation, University of Colorado at Boulder. LNCS, Springer, Heidelberg (to appear, 1998)

Stolle, R., Bradley, E.: Multimodal reasoning for automatic model construction. In: Proceedings of the Fifteenth National Conference on Artificial Intelligence, pp. 181–188. Madison, WI (1998)

Sussman, G.J., Steele, G.L.: CONSTRAINTS—a language for expressing almost hierarchical descriptions. Artificial Intelligence 14, 1–39 (1980)

Todorovski, L., Džeroski, S.: Declarative bias in equation discovery. In: Proceedings of the Fourteenth International Conference on Machine Learning, pp. 376–384. Nashville, TN (1997)

Washio, T., Motoda, H., Yuji, N.: Discovering admissible model equations from observed data based on scale-types and identity constraints. In: Proceedings of the Sixteenth International Joint Conference on Artificial Intelligence, pp. 772–779. Stockholm, Sweden (1999)

Weigend, A.S., Gershenfeld, N.S. (eds.): Time series prediction: Forecasting the future and understanding the past. Santa Fe Institute Studies in the Sciences of Complexity, Santa Fe, NM (1993)

Weld, D.S., de Kleer, J. (eds.): Readings in qualitative reasoning about physical systems. Morgan Kaufmann, San Mateo, CA (1990)

Yip, K.: KAM: A system for intelligently guiding numerical experimentation by computer. Artificial Intelligence Series. MIT Press, Cambridge (1991)

Żytkow, J.M.: Model construction: Elements of a computational mechanism. In: Proceedings of the Symposium on Artificial Intelligence and Scientific Creativity, pp. 65–71. Edinburgh, UK (1999)

Incorporating Engineering Formalisms into Automated Model Builders

Matthew Easley[1] and Elizabeth Bradley[2,*]

[1] Teledyne Scientific
Thousand Oaks, California, USA
measley@teledyne.com
[2] Department of Computer Science
University of Colorado, Boulder, Colorado, USA
lizb@cs.colorado.edu

Abstract. We present a new knowledge representation and reasoning framework for modeling nonlinear dynamic systems. The goals of this framework are to smoothly incorporate varying levels of domain knowledge and to tailor the search space and the reasoning methods accordingly. In particular, we introduce a new structure for automated model building known as a *meta-domain* which, when instantiated with domain-specific components, tailors the space of candidate models to the system at hand. We combine this abstract modeling paradigm with ideas from generalized physical networks, a meta-level representation of idealized two-terminal elements, and a hierarchy of qualitative and quantitative analysis tools, to produce dynamic modeling domains whose complexity naturally adapts to the amount of available information about the target system. Since the domain and meta-domain representation use the same type of techniques and formalisms as practicing engineers, the models produced from these frameworks are naturally communicable to their target audience.

1 Representations for Automated Model Building

System identification (SID) is the process of identifying a dynamic model of an unknown system. The challenges involved in automating this process are significant, as applications in different fields of science and engineering demand different kinds of models and modeling techniques. System identification entails two steps: *structural identification*, wherein one ascertains the general form of the model as described by an ordinary differential equation or ODE (e.g., $a\ddot{x} + b\sin(x) = 0$ for a simple pendulum), and then *parameter estimation*, in which one finds specific parameter values for the unknown coefficients that fit that model to observed data (e.g., $a = 1.0$, $b = -98.0$). For nonlinear systems, parameter estimation is difficult and structural identification is even harder;

* Supported by NSF NYI #CCR-9357740, NSF #MIP-9403223, ONR #N00014-96-1-0720, and a Packard Fellowship in Science and Engineering from the David and Lucile Packard Foundation.

S. Džeroski and L. Todorovski (Eds.): Computational Discovery, LNAI 4660, pp. 44–68, 2007.

artificial intelligence (AI) techniques can be used to automate the former, but the latter has, until recently, remained the competency of human experts.

A central problem in any automated modeling task is that the size of the search space is exponential in the number of model fragments unless severe restrictions are placed on the model-building process. One would ideally like to build black-box models without resorting to *any* domain knowledge, but the combinatorics of this method makes it impractical. "Gray-box" modeling, which uses domain knowledge to prune the search space, is much more realistic and yet still quite general, as its techniques apply in a variety of circumstances. The key to making gray-box modeling of nonlinear dynamic systems practical in an automated modeling tool is a flexible knowledge representation scheme that adapts to the problem at hand. Domain-dependent knowledge can drastically reduce the search-space size, but its applicability is fundamentally limited. The challenge in balancing these influences is to be able to determine, at every point in the reasoning procedure, what knowledge is applicable and useful.

The goal of the work described in this chapter was to develop a knowledge representation and reasoning (KRR) framework that supports automated modeling in a range of gray shades that is useful to practicing engineers. The solution described here comprises a representation that allows for different levels of subject area knowledge, a set of reasoning techniques appropriate to each level, and a control strategy that invokes the right technique at the right time (Easley & Bradley, 1999). In particular, our work is based on a new structure for automated model construction called a *meta-domain*, which combines hypotheses into ordinary differential equation models without generating overly large search spaces. Meta-domains may be used directly, or refined with subject matter knowledge to create a more specific modeling *domain*. We combined the meta-domain representation with ideas from generalized physical networks (GPN) (Sanford, 1965), a meta-level representation of idealized two-terminal elements, and traditional compositional model building (Falkenhainer & Forbus, 1991) and qualitative reasoning (Weld & de Kleer, 1990) with the intent to bridge the gap between highly specific KRR frameworks that work well in a single, limited domain (e.g., a spring/dashpot vocabulary for modeling simple mechanical systems) and abstract frameworks that rely heavily upon general mathematical formalisms at the expense of having huge search spaces, such as (Bradley & Stolle, 1996). Finally, the meta-domain representation supports dynamic modeling domains whose complexity and analysis tools naturally adapts to the available information.

To test these ideas, we have implemented two meta-domains:

- `xmission-line`, which generalizes the notion of an electrical transmission line, using an iterative template to compose models, and
- `linear-plus`, which builds models that obey fundamental linear systems properties, while also allowing for limited numbers of nonlinear terms.

We have demonstrated the effectiveness of these meta-domains by incorporating them into PRET (Bradley et al., 2001), an automatic system identification tool that constructs ordinary differential equation models of nonlinear dynamic systems. Unlike other AI modeling tools—most of which use libraries to construct models of small, well-posed problems in limited applications—PRET builds models of nonlinear systems in multiple areas and uses sensors and actuators to interact directly and automatically with the target system (Bradley & Easley, 1998). PRET takes a generate-and-test approach, using a small, powerful domain theory to build models, and then applies a body of mathematical and physical knowledge encoded in first-order logic to test those candidate ODEs against behavioral observations of the target system. The "test" phase of the generate-and-test cycle is described in detail in (Stolle & Bradley, this volume). The emphasis in the remainder of this current chapter is upon the "generate" phase.

Meta-domains effectively shrink PRET's search spaces, and hence increase its power. A PRET operator may employ a meta-domain directly, or use a more-specific domain that has been customized to fit a particular engineering application. These frameworks allow PRET to search an otherwise intractable space of possible model combinations. Furthermore, this hierarchy of domains and meta-domains also contains analysis tools which PRET may use to reduce the size of the model search. Our results demonstrate that the meta-domain representation is an effective way to construct a description that is appropriate to a wide range of points on this gray-box modeling spectrum, and thus it provides a useful bridge between general and specific modeling approaches.

One of our primary goals of this work has been to make "an engineer's tool:" one that could begin to augment—or even duplicate—the work of a human engineer. Doing this in a manner that would be acceptable and understandable to the target audience required us to incorporate the techniques and formalisms of a variety of engineering fields into the knowledge representation and reasoning (KRR) frameworks described here. We did this in a multitude of ways, from the abstract description of observations of a time series, to the underlying representation of model building primitives and the ways in which these primitive components are connected into models. As we have consistently used the tools, vocabulary, and modeling representations of the engineering disciplines, it is a straightforward process for human engineers to reason with our results.

The following section covers background material on our approach: generalized physical networks for specific model-building domains. Section 3 describes the implementation of the two meta-domains mentioned above. Section 6 describes the reasoning techniques involved in dynamic modeling domains. An example of how a meta-domain integrates smoothly with the GPN representation to form a modeling domain appears in Section 7; see Easley (2000) for other examples. Section 8 covers related component-based and compositional modeling approaches, and Section 9 summarizes the major points of this chapter.

2 Generalized Physical Networks

In the late 1950s and early 1960s, inspired by the realization that the principles underlying Newton's third law and Kirchhoff's current law were identical,[1] a

[1] Summation of {*forces, currents*} at a point is zero, respectively; both are manifestations of the conservation of energy.

researcher at the Massachusetts Institute of Technology, Henry Paynter, began combining multi-port methods from a number of engineering fields into a generalized engineering domain with prototypical components (Paynter, 1961). The basis of Paynter's work is that the behavior of an ideal two-terminal element—the "component"—may be described by a mathematical relationship between two dependent variables: generalized flow and generalized effort, where *flow(t) * effort(t) = power(t)*. This generalized physical networks paradigm is one example of a generalized modeling representation, where the pair of variables manifests differently in each domain: (*flow*, *effort*) is (*current*, *voltage*) in an electrical domain and (*force*, *velocity*) in a mechanical domain. Effort and flow are known as power variables since their product is always power. This reliance upon power relationships is both the primary advantage and the primary disadvantage of generalized physical networks: if the physical system obeys this conservation principle, then GPNs are a useful modeling tool. If not, their efficacy is questionable.

2.1 Bond Graphs and Functional Modeling

In bond graphs (Karnopp et al., 1990), another generalized representation paradigm that has seen some use in the AI modeling literature (Mosterman & Biswas, 1996), flow and effort variables are reversed: velocity is now a flow variable and force is an effort variable. The difference between GPNs and bond graphs, thus, is essentially a frame-of-reference shift. While bond graphs are a good alternative to generalized physical networks—especially if causality issues are a concern—converting them into ODE models is difficult. Functional modeling (Chittaro et al., 1994) is another alternative to GPNs; indeed, it is based upon the same "Tetrahedron of State" that underlies both GPNs and bond graphs. (See Paynter (1961) for a detailed description of relationship between the four generalized variables of the tetrahedron: effort, flow, impulse, and displacement.) Functional modeling adds another layer to the GPN/bond graph idea by describing the potential functional role that a component plays in a system, as well as the functional relationship between components. Functional roles, which describe the way generalized variables influence each other, are useful in diagnostic reasoning applications where the relationship between structure and behavior is critical.

2.2 Advantages of GPNs

The GPN representation holds many advantages for automated model building. First, its two-port nature makes it easy to incorporate sensors and actuators as integral parts of a model. For example, a sinusoidal current source often has an associated impedance that creates a loading effect on the rest of the circuit. Not only does this make the model more representative of the physical world, but it also provides a handle for an automated modeler to actually manipulate an actuator's control parameter to explore various aspects of a physical system's behavior. The use of GPNs also brings out the similarities between components and properties in different domains. Electrical resistors ($v = iR$) and mechanical dampers ($v = fB$), for instance, are physically analogous: both dissipate energy in a manner that is proportional to the operative state variable. Both of these physical components can be represented by a single GPN modeling component that incorporates a *proportional* relationship between the flow and effort variables. Two other useful GPN components instantiate *integrating* and *differentiating* relationships, as shown in Table 1; other GPN instances model flow and effort sources. See Karnopp et al. (1990) or Sanford (1965) for additional domains and components.

Table 1. Example GPN component representations

Component	Electrical	Mechanical Translation	Mechanical Rotation	Fluid-Flow
Proportional	$v = Ri$	$v = Bf$	$\omega = D\tau$	$p = Rq$
Differentiating	$i = C\frac{dv}{dt}$	$f = M\frac{dv}{dt}$	$\tau = J\frac{d\omega}{dt}$	$q = C\frac{dp}{dt}$
Integrating	$v = L\frac{di}{dt}$	$v = K\frac{df}{dt}$	$\omega = K\frac{d\tau}{dt}$	$p = I\frac{dq}{dt}$
Nonlinear	$v = -R_A i^3$	$v = -R_A f^3$	$\omega = -R_A \tau^3$	$p = -R_A q^3$

A final advantage of the GPN representation is its ability to capture behavioral analogs. Both of the networks in Figure 1, for example, can be modeled by a series proportional/integrating/differentiating GPN; knowledge that the system is electronic or mechanical would let one refine the model accordingly (to a series RLC circuit or damper-spring-mass system, respectively). The available domain knowledge, then, can be viewed as a lens that expands upon the internals of some GPN components, selectively sharpening the model *in appropriate and useful ways.*

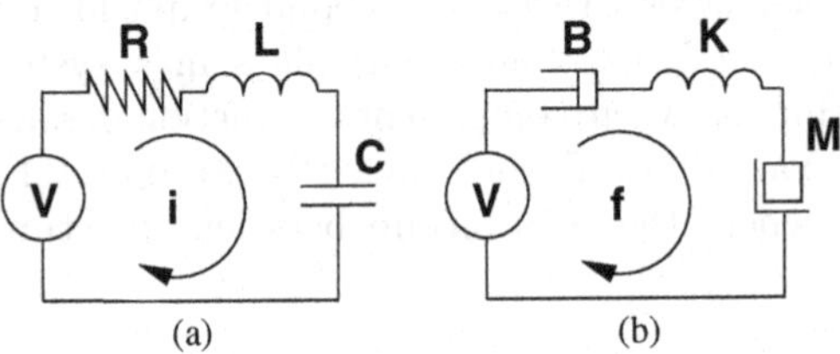

Fig. 1. Two systems that are described by the same GPN model: (a) a series RLC circuit (b) a damper-spring-mass system. **V** is a voltage source in (a) and a velocity source in (b).

2.3 Converting GPNs to ODEs

The conversion of a domain-independent GPN model into ODE form is fairly easy to automate. The powerful network-theoretic principles involved have been in the engineering vernacular for many decades , but are used here in a very different manner. Traditionally, engineers apply a tool like *modified nodal analysis*, which is based on node equations, to a known network with known parameter values in order to analyze its behavior. Our domain/meta-domain framework uses loop and node equations (formed using generalized version's of Kirchhoff's current and voltage laws) to convert a GPN network with unspecified parameter values into an ODE, also with unspecified parameter values. The advantage of doing this conversion at the end of the model-building process is that it keeps the model—and the reasoning—as abstract as possible for as long as possible.

For the example shown in Figure 1, the GPN $\rightarrow$ ODE conversion process works as follows. A generalized version of Kirchhoff's voltage law is used to form a loop equation around the network using the appropriate GPN components, in this case one proportional, one differentiating and integrating. Forming the loop equation yields the equation: $a\frac{dx}{dt}+bx+c\int xdt = df(t)$, which becomes $a\ddot{x}+b\dot{x}+cx = df'(t)$ after symbolic differentiation. The process becomes more complex when additional loop equations are required, generating another equation for every loop.

There are a variety of ways to use generalized physical networks to help automate the structural identification phase of the SID process. One could, for example, create a library of GPN components, enumerate all their possible combinations/configurations, and test each member of this succession until a valid model is found. This method is obviously impractical, as simple enumeration creates an exponential search space—a severe problem if the component library is large, as must be the case if one is attempting to model nonlinear systems.[2] The next section describes a way to alleviate this problem by incorporating powerful engineering formalisms into the reasoning framework.

3 Knowledge About Building Models

3.1 Hierarchy of Modeling Knowledge

Our method for reducing the size of the model-generation search space is to incorporate model-building heuristics and analysis tools into a carefully crafted knowledge representation and reasoning (KRR) framework. Depending upon the application area, experts use highly specific types of heuristic knowledge. Different types of problems demand different data analysis tools, for instance, and the constraints on allowable component types and connection frameworks are equally area-specific. Like these heuristics, application areas themselves vary greatly depending upon their scope. A general application area—e.g., the set of all dynamic systems—has a complex search space; a specific application like the set of conservative mechanical systems has a much smaller one. One way to reduce not only the size of the search space but also the amount of knowledge that must be encoded into the KRR framework is to organize the knowledge into

[2] Nonlinear terms are somewhat idiosyncratic, and each must be supplied as a separate library entry. This issue has not arisen in previous work on GPNs because their use has been generally confined to linear systems.

a hierarchy of generality. The inheritance tree structure of Figure 2 captures this critical system knowledge so that modeling building heuristics and analysis tools about one type of system can be reused by all of its subtypes. Detailed knowledge that is applicable to only one modeling field can then be confined to a leaf of the tree (e.g., linear viscoelastic systems). This section describes how the

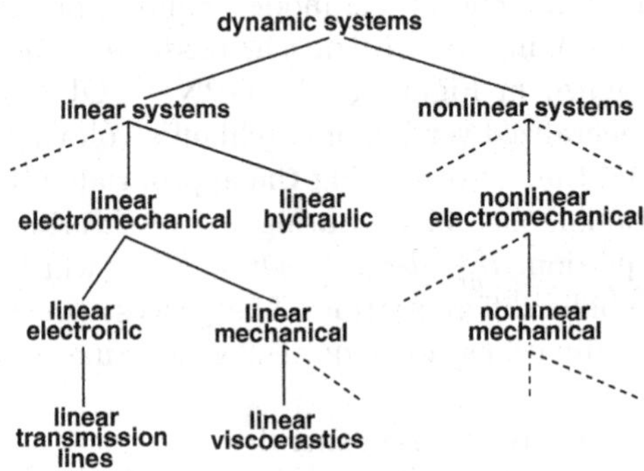

Fig. 2. A hierarchy of dynamic systems

model-building knowledge is structured inside each "node" of the tree. Section 6 describes how the structure of the tree itself is used to guide the model search process.

Knowledge within a node is encoded at two different levels: either in a meta-domain or a domain. A meta-domain is the construct that solves the search-space problem of the naïve component-based approach, where all possible combinations of possible components are generate and then tested. As implemented in our framework, meta-domains are algorithms that combine model hypotheses—either GPN components or ordinary different equation fragments—into viable ordinary differential equations. Hypotheses may either be defined by the user or may be pre-defined by a knowledge engineer for use by a more novice user. Some example hypotheses are shown in Table 2. Meta-domains also abstract knowledge about how to build models from a variety of subject areas. A thorough discussion of meta-domains follows in Sections 4 and 5.

A domain is a refinement of a meta-domain with knowledge that is specific to a particular subject area. The use of a domain requires less information from a user than a meta-domain, but at the expense of less flexibility. The relationship between domains and meta-domains is described at more length in the following section, together with some high-level issues concerning their selection and use. A specific example of how meta-domains help the PRET automated modeling tool quickly build an ODE model appears in Section 7; more examples may be found in Easley (2000).

The hierarchical representation of knowledge shown in Figure 2 is a small part of the research area of *model ontology.* One of the seminal works in the engineering sub-field of model ontology is PHYSSYS (Top & Akkermans, 1994), which is

Table 2. Example hypotheses for use in PRET. Note depending upon the meta-domain, hypotheses may either be GPN components or ordinary different equation fragments.

Hypothesis	**Description**
(<effort> (* a (integral <flow>)))	$e = a \int f dt$
(<effort> (* b (deriv <flow>)))	$e = b\dot{f}$
(<effort> (* c <flow>))	$e = cf$
(<effort> (* da (sin (* df <time>))))	$e = d_a \sin(d_f t)$

a formal ontology based upon physical system dynamics theory. PHYSSYS makes a conceptual distinction between a system's layout (network structure), physical processes underlying a behavior, and descriptive mathematical relationship. Our framework makes similar distinctions. For example, a meta-domain is used to help speed the process of determining a network structure, whereas a modeling component (provided by a user or built into a domain) captures the relationship between a physical entity and its mathematical representation. Other model ontology work is specific to hierarchies of models and/or modeling components. The Graphs of Models (Addanki et al., 1991) approach focuses on the problem of switching between different models using known assumptions in order to determine which model is appropriate. Another example is the hierarchical component library (de Vries et al., 1993), which decomposes bond-graph systems into nested subsystems and components via standard object-oriented mechanisms such as inheritence and polymorphism.

3.2 Domains and Meta-domains

Domains are constructed by application-area experts and stored in a domain-theory knowledge base. Each consists of a set of GPN component primitives and a framework for connecting those components into a model. The basic `electrical-xmission-line` domain, for example, comprises the components {`linear-resistor, linear-capacitor`}, the standard parallel and series connectors, and some codified notions of model equivalence (e.g., Thévenin (1883)).

Modeling domains are dynamic: if a domain does not contain a successful model, it automatically expands to include additional components and connections from domains higher in the hierarchy. For example, if all of the models in the initial `electrical-xmission-line` domain are rejected, the modeling domain automatically adds {`linear-inductor`} to the component set. We have constructed five specific GPN-based *modeling domains*: `mechanics`, `viscoelastics`, `electrical-xmission-line`, `linear-rotational`, and `linear-mechanics`.

Specification of state variables for these different domains—type, frames of reference, etc.—is a nontrivial design issue. In the `mechanics` domain, a body-centered inertial reference frame is assumed, together with coordinates that follow the formulation of classical mechanics (Goldstein, 1980), which assigns one coordinate to each degree of freedom, thereby allowing all equations to be written without vectors. The representations described in this chapter are designed to handle the coordinate issues associated with the remaining domains.

3.3 Choosing Domains

If a user wants to build a model of a system that does not fall in one of the existing domains, he or she can either build a new domain from scratch—a matter of making a list of components and connectors—or use one of the *meta-domains*: general frameworks that arrange hypotheses into candidate models by relying on modeling techniques that transcend individual application domains. The `xmission-line` meta-domain, for instance, generalizes the notion of building models using an iterative pattern, similar to a standard model of a transmission line, which is useful in modeling distributed parameter systems. The `linear-plus` meta-domain takes advantage of fundamental linear-systems properties that allow the linear and nonlinear components to be treated separately under certain circumstances, which dramatically reduces the model search space. Both can be used directly or customized for a specific application area.

Choosing a modeling domain for a given problem is not trivial, but it is not a difficult task for the practicing engineers who are the target audience for this work. Such a user would first look through the existing domains to see if one matched his or her problem. If none were appropriate, s/he would choose a meta-domain that matched the general properties of the modeling task. If there is a close match between the physical system's components and the model's components (i.e., it is a lumped parameter system), then `linear-plus` is appropriate; `xmission-line` is better suited to modeling distributed parameter systems. A meta-domain should be customized into a domain if:

- there exists specialized subject knowledge to assist in refining the search,
- it is to be used often, or
- to assist a novice user.

For example, the `electrical-xmission-line` domain is based on the xmission-line meta-domain. It inherits all of the meta-domain's structure, but add some specialized, built-in knowledge about transmission lines and electronics. The effect of this knowledge is to focus the search. A capacitor in parallel with two resistors, for instance, is equivalent to a single resistor in parallel with that capacitor. The `electrical-xmission-line` domain "knows" this, allowing it to avoid duplication of effort. There is significant overlap between the various domains and meta-domains; an electronic circuit can be modeled using the specific `electrical-xmission-line` domain, the `xmission-line` meta-domain, or even the `linear-plus` meta-domain. In all three cases, the model generator will eventually produce an equivalent model, but the amount of effort involved will be very different.

We chose this particular pair of meta-domains as a good initial set because they cover such a wide variety of engineering domains. We are exploring other possible meta-domains, especially for the purposes of modeling nonlinear networks. Implementation details are covered in (Easley, 2000); a modeling example that uses the `xmission-line` modeling domain appears in Section 7. Although the use, implementation, and search spaces of these meta-domains are somewhat different, their purpose is the same: to combine hypotheses into models.

4 Transmission Lines

The GPN component-based modeling representation presented in Section 2 is an excellent tool for modeling individual discrete physical components: single lumped elements with two terminals whose properties can be specified by a single state variable. In an electrical circuit, typical passive lumped elements are resistors, capacitors, inductors, etc. In mechanical systems, single GPN components describe forces like friction coefficients, stiffnesses, and masses.

However, not all physical components can be treated as lumped. In many physical systems (for example, electrical transmission lines or the cochlea of the inner ear) and physical processes (for example, thermal conduction in a rod and carrier motion in transistors), the physics cannot be described by a single variable. The process of thermal conduction in a rod, for example, is governed by heat flows that interact at a microscale throughout the volume of the rod. An accurate description of the "state" of the system, then, comprises an infinite number of variables: the temperature and thermal conduction at *every* point in the rod.

Such *distributed parameter* systems normally call for partial differential equation models, and the understanding and solution of partial differential equations is inherently far more difficult than that of ordinary differential equations with constant coefficients. A simplistic one-dimensional PDE has the form:

$$\frac{\delta u}{\delta x} + a\frac{\delta u}{\delta t} = bF(u)$$

where u is the unknown variable to be solved for (e.g., temperature or voltage), x is the dimension of interest (e.g., length down a rod or transmission line), t is time and a and b are constants. More realistic PDE for transmission lines will be of higher spatial dimension and non-linear. See Davis and Meindl (2000) for an example of modeling a distributed RLC circuit using PDEs.

However, one broadly applicable case where an ODE model does provide an accurate solution is the set of systems where a one-dimensional spatial variation assumption is valid. Under this assumption, the interesting physics of a system are confined to a finite number of physical dimensions of the system, allowing a set of ODEs to approximate a PDE. This is actually quite common in practice, as engineers expand PDEs in Fourier series and then truncate them into ODEs. For example, Edward Lorenz used analogous techniques to truncate the Navier-Stokes equations (a set of PDEs), to form the Lorenz equations (Lorenz, 1963) (a set of ODEs). The goal of the `xmission-line` meta-domain is to form a similar type of truncation by approximating a spatiotemporally distributed system using an iterative structure with a large number of identically structured lumped sections.

4.1 The Transmission-Line Meta-domain

The motivation for the `xmission-line` meta-domain was the basic engineering treatment of an electrical transmission line, wherein typical electrical parameters, such as resistance or inductance, are given in per-unit-length form. Consider, for example, the traditional network representation of a section of a power line, as shown in Figure 3. A normal model would be n such sections connected together in series. This incremental network approximation becomes more accurate as the number of sections in the model is increased. As the physical power line becomes longer, or as increased model accuracy is required, the model builder adds more elements.

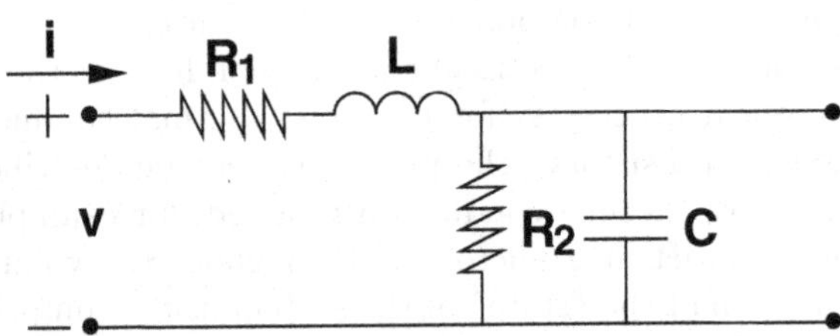

Fig. 3. The traditional incremental network model for an electrical transmission line

The transmission line example of Figure 3 may be abstracted into unspecified series and parallel impedances, as shown in Figure 4. This structure—a generalized iterative two-port network with n uniform sections, each of which has a series (A_i) and a parallel (B_i) element—is useful because different GPN components may be placed into this abstracted framework to create a large variety of transmission line-based modeling domains. In our implementation, each of these elements can contain one or more GPN components; they may also be "null" (essentially a short for the A_i and an open for the B_i).

Note that the topology of the sections is fixed in this metaphor: all the A_i contain the same network of GPN components, as do all the B_i; coefficient values within the individual elements can, of course, vary. If the user knows the internal structure of the elements *a priori*, s/he can specify a single option for each of the A_i and B_i in the `hypotheses` argument to the `find-model` call. For example, if an engineer were trying to model the traditional electrical transmission line using the incremental model of Figure 4, s/he would specify the A_i as an integrating and a proportional GPN in series, and the B_i as a differentiating and a proportional GPN in parallel. See Section 7 for an example.

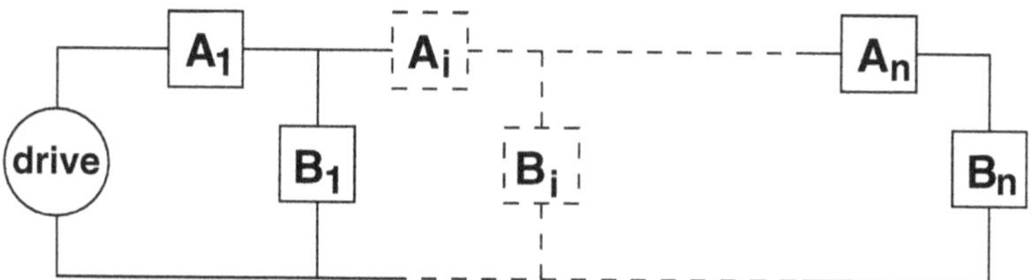

Fig. 4. The `xmission-line` meta-domain allows lumped-element GPN components to model spatially distributed systems like transmission lines, vibrating strings, and so on. The basic paradigm is an iterative structure with a variable number of sections, each of which has the same topology—a series element A_i and a parallel element B_i. The number of sections, each of which models a small piece of the continuum physics, rises with the precision of the model.

4.2 Refining a Meta-domain

Instead of using the meta-domain directly, in the manner described above, one can also refine it with subject matter knowledge to create a specific model-building domain. Building such a domain can be as simple as fixing the set of hypotheses and then renaming the meta-domain, which can be useful if one wants a modeling tool for naïve users. However, a domain can contain much more knowledge, such as which hypotheses are more likely to occur and in which combinations, or a specification of a set of data-analysis tools to help guide the model search process.

A model-building domain for use on viscoelastic problems, for instance, can easily be constructed via the `xmission-line` meta-domain by fixing the A_i as null and the B_i as an integrating and a proportional GPN—which correspond to a linear spring and a linear dashpot, respectively—connected in series. Viscoelastic data analysis tools can then be incorporated by a human expert into the model-building domain to reduce the model search by a factor of four (duplicating the functionality described in Capelo et al. (1998)). Specifying the A_i and B_i, as in these two examples, sets almost all of the structure of the model, and so the search space is small: $O(n)$, where n is the number of sections that are ultimately required to build an adequate model.

If the structure within a section is not known, the user would suggest several hypotheses; the `xmission-line` meta-domain would then try out various combinations of those components in the A_i and B_i, iterating each combination out to a predetermined depth.[3] Although this does give an exponential search space—$O(2^d n)$, if there are d possible hypotheses and n required sections—d is almost always very small in engineering practice: typically four or less. If the application really demands more than three or four components, it would be best to build a new application-specific domain, based on those components and

[3] This is currently set at five; we are investigating other values, as well as intelligent adaptation of that limit.

incorporating knowledge about how to efficiently combine them, as described in the beginning of Section 3.

The power that this knowledge representation framework brings to automated model generation is that it uses domain knowledge to tailor the search space to the problem at hand. Because resources for model testing are limited, this effectively expands the range of problems that an automated modeling tool can handle.

5 The Linear Plus Meta-domain

In many engineering fields, the word "system" implies linearity, and other types of systems must be prefaced by words such as nonlinear, dynamic, or chaotic to warn the reader that all their traditional education on "systems" is not going be valid. To the average practicing engineer, this is not a great concern; for decades, most engineering systems and techniques were thought to fall into (or at least be well approximated by) the linear realm. Even in many textbooks with chapters on nonlinear systems, the first and primary technique used to solve most nonlinear problems is to linearize them—either to restrict an operating regime so that the system appears to be linear, or to make multiple piecewise-linear models for various regimes.

There are a number of reasons engineers favor linear systems: operating behavior is predictable and "reasonable"; the number of devices/components involved is small; and the mathematical techniques and associated analysis tools are powerful. Unfortunately, the types of problems that can be treated with linear systems techniques are severely limited, as the real world is largely nonlinear. For this reason, the goal of this framework is to model all systems governed by ordinary differential equations, both linear *and* nonlinear. However, since the knowledge of linear systems is so pervasive in engineering practice, and because many of its tools are powerful enough to model many interesting systems, linear-system techniques can usefully serve as a basis for a meta-domain.

The `linear-plus` meta-domain, which instantiates standard linear systems theory, separates components into a linear and a nonlinear set, as shown in Figure 5, in order to exploit two fundamental properties of linear systems: (1) there are a polynomial number of unique nth-order linear ODEs (Brogan, 1991), and (2) linear system inputs (drive terms) appear verbatim in the resulting ODE model. The first of these properties effectively converts an exponential search space to polynomial; functionally equivalent linear networks reduce to the same Laplace transform transfer function, which allows a model generator to identify and rule out any ODEs that are equivalent to models that have already failed the test. The second property allows this meta-domain (and thus any specific domain constructed upon it) to handle a limited number of nonlinear terms (thus the name linear-*plus*) by treating them as system inputs. As long as the number of nonlinear hypotheses remains small, the search space of possible models remains tractable.

The linear part of `linear-plus` works via an analysis of canonical forms of linear systems. The fundamental idea behind this meta-domain is that many

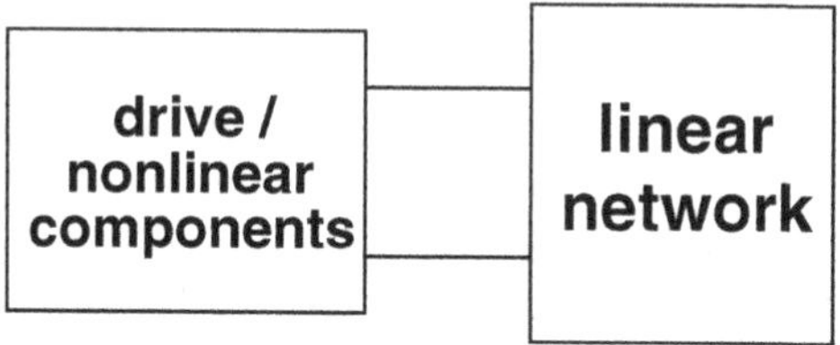

Fig. 5. PRET's **linear-plus** meta-domain is designed for systems that are basically linear, but that incorporate a few nonlinear terms and a few drive terms. Separating the linear and nonlinear/drive parts of the model reduces an exponential search space to polynomial and streamlines handling of drive terms.

different networks of linear components may be modeled by identical systems of ODEs. This allows the search space to be pruned effectively and correctly: if five network configurations all correspond to the same set of ODEs, then only one of those five should be passed from the model generator to the testing phase. Testing the other four would be a duplication of effort. In some circumstances, it is fairly simple to determine that two different networks have identical behavior and thus identical ODE models. For example, consider adding a resistor in parallel to any other resistor in the circuit. Engineers often do this in practice as they build a circuit in order to "tweak" it into some desired behavior, but it does not change the structure of the ODE model. The behavior of the circuit may change, but the circuit topology is effectively the same, which is what is important for system identification.

Delving deeper into linear systems provides a more systematic method to solve reduce the number of equivalent systems: the Laplace transform. This method converts linear ODEs to algebraic equations via a set of transform theorems; see any undergraduate linear systems textbook (e.g., Kuo (1995)) for a description of the method. Constant-coefficient linear systems may be described by a Laplace transform of the input-output transfer function, which is of the form:

$$\frac{Y(s)}{u(s)} = T(s) = \frac{\beta_m s^m + \beta_{m-1} s^{m-1} + \ldots + \beta_1 s + \beta_0}{s^n + a_{n-1} s^{n-1} + \ldots + a_1 s + a_0}, \tag{1}$$

where Y is the system output, u the input, T the transfer function, s the Laplace transform variable, and the a_is and β_is are constant coefficients. For any physical system, causality (that is, physical realizability) requires that $m \leq n$. Otherwise, the output $y(t_k)$ at time t_k would depend upon future inputs $u(t_j)$ at time t_j with $j > k$.

The primary disadvantage of the transfer-function approach is that it is limited to linear time-invariant systems only. While this analysis tool is powerful, it is not directly usable within our framework, as we have concentrated our work on time-based—not frequency-based—approaches. A more modern approach called the *state-variable format* is more compatible with the system identification approach used here. The state-variable format represents models via a state-transition matrix, and is much more general than the transfer function approach; it allows both linear and nonlinear systems—both time-invariant and time-varying systems—to be modeled in a unified manner.

There are four major categories of state-variable models, called *realizations*, each with its own particular set of advantages, disadvantages, and implications, as described in (Rowland, 1986). The realization chosen for this work is the cascade form, which is obtained by artificially splitting the transfer-function denominator polynomial into simpler functions of either first- or second-order factors, as shown in Figure 6. This is allowable because transfer functions of order three or higher may always be factored into single or quadratic terms (Brogan 1991). The advantage of splitting the denominator polynomial into smaller terms

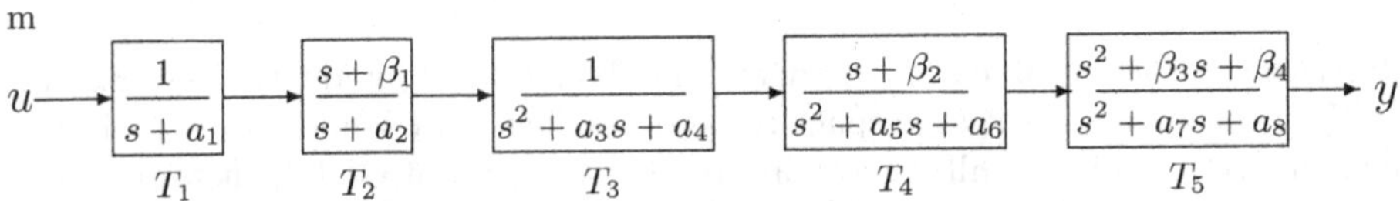

Fig. 6. The cascade realization of a Laplace transfer function. Each individual transfer function is either a simple first- or second-order factor, and the product ranges from $i = 1$ to 5 in equation 1.

is that doing so greatly reduces the number of possible models. Consider the denominator of Equation 1, which has $n + 1$ total terms. There are 2^n possible combinations of s terms, but the s^n is required, which leaves n terms to choose from. Taking all possible combinations of these s^i terms (where i varies from 0 to $n - 1$) yields an exponential number of possibilities.

The cascade realization reduces this number to polynomial. For example, a third-order polynomial only has two factorizations: one with three first-order T_i, and one with one first-order T_i and one second-order T_i. In general, the total number of factorizations for an nth order polynomial is $\lfloor n/2 \rfloor$.

for $a \leftarrow 0$ **to** $\lfloor n/2 \rfloor$

model $\leftarrow$ ($n - 2a$ first-order factors) + (a second-order factors)

Factoring the numerator is more complex: instead of two options (first-order and second-order factors, as with the denominator), there are three: zeroth-order, first-order, and second-order. Luckily, there is a linear-systems constraint that degree of the numerator polynomial must be less than or equal to that of the denominator. By choosing a set of five possible numerator-denominator combinations, the `linear-plus` meta-domain can generate all equivalent transfer functions of a given order with an $O(n)$ algorithm. See Easley (2000) for details.

Our GPN-based implementation of this meta-domain consists of a library of five two-port networks whose entries have the prototypical factorizations shown in Figure 6. The meta-domain algorithm cascades these two-port networks to give a transfer function with the desired numerator and denominator factorizations (as determined from the analysis listed above). After the `linear-plus` meta-domain generates the linear model using these transfer-function based techniques, it may add nonlinear terms by treating them as system inputs. Chua

(1969) shows in his monumental treatise on nonlinear network theory how nonlinear components may be treated as drive terms exciting a linear network. When the linear network is treated as a black box, nonlinear hypotheses such as $a \sin\theta(t)$ or drive terms such as $\beta \sin\alpha t$ appear verbatim in the resulting ODE. This approach is limited in that it does not attempt to model full networks of nonlinear components; as shown in Figure 5, nonlinear terms are external to the linear model.

This approach also loses some of the domain-specificity that a strictly component-based approach would maintain. With a state-variable approach, the direct relationship between a physical variable in a network (e.g. a voltage in a electrical circuit) and its representation in a model (e.g. v) may be lost in a set of state-variable equations. The only *component* piece that is maintained in the `linear-plus` meta-domain is the nonlinear components added to the state variable equations. Maintaining a closer linkage between physical entities and modeling representations is normally an advantage, as it allows for more domain-centered knowledge to be used in the model-building process. However, modelers in an engineering settings often switch between these state-variable and component based representations depending upon the type and complexity of the modeling task.

Canonical forms of linear systems are exactly the types of tools that an engineer would use to decompose a larger system into smaller, more-manageable sub-systems. This is important for automated model generation, as it makes the search space much more manageable as well. These canonical forms also increase the communicability of the models to practicing engineers as these are the types of models that they are trained to used.

6 Reasoning About Building Models

The technical challenge in the process of automated modeling is efficiency: the search space is huge, and so an automated modeling program must use analysis tools are possible to quickly and cheaply rule out candidate models. The PRET automated modeling tool makes a concerted effort to use domain-dependent and independent knowledge, whenever possible, in order to help guide the search. To better understand the utility of automated reasoning tools in the process, we first examine how PRET builds models.

Table 3 demonstrates how human users work with PRET to build models. Automated reasoning tools are used in all of PRET's steps. The analysis tools used for data collection and analysis are described briefly below and in more depth in Bradley and Easley (1998). Reasoning techniques used in the testing phases (qualitative and quantitative) are described elsewhere (Stolle & Bradley, this volume). Reasoning techniques used in model generation is covered in this section.

The meta-domain representation is an effective basis for dynamic modeling domains whose complexity naturally rises and falls according to the available information about the target system. The challenge in reasoning about models

Table 3. Automated modeling building steps in PRET. Automated reasoning tools are used by PRET in all of its steps.

Major Step	Subtasks	Actor
Collect Inputs	choose modeling domain	user
	select hypotheses	user (optional)
	describe observations	user (optional)
	set modeling specifications	user
	collects time-series data	user or PRET
	analyze time-series data	PRET
Generate	build model from given hypotheses	PRET
Test	qualiative testing	PRET
	quantitative testing	PRET

in the context of meta-domains is to tailor the reasoning to the knowledge level in such a way as to prune the search space to the minimum. Organizing domains into a hierarchy of generality—as was shown in Figure 2—is not enough; what is needed is a hierarchical set of analysis tools for reasoning about models and systems, as well as a means for assessing the situation and choosing which tool is appropriate.

Focused, appropriate analysis is critical to the efficiency of any automated model building process. As demonstrated in (Stolle & Bradley, this volume), invalid models can often be ruled out using purely qualitative information, rather than expensive point-by-point numerical comparisons. One challenge in designing the meta-domain modeling paradigm was to come up with a framework that supported this kind of reasoning. The key idea is that different analysis techniques are appropriate in different domains, and our solution combines a structured hierarchy of analysis tools, part of which is shown in Table 4, with a scheme that lets the component type and domain knowledge dictate which tools to use. See Easley (2000) for details of these tools.

The use and organization of these analysis tools may be customized to specific applications, just as they are in engineering disciplines. *Delay-coordinate embedding* is applicable in any discipline and lets one infer the dimension and topology of the internal system dynamics from a time series measured by a *single* output sensor. On the other hand, analysis tools for a viscoelastic system—e.g., creep testing—are highly domain-specific. Moreover, since the inputs, outputs, and reasoning processes used by these analysis tools are similar to those employed by expert engineers, PRET's results are naturally communicable. For example, engineers characterize system behaviors in the same type of language that *cell dynamics* techniques generate. A simple undamped pendulum oscillates in a *limit cycle*—a property that either a human engineer or an automated model-building tool may use to reason about a model, or to communicate with one another.

Table 4. Component type and domain knowledge dictate what analysis tools should be used to build and test models. Tools higher in this table are more general but their results can be less powerful.

System Type	Analysis Tools
	Cell dynamics (Hsu, 1987)
Nonlinear	Delay-coordinate embedding (Sauer et al., 1991)
	Nonlinear time series analysis (Bradley, 2000; Strogatz, 1994)
Linear	Linear system analysis (Reid, 1983)
Restricted linear	Domain dependent (e.g., (Capelo et al., 1998))

Reasoning about model-building proceeds in the obvious manner, given this hierarchy: if no domain knowledge about the target system is available (i.e., the true "black box" situation), then models are constructed using general reasoning techniques and analysis tools that apply to *all* ODEs—those in the top line of Table 4. This highly general approach is computationally expensive but universally applicable. If the system is known to be linear, the extensive and powerful repertoire of linear analysis tools developed over the last several decades makes the model building and testing tasks far less imposing. Moreover, system inputs (drive terms) in linear systems appear verbatim in the resulting system ODE, which makes input/output analysis much easier. In more-restricted domains, analysis tools are even more specific and powerful. In viscoelastics, for example, three qualitative properties of a "strain test" reduce the search space of possible models to linear (Capelo et al., 1998).

Given all of this representation and reasoning machinery, PRET's model-generation phase builds models as follows. First, a candidate model is constructed using basic GPN components and connectors, which are either supplied by the user or built into a particular domain or meta-domain, as described in Section 3.2. This process is guided by the analysis tools in Table 4. If the system is nonlinear, for example, the cohort of nonlinear tools is applied to the sensor data to determine the dimension d of the dynamics; this fact allows the generation phase of an automated system identification tool to automatically disregard all models of order $< d$. Other nonlinear analysis techniques yield similar search-space reductions. If the system is linear, many more tools apply; these tools are cheaper and more powerful than the nonlinear tools, and so the hierarchy guides the testing phase to use the former before the latter. Knowledge that the target system is oscillating, for example, not only constrains any autonomous linear model to be of least second order, but also implies some constraints on its coefficients; this reasoning is purely symbolic and hence very inexpensive as well as broadly powerful.

The generate and test phases are interleaved: if the generate phase's first-cut search space does not contain a model that matches the observed system behavior, the modeling domain dynamically expands to include more-esoteric components. For example, as described in Section 3, the `electrical-xmission-line` domain begins with the components {`linear-resistor`, `linear-capacitor`}. If all models in this search space are rejected, the model generator automatically expands the domain to include the component type `linear-inductor`. The intuition captured by this notion of "layered" domains is that inductors are much less common in practical engineering systems than capacitors. Expanding domains beyond linear components is more difficult, since the number of possible component types increases dramatically and any ordering scheme necessarily becomes somewhat *ad hoc.*

In many engineering domains, however, there exist well-defined categorization schemes that help codify this procedure. Tribology texts, for example,[4] specify different kinds of friction for different kinds of ball bearings, as well as some notions about which of those are common and should be tried first, and which are rare (Halling, 1978). The meta-domain paradigm combined with GPN-modeling lets an expert user—a "domain builder"—encode this kind of information quickly and easily, and allows any reasoning tool that uses that meta-domain to exploit that knowledge.

After the model generator creates a GPN network using the relevant domain knowledge and the model tester successfully verifies it against the observations of the target system, the final step in the model-building process is to convert that GPN into ODE format, as described in Section 2.3.

7 Modeling Water Resource Systems

Water resource systems are made up of streams, dams, reservoirs, wells, aquifers, etc. In order to design, build, and/or manage these systems, engineers must model the relationships between the inputs (e.g., rainfall), the state variables (e.g., reservoir levels), and the outputs (e.g., the flow to some farmer's irrigation ditch). To do this in a truly accurate fashion requires partial differential equations (PDEs) because the physics of fluids involves multiple independent variables—not just time—and an infinite number of state variables. PDEs are extremely hard to work with, however, so the state of the art in the water resource engineering field falls far short of that. Most existing water resource applications, such as river-dam or well-water management systems, use rule-based or statistical models.

ODE models, which capture the dynamics more accurately than these simple models but are not as difficult to handle as PDEs, are a good compromise between these two extremes, and the water resource community has begun to take this approach (Bredehoeft et al., 1966; Chin, 2000). In this section, we show how meta-domains allow the PRET automatic modeling tool to duplicate some of these research results and model the effects of sinusoidal pressure fluctuation in an aquifer on the level of water in a well that penetrates that aquifer, as shown in Figure 7. This example is a particularly good demonstration of how domain knowledge and the structure inherited from the meta-domain let the model generator build systems without creating an overwhelming number of models.

---

[4] Tribology is the science of surface contact.

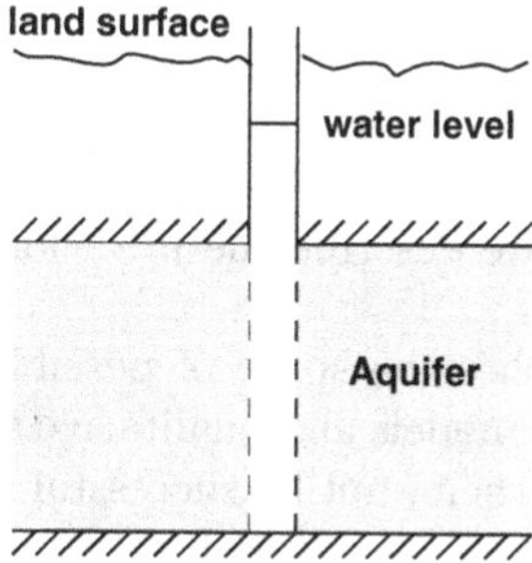

Fig. 7. An idealized representation of an open well penetrating an artesian aquifer. The motion of the water level in the well is controlled by sinusoidal fluctuations of the pressure in the aquifer.

This example also demonstrates how meta-domains allow automated modeling tools to generate a models in the language of engineering practitioners. PRET's use of the meta-domain and GPN constructs makes the resulting models—and the associated reasoning concepts and implications—immediately recognizable to most engineers.

```
(find-model
  (domain xmission-line)
  (state-variables (<well-flow> <flow>)) ...
  (hypotheses
    (<effort> (* c (integral <flow>)))
    (<effort> (* l (deriv <flow>)))
    (<effort> (* r <flow>)))
  (drive (<effort> (* da (sin (* df <time>)))))
  (observations
    (numeric (<time> <well-flow> (deriv <well-flow>))
             ((0 4.0 0.5) (0.1 4.25 0.6) ...)))
  (specification
    (<well-flow> absolute-resolution 0.1 (-infinity infinity))) )
```

Fig. 8. A `find-model` call fragment for the well/aquifer example of Figure 7. See text of this section for an explanation of the predictes.

The user describes the modeling problem to PRET via a `find-model` call, a fragment of which is shown in Figure 8. The first line specifies the meta-domain (`xmission-line`) and instantiates the relevant domain theory: in this case the template of an iterative network structure. The next line informs PRET of the relationship between the domain specific (`<well-flow>`) and generalized (`<flow>`) state variables. The following lines of the call specify some suggested model fragments—`hypotheses` and `drives`. These describe different possible aspects of the physics involved in the system (in this case, linear capacitance, inertia, and resistive relationships, and a sinusoidal drive term that represents a periodic fluctuations within the aquifer). `observations` are facts about the system and may be measured automatically via sensors and/or interpreted by the user. In this call, the `observations` contain a time-series measurement of well-flow. `specifications` are describe the precision at which the model must match the given observations.

PRET automatically searches the space of possible models, using the `xmission-line` template to build models and qualitative and quantitative techniques (Bradley et al., 2001) to test them, until a successful model is found. The testing procedure and associated issues are covered in (Stolle & Bradley, this volume). The result is shown in Figure 9. A perfect model of an infinite-dimensional

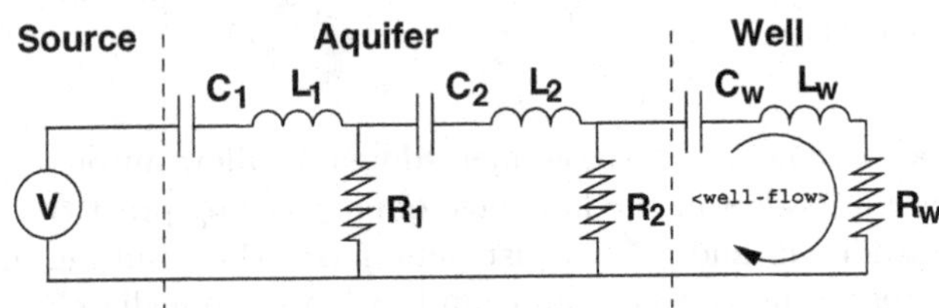

Fig. 9. PRET's model of the well/aquifer system. The drive, $V = d_a \sin d_f t$, simulates a sinusoidal pressure fluctuation in the aquifer; the arrow labeled `<well-flow>` in the network corresponds to the water flow into and out of the well. Note how the `xmission-line` meta-domain naturally incorporates the load (the well) into the model.

system, of course, requires an infinite number of discrete sections, but one can construct approximations using only a few sections; the fidelity of the match rises with the number of sections. In this case, the meta-domain based model-building framework used two sections to model the aquifer and one to model the well.

Using a meta-domain in an automated model-builder is advantageous as it avoids duplication of effort. Connecting arbitrary components in parallel and series creates an exponential number of models, many of which are mathematically equivalent (cf., Thévenin and Norton equivalents, in network theory). Here, the meta-domain contains the knowledge about how to compose the components into a series of viable networks. Note also how the meta-domain uses domain-independent terms such as `<effort>` and `<flow>`, rather than domain-specific ones like `<force>`, or `<water-flow>`; this allows the meta-domain to be applied to a variety of model-building fields with little or no customization. Using GPN components also facilitates a simple and natural expression of the drive term, which is attached directly to the iterative part of the network: its effects automatically become part of the model, just as they do in real systems. Finally, the `xmission-line` is actually more general than the example shown here; it has met with success in modeling problems from a variety of application areas, from electrical to viscoelastic systems (Easley, 2000).

Assessing the success of a model is a difficult question in any field. PRET takes a strict engineering approach to this question. Its goal is to find an ODE system that matches the given observations to within the user-specified precision. It is, of course, possible to supply several sets of observations to PRET at the same time, in which case it will try to match them all.

8 Related Work

In the qualitative reasoning/physics research that is most closely related to PRET's domains and meta-domains, ODE models are built by evaluating time-series data using qualitative-reasoning techniques, and then using parameter estimation to match the resulting model with a given observed system (Capelo et al., 1998). That approach differs from the techniques presented in this paper in that it selects models from a set of pre-enumerated solutions in a very specific domain (linear viscoelastics). The machine-discovery community has also considered representations for automated modeling. LAGRAMGE (Todorovski & Džeroski, 1997), for instance, is similar to PRET in that it builds ODE models of dynamic systems, but it uses a grammar-based representation to compose models, rather than a GPN-based representation. The LAGRAMGE authors have extended their system to model population dynamics (Todorovski, this volume); the associated formalism encodes high-level domain knowledge (e.g., about population growth and decay rates) into context-free grammars, which the LAGRAMGE program then uses to automate equation discovery.

Other formalisms and techniques have been applied to these problems as well. Koza (1992) has been successful in automating the construction of network-based models of time-series data in a variety of domains, ranging from electrical circuits and to chemical reactions. Their approach involves establishing a set of components and connection for the application problem, simulating the behavior of an initial network of those components, defining a fitness measure between the simulated network and the observed behavior, and applying genetic programmming to "breed" a population of networks with higher fitness measures. ODEs are the best way to capture the behavior of systems that are nonlinear and dynamic, however, so PRET relies on them and does not use more-general techniques like genetic programming, neural nets, regression rules, etc. Like the computational discovery work of Schabacher et al. (this volume) and Saito & Langley (this volume), PRET makes direct contact with the applicable domain theory, and leverages that information in the model-building process. Again, the theory and methods of the two approaches are different; PRET's domain is the general mathematics of ODEs rather than the specifics of biological processes. Many of the research issues are similar, though: how best to combine concrete data and abstract models, how to communicate the results effectively to domain experts, etc. Like Garret et al. (this volume), PRET relies on mathematical logic to capture domain knowledge in a declarative form. Like Washio & Motoda (this volume), PRET clearly separates domain-specific facts and general knowledge, making the priorities and connections explicit, and expressing each in a manner that is appropriate for their use. Amsterdam's automated model construction tool Amsterdam (1992) uses a similar underlying component representation (bond graphs) and is applicable to multiple physical domains. However, it is also somewhat limited; it can only model linear systems of order two or less. The meta-domain framework described in this paper is much more general; it works on linear *and nonlinear* continuous-time ODEs in a variety of domains, and it uses *dynamic* model generation. A great deal of other automated modeling work exists, but space precludes a lengthy discussion; see Forbus (1997) for a review.

9 Conclusion

Model-building domains and meta-domains, coupled with generalized physical networks and a hierarchy of qualitative and quantitative reasoning tools that relate observed physical behavior and model form, provide the flexibility required for gray-box modeling of nonlinear dynamic systems. The meta-domains introduced in this chapter, `linear-plus` and `xmission-line`, use modeling techniques that transcend individual application fields to create rich and yet tractable model search spaces. This framework is flexible as well as powerful; one can use meta-domains directly or customize them to fit a specific engineering application, and they are designed to adapt smoothly to varying amounts and levels of domain knowledge. This approach allows the majority of the reasoning involved in the modeling-building process to proceed at a highly abstract level; only at the end of the process must the component-based representation be converted into ODE form. This type of reasoning, wherein the modeling tool applies the correct knowledge at the proper time and in the proper manner, accurately reflects the abstraction levels and reasoning processes used effectively by human engineers during the modeling procedure.

Any tool that effectively automates a coherent and useful part of the modeling art is of obvious practical importance in science and engineering: as a corroborator of existing models and designs, as a medium within which to instruct newcomers, and as an intelligent assistant. Furthermore, automated modeling tools are becoming more and more important in the context of building flexible autonomous devices which often must have accurate models of their *own* dynamics. Using existing engineering formalisms in automated modeling tools is exceedingly important, as human engineers will have to work smoothly with automated tools in order to design, build, and test the new kinds of artifacts demanded by modern engineering applications. Since the domain and meta-domain paradigms described in this chapter use the same type of techniques and formalisms as practicing engineers, the models that they produce are easily communicated to their target audience.

Our current research efforts are focused more on using PRET as a tool rather than improving its capabilities and algorithms. One focus of this work is to use PRET to deduce information about paleoclimate dynamics from radioisotope dating data. A number of other interesting future work projects, from automated modeling of hybrid systems to closer integration of the parameter estimation process with model generation process, are described in (Easley, 2000).

Acknowledgments: Apollo Hogan, Joe Iwanski, Brian LaMacchia, and Reinhard Stolle also contributed code and/or ideas to this project.

References

Addanki, S., Cremonini, R., Penberthy, J.S.: Graphs of models. Artificial Intelligence 51, 145–178 (1991)

Amsterdam, J.: Automated qualitative modeling of dynamic physical systems. Tech. Report 1412. MIT Artificial Intelligence Laboratory, Cambridge (1992)

Bradley, E.: Time-series analysis. In: Berthold, M., Hand, D. (eds.) Intelligent data analysis: An introduction, Springer, Heidelberg (2000)

Bradley, E., Easley, M.: Reasoning about sensor data for automated system identification. Intelligent Data Analysis 2, 123–138 (1998)

Bradley, E., Easley, M., Stolle, R.: Reasoning about nonlinear system identification. Artificial Intelligence 133, 139–188 (2001)

Bradley, E., Stolle, R.: Automatic construction of accurate models of physical systems. Annals of Mathematics and Artificial Intelligence 17, 1–28 (1996)

Bredehoeft, J., Cooper, H., Papadopulos, I.: Inertial and storage effects in well-aquifer systems. Water Resource Research 2, 697–707 (1966)

Brogan, W.: Modern control theory, 3rd edn. Prentice-Hall, Englewood, NJ (1991)

Capelo, A.C., Ironi, L., Tentoni, S.: Automated mathematical modeling from experimental data: An application to material science. IEEE Transactions on Systems, Man and Cybernetics - Part C 28, 356–370 (1998)

Chin, D.: Water-resource engineering. Prentice Hall, Englewood, NJ (2000)

Chittaro, L., Tasso, C., Toppano, E.: Putting functional knowledge on firmer ground. Applied Artificial Intelligence 8, 239–258 (1994)

Chua, L.O.: Introduction to nonlinear network theory. McGraw-Hill, New York (1969)

Davis, J., Meindl, J.: Compact distributed rlc interconnect models. IEEE Transactions on Electronic Devices 47, 2078–2087 (2000)

de Vries, T.J., Breedveld, P., Meindertsma, P.: Polymorphic modelling of engineering systems. In: Proceedings of the First International Conference on Bond Graph Modeling, La Jolla, CA, pp. 17–22 (1993)

Easley, M.: Automating input-output modeling of dynamic physical systems. Doctoral dissertation, University of Colorado at Boulder (2000)

Easley, M., Bradley, E.: Generalized physical networks for automated model building. In: Proceedings of the Sixteenth International Joint Conference on Artificial Intelligence, Stockholm, Sweden, pp. 1047–1052 (1999)

Falkenhainer, B., Forbus, K.D.: Compositional modeling: Finding the right model for the job. Artificial Intelligence 51, 95–143 (1991)

Forbus, K.D.: Qualitative reasoning. In: Tucker, J.A. (ed.) CRC computer science and engineering handbook, CRC Press, Boca Raton, FL (1997)

Goldstein, H.: Classical mechanics. Addison-Wesley, Reading, MA (1980)

Halling, J. (ed.): Principles of tribology. MacMillan, Hampshire, UK (1978)

Hsu, C.S.: Cell-to-cell mapping. Springer, Heidelberg (1987)

Karnopp, D.C., Margolis, D., Rosenberg, R.C.: System dynamics: A unified approach, 2nd edn. John Wiley, New York (1990)

Koza, J.R.: Genetic programming: on the programming of computers by means of natural selection. MIT Press, Cambridge, MA (1992)
Kuo, B.C.: Automatic control systems, 7th edn. Prentice-Hall, Englewood, NJ (1995)
Lorenz, E.N.: Deterministic nonperiodic flow. Journal of the Atmospheric Sciences 20, 130–141 (1963)
Mosterman, P.J., Biswas, G.: A formal hybrid modeling scheme for handling discontinuities in physical system models. In: Proceedings of the Thirteenth National Conference on Artificial Intelligence, Portland, OR, pp. 985–990 (1996)
Paynter, H.M.: Analysis and design of engineering systems. MIT Press, Cambridge, MA (1961)
Reid, J.: Linear system fundamentals. McGraw-Hill, New York (1983)
Rowland, J.: Linear control systems. John Wiley & Sons, New York (1986)
Sanford, R.S.: Physical networks. Prentice Hall, Englewood, NJ (1965)
Sauer, T., Yorke, J.A., Casdagli, M.: Embedology. Journal of Statistical Physics 65, 579–616 (1991)
Strogatz, S.H.: Nonlinear dynamics and chaos. Addison-Wesley, Reading, MA (1994)
Thévenin, L.: Extension de la loi d'Ohm aux circuits électromoteurs complexes [Extensions of Ohm's law to complex electromotive circuits]. Annales Télégraphiques 97, 159–161 (1883)
Todorovski, L., Džeroski, S.: Declarative bias in equation discovery. In: Proceedings of the Fourteenth International Conference on Machine Learning, Nashville, TN, pp. 376–384 (1997)
Top, J., Akkermans, L.: Tasks and ontologies in engineering modelling. International Journal of Human-Computer Studies 41, 585–617 (1994)
Weld, D.S., de Kleer, J. (eds.): Readings in qualitative reasoning about physical systems. Morgan Kaufmann, San Mateo, CA (1990)

Integrating Domain Knowledge in Equation Discovery

Ljupčo Todorovski and Sašo Džeroski

Department of Knowledge Technologies
Jožef Stefan Institute, Ljubljana, Slovenia
Ljupco.Todorovski@ijs.si, Saso.Dzeroski@ijs.si

Abstract. In this chapter, we focus on the equation discovery task, i.e., the task of inducing models based on algebraic and ordinary differential equations from measured and observed data. We propose a methodology for integrating domain knowledge in the process of equation discovery. The proposed methodology transforms the available domain knowledge to a grammar specifying the space of candidate equation-based models. We show here how various aspects of knowledge about modeling dynamic systems in a particular domain of interest can be transformed to grammars. Thereafter, the equation discovery method LAGRAMGE can search through the space of models specified by the grammar and find ones that fit measured data well. We illustrate the utility of the proposed methodology on three modeling tasks from the domain of Environmental sciences. All three tasks involve establishing models of real-world systems from noisy measurement data.

1 Introduction

Scientists and engineers build mathematical models to analyze and better understand the behavior of real-world systems. Establishing an acceptable model for an observed system is a very difficult task that occupies a major portion of the work of the mathematical modeler. It involves observations and measurements of the system behavior under various conditions, selecting a set of variables that are important for modeling, and formulating the model itself. This chapter addresses the task of automated modeling, i.e., the task of formulating a model from the observed behavior of the selected system variables.

There are at least three properties of mathematical models that make them an omnipresent analytic tool. The first is the *data integration* aspect — mathematical models are able to integrate large collections of observations and measurements into a single entity. Second, models allow for simulation and *prediction* of the future behavior of the observed system under varying conditions. Finally, mathematical models can provide *explanations*, i.e., reveal the structure of processes and phenomena that govern the behavior of the observed system.

Although the aforementioned properties of models are equally important to scientists and engineers, most of the current automated modeling approaches based on machine learning and data mining methods are mainly concerned with

S. Džeroski and L. Todorovski (Eds.): Computational Discovery, LNAI 4660, pp. 69–97, 2007.

the first two: data integration and accurate prediction of future behavior. In this chapter, we focus on the third, explanatory aspect of induced models. We address two issues that are related to the explanatory aspect. The first is the issue of modeling notations and formalisms used by scientists and engineers. While machine learning and data mining methods focus on inducing models based on notations introduced by researchers in these areas (such as decision trees and Bayesian networks), we induce models based on algebraic and differential equations, i.e., standard notations, already established in many scientific and engineering domains. The second is the issue of integrating knowledge from the specific domain of interest in the induction process. This is in analogy with the manual modeling process — to obtain an explanatory *white-box* model of an observed system, scientists and engineers rely on and make use of this knowledge.

We present a methodology for integrating domain-specific knowledge in the process of model induction. The methodology allows integration of several different types of knowledge. The first type includes knowledge about basic processes that govern the behavior of systems in the particular domain and the equation templates that are typically used by domain experts for modeling these processes. The second type includes knowledge about models already established and used in the domain. The method can take into account a fully specified initial model, or a partially specified model of the observed system. We relate the integration of knowledge in the induction process to the notion of inductive language bias, which defines the space of candidate model structures considered by the induction method. We show that the three different types of domain-specific knowledge mentioned above can be transformed to inductive language bias, i.e., a specification of the space of models. We illustrate the utility of the presented methodology on three practical tasks of establishing models of real-world systems from measurement data in the domain of Environmental sciences.

The chapter is organized as follows. In Section 2, we start with a discussion of the relation between domain-specific knowledge and language bias in induction methods. In the following Section 3, we introduce the equation discovery task, i.e., the task of inducing equation-based models and propose methods for specifying language bias for equation discovery. Section 4 illustrates how to use the knowledge about basic processes that govern the behavior of population dynamics systems for establishing explanatory models thereof. In Section 5, we focus on the integration of (full or partial) initial model specifications in the model induction process. Section 6 discusses related research. Finally, in Section 7 we summarize and propose directions for further research.

2 Domain-Specific Knowledge and Language Bias

Many studies of machine learning methods and especially their applications to real-world problems show the importance of background knowledge for the quality of the induced models. Pazzani and Kibler (1992) show that the use of domain-specific knowledge improves the predictive performance of induced models on test examples unseen in the induction phase. Domain knowledge is also important for the acceptance of the induced concepts by human experts. As Pazzani et al. (2001) show, the consistency of the induced concepts with existing knowledge in the domain of interest is an important factor that influences the opinion of human experts about the credibility of the concepts and their acceptability for further use. This is especially true in complex scientific and engineering domains, where a vast amount of knowledge is being systematically collected and well documented.

Although these results are well known, most state-of-the-art machine learning methods do not allow for *explicit* integration of domain knowledge in the induction process. Knowledge is usually *implicitly* involved in the phases that precede or follow the induction process, that is the data preparation or preprocessing phase, when the set of variables important for modeling of the observed phenomena are identified, or in the model interpretation phase. A notable exception are learning methods developed within the area of inductive logic programming (ILP) (Lavrač & Džeroski, 1994). The notion of (background) knowledge and its integration in the induction process is made explicit there and background knowledge is a part of the learning task specification. ILP methods deal with the induction of first-order logic programs (or theories) from examples and domain knowledge. Background knowledge predicates allow user to integrate knowledge from the domain under study in ILP. They define concepts (or building blocks) that may be used in the induced theories. Note, however, that background knowledge predicates specify building blocks only and do not specify how to combine them into proper programs or theories.

Another way to integrate domain knowledge in the induction process is through the use of inductive bias, which refers to any kind of mechanism or preference used by the induction algorithm to choose among candidate hypotheses (Utgoff, 1986; Mitchell, 1991). The preference usually goes beyond the mere consistency (or degree-of-fit) of the hypothesis with the training examples and determines in which region of the candidate hypotheses space we are more likely to find a solution. Thus, the term bias is related to domain knowledge, since knowledge can help constrain the space of candidate hypotheses. Nédellec et al. (1996) define three types of inductive bias. The first type, language bias, is used to specify the space of hypotheses considered in the induction process. In our case, language bias would specify the space of candidate equation-based models. Note that language bias does not only specify a set of building blocks for establishing models (as background knowledge predicates do), but it also specifies recipes for combining them into proper candidate models (or hypotheses). The second type is search bias, which specifies the order in which the hypotheses are considered during the induction process. The third type is validation bias that specifies the acceptance or stopping criteria for the induction process.

Depending on the kind of formalism available for specifying the bias for an induction method, it may be non-declarative (built-in), parametrized, or declarative (Nédellec et al., 1996). Methods for inducing decision trees (Quinlan, 1993) are typical examples of methods with non-declarative or parametrized language bias, since they explore the fixed hypothesis space of decision trees using variables from the given data set. Parametrized bias lets the user influence the set of candidate hypotheses by setting some of its parameters, such as the depth (or complexity) of the induced decision trees. On the other hand, declarative language bias provides the user with complete control over the space of candidate hypotheses. Thus, declarative language bias provides a powerful mechanism for integrating domain knowledge in the process of induction.

Different formalisms can be used for specifying the space of candidate hypotheses. Nédellec et al. (1996) provide an overview of declarative bias formalisms used in ILP. Note, however, that these formalisms are developed for concepts and hypotheses expressed in first-order logic and are not directly applicable to the task of building equation-based models. In the next section, we propose a declarative bias formalism for equation discovery that will allow for integrating different aspects of domain-specific modeling knowledge in the process of inducing models from data.

3 Equation Discovery and Language Bias

Before we propose an appropriate declarative bias formalism for equation-based models, we define the task of inducing equations from data. The task is usually referred to as the task of equation discovery. Equation discovery is the area of machine learning that develops methods for automated discovery of quantitative laws, expressed in the form of equations, in collections of measured data (Langley et al., 1987; Langley & Żytkow, 1989; Džeroski & Todorovski, 1995; Washio & Motoda, 1997). In the rest of this section, we then provide an overview of typical language biases used in existing equation discovery methods, propose a declarative bias formalism for equation discovery, and provide an example bias based on domain knowledge for modeling population dynamics.

3.1 Task Definition

The equation discovery task can be formalized as follow. Given:

- a set of (numeric) variables V that are important for modeling the observed system, where $V = S \cup E$, $S = \{s_1, s_2, \ldots s_n\}$ is a set of *system* (or dependent) variables that we want to explain (or be able to predict) with the induced model, while $E = \{e_1, e_2, \ldots e_m\}$ is a set of *exogenous* (or input/output) variables that are not explained by the induced model, but may be used in the model equations;
- a designated time variable t, in cases when we deal with a task of modeling a dynamic system;
- one or more tables with measured/observed values of the variables from V (in consecutive time points or independent experiments),

find a set of n equations, one for each variable in S, that can explain and accurately predict the values of the system variables and take one of the two forms:

- differential $\frac{d}{dt} s_i = f_i(s_1, s_2, \ldots s_n, e_1, e_2, \ldots e_m)$ or
- algebraic $s_i = f_i(s_1, s_2, \ldots s_{i-1}, s_{i+1}, \ldots s_n, e_1, e_2, \ldots e_m)$.

Table 1. Two example context free grammars that specify distinct classes of arithmetical expressions. The one on the left-hand side is the "universal" grammar that specifies the space of all arithmetical expressions that can be built using the four basic arithmetical operators (and parentheses). The grammar on the right-hand side specifies the space of polynomials.

$E \to E + F \mid E - F$ $F \to F * T \mid F/T$ $T \to (E) \mid V \mid const$	$P \to P + const * T \mid const * T$ $T \to T * V \mid V$

3.2 A Language Bias Formalism for Equation Discovery

Once provided with the task definition, we can identify the language bias of an equation discovery method as the space or class of candidate arithmetical expressions f_i that can appear on the right-hand sides of the induced model equations. Thus, the declarative language bias formalism for equation discovery should allow us to specify classes of arithmetical expressions. We argue here that context free grammars (Hopcroft & Ullman, 1979) provide an ideal framework for this purpose. Chomsky (1956) introduced context free grammars as formal models of natural or synthetic languages and since then they have became a standard formalism for specifying the syntax of programming languages.

Table 1 provides two example grammars that specify two distinct classes of arithmetical expressions. The grammar on the left-hand side of the table specifies the space of all arithmetical expressions that can be built using the four basic arithmetical operators and parentheses. Formally, a context free grammar consists of a finite set of variables, usually referred to as *nonterminals* or syntactic categories, each of them representing a subclass of subexpressions or phrases in the language represented by the grammar.

In the particular example, the nonterminals E, F, T, and V refer to arithmetical expressions, factors, terms, and variables, respectively. The class of factors F represents arithmetical expressions that are built using multiplication and division only[1]. Terms represent atomic arithmetical expressions that are further combined into factors. A term T can be either (1) an arbitrary expression E in parentheses, (2) a variable V, or (3) a constant parameter. We specify these three classes of terms using three grammar *productions* (also named rewrite rules). The first rule is formalized as $T \to (E)$, the second as $T \to V$, and the third as $T \to const$. In short, this is written as $T \to (E) \mid V \mid const$, where the symbol "|" separates alternative productions for the same nonterminal.

The symbol *const* is a *terminal* grammar symbol that represents a constant parameter in the arithmetical expression with an arbitrary value, which is usually fitted against training data in the process of equation discovery. The terminals are primitive grammar symbols that can not be further rewritten, i.e., no

[1] Note that the organization of the grammar follows the relative priority of the basic arithmetic operators—multiplication and division are always performed before addition and subtraction.

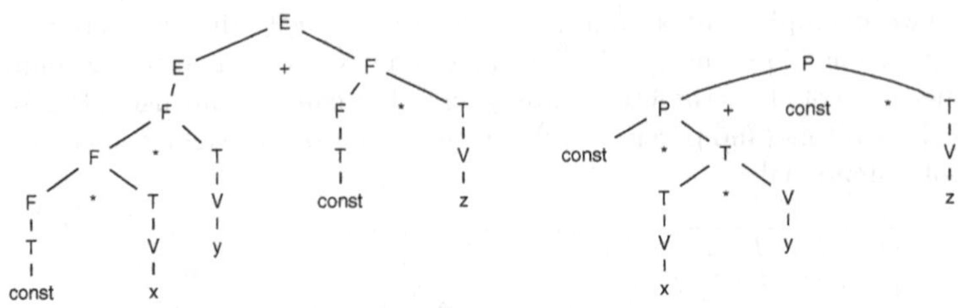

Fig. 1. Two parse trees that derive the same polynomial expression $const * x * y + const * z$. The one on the left-hand side corresponds to the universal grammar, while the parse tree on the right-hand side corresponds to the polynomial grammar from Table 1.

productions are affiliated with them. Other terminals in the example grammar are "+", "-", "*", "/", "(", and ")".

Similarly, the grammar on the right-hand side of Table 1 specifies the space of polynomials over the variables in V. Productions for the nonterminal P specify that each polynomial is a linear combination of one or more terms T multiplied by constants. Each term, in turn, is a multiplication of one or more variables from V. Note that the productions for the nonterminal V depend on the particular equation discovery task at hand. Typically, a single production $V \rightarrow v$ will be added to the grammar for each observed/measured variable v. For example, if we deal with an equation discovery task where we measure three variables x, y, and z, the corresponding productions will be $V \rightarrow x \mid y \mid z$.

Given a context free grammar, we can check whether a certain expression (string) belongs to the language defined by the grammar (parse task) or generate expressions that belong to the language (generate task). For both purposes, we use the notion of a parse tree, which describes the way a certain expression can be derived using the grammar productions. Figure 1 presents two example parse trees that derive the same (polynomial) expression $const * x * y + const * z$. The left-hand tree presents the derivation of this expression using the universal grammar, while the one on the right-hand side of the figure presents its derivation using the polynomial grammar. Each internal node in a parse tree corresponds to a nonterminal symbol N from the grammar. Its children nodes $n_1, n_2, \ldots n_k$, ordered from left to right, always correspond to a grammar production $N \rightarrow n_1\ n_2 \ldots n_k$. The leaf nodes of a parse tree are terminal symbols: reading them from left to right, we get the expression derived with the parse tree.

Most equation discovery methods would adopt one of the language biases based on the grammars from Table 1. While the BACON (Langley et al., 1987) and SDS (Washio & Motoda, 1997) methods for equation discovery can induce equations based on arbitrary arithmetical expressions (i.e., use the universal grammar as a built-in bias), FAHRENHEIT (Zembowicz & Żytkow, 1992) and LAGRANGE (Džeroski & Todorovski, 1995) limit their scope to inducing polynomial equations. Note that the language biases of the aforementioned equation discovery methods are usually non-declarative and often take the form of a relatively small (pre-defined or built-in) class of possible equations. The user is allowed to influence the built-in bias using a number of parameters. LAGRANGE would let user specify the maximal degree of polynomials or maximal number of multiplicative terms in the polynomial (Džeroski & Todorovski, 1995). Additionally, Zembowicz and Żytkow (1992) let the user specify functions that can be used to transform the variables before they are incorporated in the polynomials.

Another type of language constraints used for equation discovery is based on information about the measurement units of the observed system variables. COPER (Kokar, 1986) uses dimensional analysis theory (Giordano et al., 1997) in order to constrain the space of equations to those that properly combine variables and terms taking care about the compatibility of their measurement units. The SDS method (Washio & Motoda, 1997) extends this approach to cases in which the exact measurement units of the system variables are not known. In such cases, SDS employs knowledge about the type of the measurement scale for each system variable, which is combined with knowledge from measurement theory to constrain the space of possible equations.

Note, however, that knowledge about measurement units or the scale types thereof is domain independent. Experts from a specific domain of interest can usually provide much more modeling knowledge about the system at hand than merely enumerating the measurement units of the system variables. Many textbooks on mathematical modeling give comprehensive overviews of the modeling knowledge for specific domains, such as biology (Murray, 1993) or biochemistry (Voit, 2000). In the next section, we will show how this kind of knowledge can be transformed to a grammar that provides an appropriate knowledge-based bias for equation discovery in the domain of population dynamics.

3.3 An Example from Modeling Population Dynamics

The domain of population dynamics falls within the field of population ecology, which studies the structure and dynamics of populations. A population is a group of individuals of the same species that inhabit a common environment. More specifically, we consider modeling the dynamics of populations, especially how their concentrations change through time (Murray, 1993). Here, we focus on population dynamics in aquatic systems, where we are mainly concerned with interactions between dissolved inorganic nutrients (e.g., nitrogen and phosphorus), primary producers (or plants, such as phytoplankton and algae), and secondary producers (or animals, such as zooplankton and fish).

Models of population dynamics are based on the mass conservation principle, where influences of different interactions and processes on a single variable are summed up. Following that principle, the change of a primary producer population can be modeled as:

$$\frac{dPP}{dt} = growth(PP) - grazing(SP, PP) - mortality(PP), \tag{1}$$

where the *growth* and *mortality* are expressions modeling the influence of growth and mortality processes on the concentration of the primary producer *PP*, while

Table 2. An example language bias (context free grammar) for equation discovery based on knowledge about modeling population dynamics in aquatic ecosystems

```
double monod(double v, double c) return v / (v + c);
```

$$PPChange \rightarrow PPGrowth - Grazing - PPMortality$$

$$PPGrowth \rightarrow const * V_{PP} \mid const * Limitation * V_{PP}$$
$$Limitation \rightarrow V_{Nutrient} \mid \mathtt{monod}(V_{Nutrient}, const)$$

$$Grazing \rightarrow const * V_{PP} * V_{SP}$$

$$PPMortality \rightarrow const * V_{PP}$$

$$V_{PP} \rightarrow phytoplankton$$
$$V_{SP} \rightarrow zooplankton$$
$$V_{Nutrient} \rightarrow phosphorus \mid nitrogen$$

the *grazing* expression models the influence of secondary producer (zooplankton) grazing on *PP*. Note that the growth process positively influences the change of primary producer concentration, while mortality and grazing influence it negatively. Murray (1993) enumerates different expressions typically used by ecologists to model these processes. The grammar presented in Table 2 captures this kind of knowledge about modeling population dynamics.

The first production in the grammar follows the mass conservation equation template (1) and introduces three nonterminals *PPGrowth*, *Grazing*, and *PPMortality*, one for each of the processes of growth, grazing, and mortality. Productions for these three nonterminals, further in the grammar, define alternative models for these processes. The productions for the growth process define unlimited and limited growth. The first model assumes unlimited exponential growth of the primary producer population in the absence of interactions with secondary producers. The second model corresponds to the more realistic situation where the growth of the primary producer population is limited by the supply of nutrients. The nonterminal *Limitation* defines two alternative limitation terms. The first assumes unlimited capacity of the primary producer for nutrient absorption. The second (Monod) term is used when the absorption capacity of the primary producer saturates to a certain value, no matter how plentiful the nutrient supply is. Ecologists use different expressions for modeling saturation of the absorption capacity — one possibility is Monod (also known as Michaelis-Menten's) term defined as:

$$\mathtt{monod}(V_{Nutrient}, const) = \frac{V_{Nutrient}}{V_{Nutrient} + const},$$

where the constant parameter is inversely proportional to the saturation rate. The graphs in Figure 2 depict the difference between the unsaturated and saturated model of the nutrient absorption capacity.

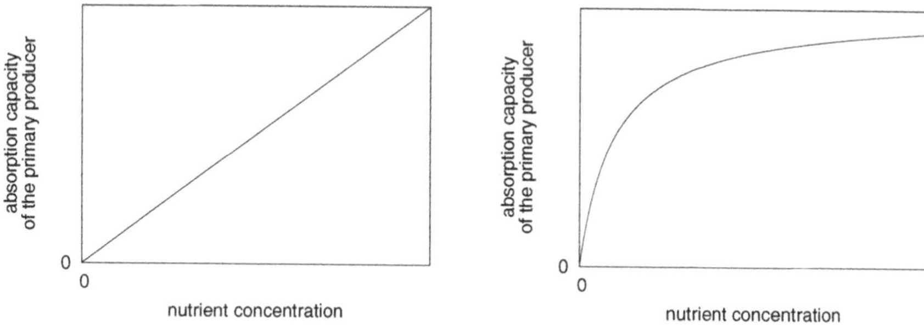

Fig. 2. A comparison of two alternative terms used to model the limitation of primary producer growth due to limited supply of nutrients. The one on the left-hand side assumes unlimited nutrient absorption capacity, while the model on the right-hand side corresponds to situations where the absorption capacity saturates to a certain value.

The remaining productions in the grammar provide models for the other two processes of grazing and mortality. Note that the grammar includes a single model for each process. One can add modeling alternatives for these two processes by adding alternative productions to the appropriate nonterminal symbol.

3.4 LAGRAMGE

Once provided with domain knowledge encoded in the form of a grammar and measurement data about the variables of the observed system, we can employ the LAGRAMGE induction method to perform search through the space of candidate equations defined by the grammar. LAGRAMGE follows a beam strategy to search through the space of parse trees generated by the grammar in the order from simplest (shallowest) trees to more complex (deeper) ones (Todorovski & Džeroski, 1997). In addition to the grammar that constrains the space of candidate models, the user can specify a maximal depth of the parse tree and further limit the complexity of the induced equations.

Each parse tree generated by the grammar is evaluated with respect to its fit to the provided measurement data. To this end, LAGRAMGE first fits the values of the constant parameters against data using a nonlinear least-squares algorithm (Bunch et al., 1993) that carries out second-order gradient descent through the parameter space. To avoid entrapment in local minima, the procedure involves multiple restarts with different initial parameter values. Once LAGRAMGE finds the optimal parameter values, it measures the discrepancy between the observed values of the system variables and the values predicted by the equation based on the parse tree and the optimal parameter values. Note that prediction in the case of dynamic systems requires simulating the model over the full time span. To improve the convergence rate of the gradient search, we employ *teacher forcing* simulation (Williams & Zipser, 1989), which finds parameters that best predict observations at the next time point based solely on those at the present time point. The discrepancy is measured as mean squared error (MSE) and this measure is used as a heuristic function that guides the search. An alternative heuristic function MDL (which stands for minimal description length) takes into account the complexity of the parse tree and introduces a preference towards simpler models (Todorovski & Džeroski, 1997). At the end of the beam search procedure, LAGRAMGE returns a user specified number of models with the lowest values of the heuristic function selected by the user (among the two choices — MSE and MDL).

Todorovski and Džeroski (1997) show that LAGRAMGE is capable of integrating many different types of domain knowledge through grammars. For example, knowledge about the measurement units of system variables has been used to build a grammar for modeling a mechanical pole on cart system. In another example, knowledge about the basic processes that govern population dynamics has been used for automated modeling of phytoplankton growth in Lake Glumsø in Denmark from a sparse and noisy set of real-world measurements (Todorovski et al., 1998). A drawback of LAGRAMGE is that it is difficult for domain experts to express or encode their knowledge in the form of a grammar. Furthermore knowledge-based grammars such as the one from Table 2 are usually task specific, i.e., a grammar built for modeling one system (e.g., phytoplankton growth in Lake Glumsø) cannot be reused for modeling other systems from the same domain (e.g., population dynamics in another lake with a slightly different set of observed variables). These limitations can be addressed by the transformational approach, which is the topic we focus on in the rest of this chapter. In the next two sections, we show that different types of higher level domain knowledge encoded in various formalisms can be easily transformed to grammars and thus integrated in the process of equation discovery.

4 A Language Bias Formulation of Process-Based Modeling Knowledge

This section presents a flexible and high-level formalism for encoding domain knowledge that is more accessible to domain experts than the grammar-based formalism discussed above. The formalism organizes knowledge in a taxonomy of process classes, each of which represents an important class of basic processes that influence the behavior of entities in the domain of study. For each process class, a number of alternative equation models, usually used by modeling experts in the domain, can be specified. The formalism also encodes knowledge on how to combine the models of individual basic processes into a single model of the whole observed system. We illustrate the use of the formalism by encoding knowledge on modeling population dynamics.

We designed the formalism for knowledge representation so that the resulting library of knowledge for a particular domain is independent of the modeling task at hand. Provided with such a library, the user only needs to specify the modeling task by stating the types of the observed variables along with a list of processes that are most likely to influence the system behavior. While the library of domain-specific knowledge should be provided by a modeling expert with extensive modeling experience, the task specification can written by a naïve user who is familiar with the domain but does not have much modeling expertise.

4.1 Process-Based Knowledge for Modeling Population Dynamics

In this section, we reconsider the population dynamics domain introduced in Section 3.3. As we already stated there, population dynamics studies the structure and dynamics of species sharing a common environment and is mainly concerned with the study of interactions between different species. While Section 3.3 focuses on the absorption of inorganic nutrients by the primary producer species (plants) and its effect on the change of primary producer concentration only, we extend our focus here to other types of interactions and their effect on the concentrations of all the species involved.

Research in the area of population dynamics modeling was pioneered by Lotka and Volterra and their well-known Volterra-Lotka model of predator-prey interaction (Murray, 1993), which can be schematized as:

$$\dot{N} = \mathit{growth\_rate}(N) - \mathit{feeds\_on}(P, N)$$
$$\dot{P} = \mathit{feeds\_on}(P, N) - \mathit{decay\_rate}(P),$$

where P and N denote the concentrations of predator and prey, respectively. The model combines the influences of three fundamental processes: prey growth, predation, and predator decay. Each of the processes is modeled following one of the following three assumptions. The first assumption is that the growth rate of the prey population in the absence of predation is proportional to its density, i.e., $\mathit{growth\_rate}(N) = aN$. This means that the growth of the population is exponential and unlimited, which is unrealistic in many cases. Natural environments often have a limited carrying capacity for the species. In such cases, one can use the alternative logistic growth model $\mathit{growth\_rate}(N) = aN(1 - N/K)$, where the constant parameter K determines the carrying capacity of the environment for species N.

The second assumption made in the simple Volterra-Lotka model is that the predation rate is proportional to the densities of both the predator and the prey populations, i.e., $\mathit{feeds\_on}(P, N) = bPN$. In analogy with growth, this means that the predation capacity grows exponentially and is unlimited. Again, in some cases the predators have limited predation capacity. When the prey population density is small the predation rate is proportional to it, but when the prey population becomes abundant, the predation capacity saturates to a certain limit value. Several different terms can be used to model the predator saturation response to the increase of prey density (Murray, 1993):

$$(a)\ P\frac{N}{N+B}; \quad (b)\ P\frac{N^2}{N^2+B}; \quad (c)\ P(1 - e^{-BN}),$$

where B is the constant that determines the saturation rate.

Table 3. A taxonomy of process classes encoding knowledge about alternative models of the predation process

```
process class Feeds_on(Population p, Populations ns)
   condition p ∉ ns
   expression p * ∏n∈ns Saturation(p, n)

process class Saturation(Population p, Population n)
process class Type_0 is Saturation
   expression n
process class Type_A is Saturation
   expression n / (n + const(saturation_rate,0,Inf))
process class Type_B is Saturation
   expression n * n / (n * n + const(saturation_rate,0,Inf))
process class Type_C is Saturation
   expression 1 - exp(const(saturation_rate,0,Inf) * n)
```

Table 4. A combining scheme specifies how to combine the models of individual population dynamics processes into a model of the entire system

```
combining scheme Population_dynamics(Population p)
   ṗ = + Growth(p) - Decay(p)
      + ∑food const(_,0,Inf) * Feeds_on(p, food)
      - ∑predator const(_,0,Inf) * Feeds_on(predator, p)
```

The third assumption is that the predator decay rate is proportional to its concentration, which leads to exponential decay in the absence of interactions with other species. Relaxing the three assumptions made in the simple Volterra-Lotka model, we can build different, more complex and more realistic models of predator-prey population dynamics.

The modeling knowledge about the different modeling alternatives for the predation process described above can be encoded as shown in Table 3. The `Feeds_on` process class refers to predation processes, while `Saturation` refers to different saturation models. The first line specifies that each predation process relates a single predator population `p` with one or more prey populations `ns`. The next line specifies the constraint that `p` can not predate on itself (i.e., specifying non-cannibalistic behavior). The last line of the `Feeds_on` process declaration specifies the expression template that is used to model the influence of the predation process. The expression defines this influence as a product of predator concentration `p` and the influences of saturation processes for each prey population `n` $\in$ `ns`. Similarly, the saturation process specifies alternative expressions for modeling the limited (or unlimited) predation capacity of `p` on `n`. Each modeling alternative is encoded as a subclass of the `Saturation` process class.

The second part of the process-based knowledge encodes a recipe for combining the models of individual processes into a model of the entire observed system. The combining scheme presented in Table 4 provides such a recipe for

Table 5. A formal specification of predator-prey interaction between two species. Given this specification and the library of knowledge presented in Tables 3 and 4, one can reconstruct all candidate models of predator-prey interactions between two species.

```
system variable Population predator, prey

process Growth(prey)
process Feeds_on(predator, prey)
process Decay(predator)
```

building models of population dynamics. The recipe is based on the "*conservation of mass*" principle, explained earlier in Section 3.3. It specifies that the change of a population concentration is positively influenced by the process of population growth `Growth(p)` processes and negatively influenced by population decay `Decay(p)`. Similarly, `Feeds_on` processes can positively or negatively influence the change of `p`, depending on the role of the species in the predator-prey interaction. The predation processes that involve `p` as a predator positively influence the change of `p`, while the processes where `p` is involved as a prey negatively influence the change of `p`.

Given this kind of knowledge encoded in a library, one can specify the simple Volterra-Lotka model as illustrated in Table 5. The task specification contains declarations of the two variables involved along with the set of three processes of growth, decay, and predation. An automated modeling system, as the one presented in (Todorovski, 2003), can transform this specification to the candidate model equations. Note that this abstract specification accounts for more than one model, since there are many alternative models of predator saturation that can be used for the `Feeds_on` process. In such cases, the system would transform the specification to a grammar enumerating all possible models of predator-prey interaction between two species. We will not provide further details about the transformation method, that can be found in (Todorovski, 2003), but rather illustrate its use on a practical task of modeling population dynamics.

4.2 Modeling Algal Growth in the Lagoon of Venice

The Lagoon of Venice measures 550 km^2, but is very shallow, with an average depth of less than 1 meter. It is heavily influenced by anthropogenic inflow of nutrients – 7 mio kg/year of nitrogen and 1.4 mio kg/year of phosphorus (Bendoricchio et al., 1994). These (mainly nitrogen) loads are above the Lagoon's admissible trophic limit and generate its dystrophic behavior, which is characterized by excessive growth of algae, mainly Ulva rigida. Coffaro et al. (1993) present four sets of measured data available for modeling the growth of algae in the Lagoon. The data was sampled weekly for slightly more than one year at four different locations in the Lagoon. Location 0 was sampled in 1985/86, locations 1, 2, and 3 in 1990/91. The sampled quantities are nitrogen in ammonia NH_3, nitrogen in nitrate NO_3, phosphorus in orthophosphate PO_4 (all in [μg/l]),

Table 6. A task specification used for modeling algae growth in the Lagoon of Venice

```
exogenous variable Inorganic temp, DO, NH3, NO3, PO4
system variable Population biomass

process Growth(biomass) biomass_growth
process Decay(biomass) biomass_decay
process Feeds_on(biomass, *) biomass_grazing
```

dissolved oxygen *DO* (in percentage of saturation), temperature T ([degrees C]), and algal biomass B (dry weight in [g/m^2]).

In a previous study, Kompare and Džeroski (1995) apply the equation discovery method GoldHorn (Križman, 1998) to the task of modeling algal growth in the Lagoon of Venice. Since GoldHorn could not find an accurate model based on the set of measured variables, two additional variables were calculated and added to this set — growth and mortality rates, which are known quantities in ecological modeling and were calculated according to the simplified version of an existing model of algal growth in the lagoon proposed by Coffaro et al. (1993). From the extended set of system variables and data measured at Location 0, GoldHorn discovered a difference equation for predicting biomass that, due to the large measurement errors (estimated at the level of 20-50%), does not fit the data perfectly, but still predicts most of the peaks and crashes of the biomass concentration correctly (Kompare & Džeroski, 1995). Although the equation model involves the mortality rate, as calculated by domain experts, the model itself is still a black-box model that does not reveal the limiting factors for the biomass growth in the lagoon.

The task of modeling algal growth in the Lagoon of Venice from Table 6 specifies the types of the observed system variables and the processes that are important for the growth of the biomass (algae) in the lagoon. Note that the specification of the `biomass_grazing` process leaves the nutrient parameter of the `Feeds_on` process class unspecified (denoted using the symbol `*`). Since ecologists did not know the limiting factors for the biomass growth, they let Lagramge search for the model that would reveal them. The search space consists of 6248 candidate models with generic constant parameters.

The model induced from Location 0 data reveals that the limiting factors for biomass growth in the lagoon are dissolved oxygen (`DO`) and nitrogen in nitrate (`NO3`). The model induced from Location 2 data reveals that the limiting factors for the biomass growth are temperature (`temp`), dissolved oxygen (`DO`), and nitrogen in ammonia (`NH3`). Although the two models are not completely consistent, they both identify dissolved oxygen and nitrogen based nutrients to be limiting factors for biomass growth. The differences between the two models may be due to the fact that the measurements were taken during two different time periods.

In the experiments with the data measured at the other two locations (1 and 3), Lagramge did not find an accurate model of biomass growth. Note that these results still compare favorably with the results obtained by GoldHorn, which discovered an acceptable model for Location 0 only.

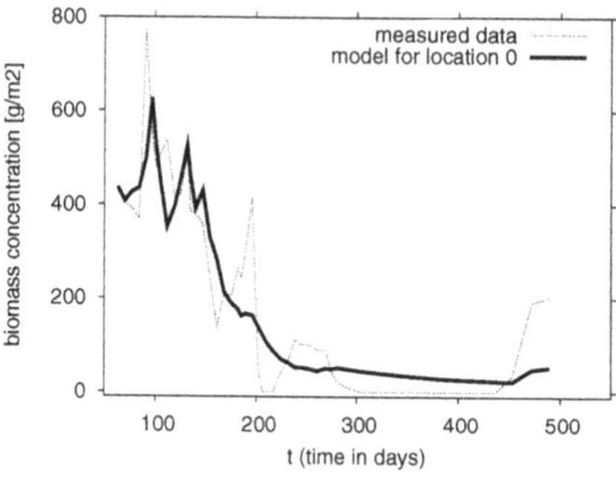

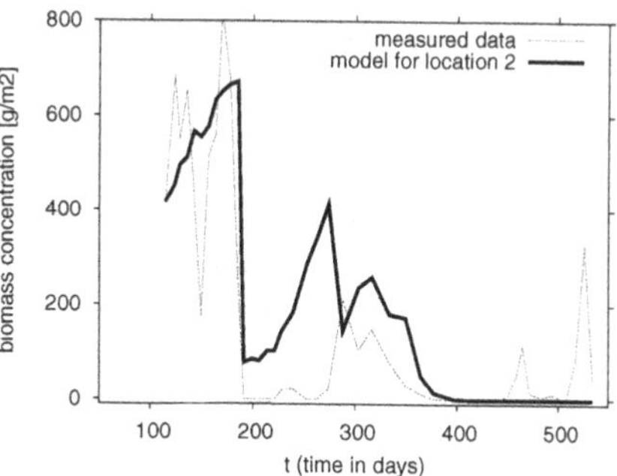

Fig. 3. Simulations of the two models of biomass growth in the Lagoon of Venice, discovered by LAGRAMGE, compared to the measured biomass concentration (left-hand side: Location 0, right-hand side: Location 2)

Figure 3 compares the measured and simulated values of the biomass for both models. We ran long-term simulations of the models from the initial value of the biomass without restarting the simulation process at each measurement point. For values of all other system variables needed during the simulation, we used the measurement at the nearest time point in the past. As in the GOLDHORN experiments, due to the high measurement errors of the order 20-50%, the models discovered by LAGRAMGE did not fit the measured data perfectly. However, they correctly predict most of the peaks and crashes of the biomass concentration. These events are more important to ecologists than the degree of fit. Note however another important advantage of these models over the one discovered by GOLDHORN. While the GOLDHORN model is black-box, the models discovered by LAGRAMGE identify the most influential limiting factors for biomass growth in the Lagoon of Venice.

5 Language Bias Based on Existing Models

Another type of domain-specific knowledge that is typically neglected by equation discovery methods is contained in the existing models already established in the domain of interest. Rather than starting the search with an existing model, current equation discovery methods start their search from scratch. In contrast, theory revision methods (Ourston & Mooney, 1994; Wrobel, 1996) start with an existing theory and use heuristic search to revise it in order to improve its fit to data. However, research on theory revision has been mainly concerned with the revision of models expressed in propositional or first-order logic. Therefore, existing methods for theory revision are not directly applicable to the task of revising models based on equations.

In this section, we present a flexible, grammar-based methodology for revising equation-based models. To support the revision of existing models, we first transform the given model into an initial grammar that can be used to derive the given model only. The nonterminals in the grammar and their productions reflect the structure of the initial model. Next, we extend the initial grammar with alternative productions that specify the possible modeling alternatives. The modeling alternatives can be specified by a domain expert or can be determined from modeling knowledge about the domain at hand. The extended grammar built in this manner specifies the space of possible revisions of the initial model. In the last step, we employ LAGRAMGE to search through the space of possible revisions and find those that fit the data better than the initial model.

5.1 Revising the CASA Model

The CASA model, developed by Potter and Klooster (1997) at NASA Ames, accounts for the global production and absorption of biogenic trace gases in the Earth atmosphere. It also allows us to predict changes in the geographic patterns of major vegetation types (e.g., grasslands, forest, tundra, and desert) on the land. CASA predicts annual global fluxes in trace gas production as a function of surface temperature, moisture levels, soil properties, and global satellite observations of the land surface. It operates on gridded input at different levels of resolution, but typical usage involves grid cells that are eight kilometers square, which matches the resolution for satellite observations of the land surface.

The overall CASA model is quite complex, involving many variables and equations. We decided to focus on one portion that is concerned with modeling net production of carbon by terrestial plants (NPPc). Table 7 presents the NPPc portion of the CASA model. The model predicts the NPPc quantity as the product of two unobservable variables, the photosynthetic efficiency, E, at a site (grid cell) and the solar energy intercepted, $IPAR$, at that site.

Photosynthetic efficiency is in turn calculated as the product of the maximum efficiency (0.56) and three stress factors that reduce this efficiency. The stress term *T2* takes into account the difference between the optimum temperature, *topt*, and actual temperature, *tempc*, for a site. The stress factor *T1* involves the nearness of *topt* to a global optimum for all sites. The third term, W, represents stress that results from lack of moisture as reflected by *eet*, the estimated water loss due to evaporation and transpiration, and PET, the water loss due to these processes given an unlimited water supply. In turn, PET is defined in terms of the annual heat index, *ahi*, for a site, and *pet_tw_m*, a modifier on PET to account for day length at differing locations and times of year.

The energy intercepted from the sun, $IPAR$, is computed as the product of *FPAR_FAS*, the fraction of energy absorbed through photo-synthesis for a given vegetation type, *monthly_solar*, the average radiation for a given month, and *SOL_CONV*, the number of days in that month. *FPAR_FAS* is a function of *fas_ndvi*, which indicates overall greenness at a site as observed from space, and *srdiff*, an intrinsic property that takes on different numeric values for different vegetation types.

Of the variables we have mentioned, *NPPc*, *tempc*, *ahi*, *monthly_solar*, *SOL_CONV*, and *fas_ndvi*, are observable. Two additional terms, *eet* and *pet_tw_m*, are defined elsewhere in the model, but we assume their definitions are correct and thus we can treat them as observables. The remaining variables are unobservable

Table 7. The NPPc portion of the CASA model that accounts for the net production of carbon by terrestial plants

$$
\begin{array}{l}
NPPc = \max(0, E \cdot IPAR) \\
\quad E = 0.56 \cdot T1 \cdot T2 \cdot W \\
\quad\quad T1 = 0.8 + 0.02 \cdot topt - 0.0005 \cdot topt^2 \\
\quad\quad T2 = 1.1814/((1 + \exp(0.2 \cdot (TDIFF - 10))) \cdot (1 + \exp(0.3 \cdot (-TDIFF - 10)))) \\
\quad\quad\quad TDIFF = topt - tempc \\
\quad\quad W = 0.5 + 0.5 \cdot eet/PET \\
\quad\quad\quad PET = 1.6 \cdot (10 \cdot \max(tempc, 0)/ahi)^A \cdot pet\_tw\_m \\
\quad\quad\quad\quad A = 0.000000675 \cdot ahi^3 - 0.0000771 \cdot ahi^2 + 0.01792 \cdot ahi + 0.49239 \\
\quad IPAR = FPAR\_FAS \cdot monthly\_solar \cdot SOL\_CONV \cdot 0.5 \\
\quad\quad FPAR\_FAS = \min((SR\_FAS - 1.08)/srdiff, 0.95) \\
\quad\quad\quad SR\_FAS = (1 + fas\_ndvi/1000)/(1 - fas\_ndvi/1000) \\
\quad\quad SOL\_CONV = 0.0864 \cdot days\_per\_month
\end{array}
$$

and must be computed from the others using their definitions. This portion of the model also contains a number of numeric parameters, as shown in Table 7.

The ecologists that developed the CASA model pointed out which parts of the initial CASA-NPPc model are likely candidates for revision. Their confidence in the equations used to calculate the values of the four intermediate variables E, $T1$, $T2$, and SR_FAS is low, so they considered them to be "weak" parts of the model. Thus, we focused our revision efforts to the equations corresponding to these four intermediate variables. The data set for revising the CASA-NPPc model contains measurements of the observed variables at 303 sites. We assess the quality of the revised models using root mean squared error evaluated on training data and cross-validated. The RMSE of the initial model on the training data set is 465.213.

Grammars Used for Revision. A set of equations defining a target variable through a number of intermediate variables can easily be turned into a grammar. The staring nonterminal symbol represents the dependent variable *NPPc*, while other nonterminals correspond to the intermediate variables that appear on the left-hand sides of the equations in the initial model. Each nonterminal has a single production that is obtained by replacing the equality ("=") sign in the equation with "→". In such a grammar, the terminals denote observed variables, the model's constant parameters, and the arithmetical operators involved in the equations. Such a grammar generates a single model which is exactly the initial CASA-NPPc model from Table 7.

The grammar obtained from the initial model lets us specify an arbitrary number of alternative models for any intermediate variable by adding productions to the corresponding nonterminal. These additional productions specify alternative modeling choices, only one of which will eventually be chosen to revise the initial model. In general, there are two classes of alternative productions. The productions from the first class replace one or more constant parameter values with a generic constant parameter *const*. In our experiments with the CASA-NPPc

Table 8. Grammar productions specifying modeling alternatives that can be used to revise the initial CASA-NPPc model

`Ec-100`	$E \rightarrow const[0:1.12] \cdot T1 \cdot T2 \cdot W$
`Es-exp`	$E \rightarrow const[0:1.12] \cdot T1^{const[0:]} \cdot T2^{const[0:]} \cdot W^{const[0:]}$
`T1c-100`	$T1 \rightarrow const[0:1.6] + const[0:0.04] \cdot topt$ $- const[0:0.001] \cdot topt^2$
`T1s-poly`	$T1 \rightarrow const \mid const + (T1) * topt$
`T2c-100`	$T2 \rightarrow const[0:2.3628]/((1+\exp(const[0:0.4]$ $\cdot(TDIFF - const[0:20]))) \cdot (1+\exp(const[0:0.6]$ $\cdot(-TDIFF - const[0:20]))))$
`T2s-poly`	$T2 \rightarrow const \mid const + (T2) \cdot TDIFF$
`SR_FASc-25`	$SR\_FAS \rightarrow (1 + fas\_ndvi/const[750:1250])$ $/(1 - fas\_ndvi/const[750:1250])$

model, we use alternative productions that allow for a 100% relative change of the initial value of a constant parameter. This can be specified by replacing the fixed value constant parameter v with a terminal symbol $const[0 : 2 \cdot v]$. Thus, the lower bound for the newly introduced constant parameter is set to $v - 100\% \cdot v = 0$, while the upper bound is set to $v + 100\% \cdot v = 2 \cdot v$. Productions in the second class are slightly more complex and allow for *structural* revision of the model.

As we pointed out above, the focus of the model revision is on the equations corresponding to the four intermediate variables E, $T1$, $T2$, and SR_FAS. Table 8 presents the modeling alternatives we used to revise the equations for these four variables.

Alternative productions for E

`Ec-100` allows refitting the value of the constant parameter in the equation for E.

`Es-exp` enables structural revision of the E equation. It replaces the initial $T1 \cdot T2 \cdot W$ product with an expression that allows for arbitrary exponents of the three multiplied variables (i.e., $T1^{e_1} \cdot T2^{e_2} \cdot W^{e_3}$). The obtained values of the exponents in the revised model would then correspond to the relative magnitude of the influences of $T1$, $T2$, and W on E.

Alternative productions for T1

`T1c-100` allows refitting the values of the constant parameters in the *T1* equation.

`T1s-poly` replaces the initial second degree polynomial for calculating *T1* with an arbitrary degree polynomial of the same variable *topt*.

Alternative productions for T2

`T2c-100` enables 100% relative change of the constant parameter values in the *T2* equation.

`T2s-poly` replaces the initial equation for *T2* with an arbitrary degree polynomial of the variable *TDIFF*.

Table 9. The root squared mean error (RMSE) of the revised models and the percentage of relative error reduction (RER) over the initial CASA-NPPc model. We evaluated RMSE of the revisions on training data (training - the left-hand side of the table) and using 30-fold cross-validation (the right-hand side of the table).

	training		cross-validation	
alternative production(s)	RMSE	RER(%)	RMSE	RER(%)
`Ec-100`	458.626	1.42	460.500	1.01
`Es-exp`	443.029	4.77	443.032	4.77
`T1c-100`	458.301	1.49	460.799	0.95
`T1s-poly`	450.265	3.21	457.370	1.69
`T2c-100`	457.018	1.76	459.633	1.20
`T2s-poly`	450.972	3.06	461.642	0.77
`SR_FASc-25`	453.157	2.59	455.281	2.13
`ALL`	414.739	10.85	423.684	8.93

Alternative productions for SR_FAS

`SR_FASc-25` allows for a 25% relative change of the constant parameter values in the *SR_FAS* equation. We used 25% instead of usual 100% to avoid values of the constant parameters below 750, which would cause singularity (division by zero) problems when applying the equation.

Note that we can add an arbitrary combination of these alternative productions to the initial grammar. If all of them are added at the same time, then LAGRAMGE will search for the most beneficial combination of revisions. In the latter case, LAGRAMGE searches the space of 384 possible revisions of the original CASA-NPPc model (note that we limit the parse tree depth so that the maximal degree of the polynomials corresponding to $T1$ and $T2$ is five).

Results. Table 9 summarizes the results of the revision. When we allow only a single of the seven alternatives, revising the structure of the E equation leads to the largest (almost 5%) reduction of the initial model error. However, we obtain the largest error reduction of almost 11% on the training data and 9% when cross-validated with a combination of revision alternatives. The optimal combination of revisions is `Es-exp`, `T1c-100`, `T2s-poly`, and `SR_FASc-25` , which leads to the model presented in Table 10. It is worth noticing that the reductions obtained with single alternative productions nearly add up to the error reduction obtained with the optimal combination of revisions.

The most surprising revision result proposes that the water stress factor W does not influence the photosynthetic efficiency E, i.e., $E = 0.402 \cdot T1^{0.624} \cdot T2^{0.215} \cdot W^0$. Ecologists that developed the CASA model suggested that this surprising result might be due to the fact that the influence of the water stress factor on E is already being captured by the satellite measurements of the relative greenness, *fas_ndvi*, and thus, W becomes obsolete in the E equation.

Furthermore, the revised model replaces the initial second-degree polynomial for calculating *T1* with a linear equation. The structural revision `T2s-poly` replaced the complex initial equation structure for calculating *T2* with a relatively

Table 10. The best revision of the initial CASA-NPPc model. The parts that were left intact in the revision process are printed in gray.

$$NPPc = \max(0, E \cdot IPAR)$$
$$E = 0.402 \cdot T1^{0.624} \cdot T2^{0.215} \cdot W^0$$
$$T1 = 0.680 + 0.270 \cdot topt - 0 \cdot topt^2$$
$$T2 = 0.162 + 0.0122 \cdot TDIFF + 0.0206 \cdot TDIFF^2 - 0.000416 \cdot TDIFF^3 - 0.0000808 \cdot TDIFF^4 + 0.000000184 \cdot TDIFF^5$$
$$TDIFF = topt - tempc$$
$$W = 0.5 + 0.5 \cdot eet/PET$$
$$PET = 1.6 \cdot (10 \cdot \max(tempc, 0)/ahi)^A \cdot pet\_tw\_m$$
$$A = 0.000000675 \cdot ahi^3 - 0.0000771 \cdot ahi^2 + 0.01792 \cdot ahi + 0.49239$$
$$IPAR = FPAR\_FAS \cdot monthly\_solar \cdot SOL\_CONV \cdot 0.5$$
$$FPAR\_FAS = \min((SR\_FAS - 1.08)/srdiff, 0.95)$$
$$SR\_FAS = (1 + fas\_ndvi/750)/(1 - fas\_ndvi/750)$$
$$SOL\_CONV = 0.0864 \cdot days\_per\_month$$

simple fifth degree polynomial. While the initial form of the *T2* equation is fairly well grounded in first principles of plant physiology, it has not been extensively verified from field measurements. Therefore, both empirical improvements are considered plausible by the ecologists.

5.2 Completing a Partial Model of Water Level Change in the Ringkøbing Fjord

In the last series of experiments, we illustrate that the proposed methodology can be also used for completing a partial model specification. In such a case, human experts specify only some parts of the model structure and leaves others unspecified or partly specified. The goal is then to determine both the structure and constant parameter values of the unspecified parts.

An example of such a task is modeling water level variation in Ringkøbing fjord, a shallow estuary located at the Danish west coast, where it experiences mainly easterly and westerly winds.[2] Wind forcing causes large short term variation of the water level (h) measured at the gate between the estuary and the North Sea. Domain experts specified the following partial model for the temporal variation of the water level in the estuary:

$$\dot{h} = \frac{f(a)}{A}(h_{sea} - h + h_0) + \frac{Q_f}{A} + g(W_{vel}, W_{dir}). \quad (2)$$

The water level response to the wind forcing depends on both wind speed (variable W_{vel}, measured in [m/s]) and direction (W_{dir}, measured in degrees) and is

[2] The task was used as an exercise within a post-graduate course on modeling dynamic systems organized in 2000. Since the Web page of the course is no longer available, we cannot provide a proper reference to the original task specification. Note also that we could not consult domain experts and therefore could not obtain expert comments on the induced models.

Table 11. The grammar for modeling water level variation in the Rinkøbing fjord that follows the partial model specification from Equation 2

$WaterLevelChange \rightarrow$	$(F/A) * SaltWaterDrive + FreshWaterFlow + G$
$SaltWaterDrive \rightarrow$	$(h_{sea} - h + const)$
$FreshWaterFlow \rightarrow$	Q_f/A

modeled by an unknown function g. Apart from wind forcing, fresh water supply (Q_f, measured in [m^3/s]) influences the water level change. When the gate is closed, fresh water is accumulated in the estuary causing a water level rise of Q_f/A, where A is the surface area of the estuary measured in squared meters. During periods when the gate is open, the stored fresh water is emptied in the North Sea. The gate is also opened in order to maintain sufficient water level in the estuary, in which case the water rise is driven by the difference between the water level in the open sea (variable h_{sea}, measured in meters), the water level in the estuary (h, measured in meters), and the constant parameter (h_0). The flow is restricted by the friction of the flow, modeled by an unknown function f of the number of gate parts being open (a). Namely, the gate consists of 14 parts and allows for opening some parts and closing others. The value of A is not directly observed, but a function that calculates A on the basis of h is provided, so A can be also treated as an observed variable.

Given the partial model specification and measurements of the observed variables, the task of model completion is to find the structure and parameters of the unknown functions f and g. The data set contains hourly measurements of all the observed variables within the period from 1st of January to 10th of December 1999.

Grammars for Completion. In order to apply our methodology to the task of model completion, we first recode the partial model specification into a grammar. The grammar presented in Table 11 follows the partial model formula along with the explanations of its constituent terms. The grammar contains two nonterminal symbols F and G that correspond to the unknown functions f and g, respectively.

In the second step, we add productions for F and G to the grammar. They specify modeling alternatives for the completion task. In absence of domain-specific knowledge, we make use of simple polynomial models as presented in Table 12. The first two modeling alternatives, `F0` and `G0`, are the simplest possible models, i.e., constants. The next two, `F1` and `G1`, are using polynomials of the appropriate system variables. Finally, we used an additional modeling alternative for the g function, `G2`, that replaces the wind direction value (that represents angle) with its sine and cosine transformation, respectively.

Results. Table 13 summarizes the results of the model completion and provides a comparison with the black-box modeling case where no knowledge about model structure was used (`Polynomial`). The best cross-validated performance is gained using the partial model specification provided by the experts in combination with

Table 12. Grammar productions specifying modeling alternatives that can be used to complete the model of water level variation in the Ringkøbing fjord

F0	$F \rightarrow const$
F1	$F \rightarrow F + const * T_F \mid const * T_F$
	$T_F \rightarrow T_F * V_F \mid V_F$
	$V_F \rightarrow a$
G0	$G \rightarrow const$
G1	$G \rightarrow G + const * T_F \mid const * T_F$
	$T_G \rightarrow T_G * V_G \mid V_G$
	$V_G \rightarrow W_{vel} \mid W_{dir}$
G2	$V_G \rightarrow W_{vel} \mid \sin(W_{dir}) \mid \cos(W_{dir})$

Table 13. The root mean squared errors (RMSE, estimated on both training data and using 10-fold cross-validation) of the four water level variation models induced by LAGRAMGE with (three first rows) and without (last row) using the partial model specification provided by the domain experts. The last column gives number of candidate models (#CMS) considered during the search.

task	training RMSE	cross-validated RMSE	#CMS
F0 + G0	0.0848	0.106	1
F1 + G1	0.0655	0.0931	378
F1 + G2	0.0585	0.0903	2184
Polynomial	0.0556	2.389	2801

the F_1 and G_2 modeling alternatives for the unspecified parts of the structure. The graph on the left-hand side of Figure 4 shows the simulation of this model compared to the measured water level in the Ringkøbing fjord. We ran a long-term simulation of the model from the initial value of the water level without restarting the simulation process at any measurement point. For the values of all other system variables needed during the simulation, we used the measurements at the nearest time point in the past.

Note that the model follows the general pattern of water level variation. The long-term simulation of the model, however, fails to precisely capture the short-term (hour) changes of the water level. To test the short-term prediction power of the model, we performed two additional simulations, which we restarted with the true measured water level values at every hour and at every day (24 hours). Table 14 presents the results of this analysis. They show that the model is suitable for short-term prediction of the water level in the Ringkøbing fjord.

Since the model induced by LAGRAMGE follows the partial structure specification provided by the human experts, further analysis can be performed. For example, we can compare the influence of the gate opening (modeled by $f(a)(h_{\text{sea}} - h + h_0)/A$) with the effect of the wind (modeled by $g(W_{\text{vel}}, W_{\text{dir}})$).

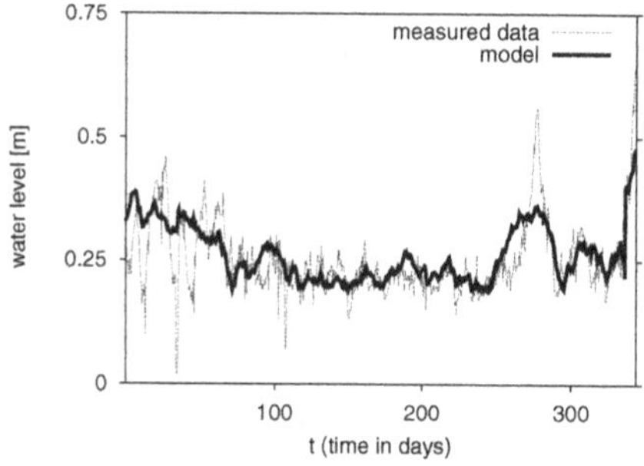

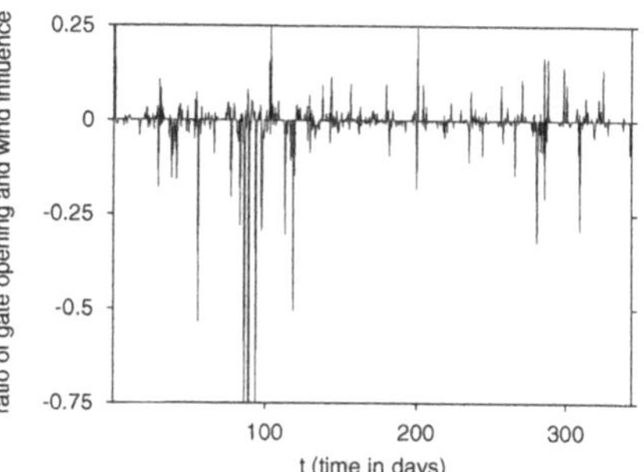

Fig. 4. Simulation of the best water level variation model induced by LAGRAMGE compared to the measured water level (left-hand side) and analysis of the influence of gate opening relative to wind forcing as modeled by LAGRAMGE (right-hand side)

Table 14. The RMSE and correlation coefficient (r) for the short-term (one hour and one day) prediction of the water level in the Ringkøbing fjord compared to the RMSE and r^2 of the simulation over the whole observation period

prediction/simulation period	RMSE	r
one hour	0.0168	0.976
one day	0.0425	0.845
whole observation period	0.0585	0.659

The graph on the right-hand side of Figure 4 shows the ratio of the gate opening and the wind influences on the water level change in the Ringkøbing fjord. The low magnitude of the ratio shows that the influence of the wind prevails over the influence of the gate opening most of the time. The only exceptions occur in the period from 80 to 100 days from the beginning of the measurement, that is, the end of March and beginning of April 1999.

Finally, note that the polynomial model of the water level variation ignores the partial specification, but performs best on the training data. However, the model's small RMSE is due to the obvious overfitting of the training data, since the cross-validated RMSE of the same model (2.389) is much larger than the cross-validated RMSE of the models that follow the partial model specification. This result confirms the importance of integrating available knowledge in the process of model induction.

6 Related Work

The work presented in this chapter is mainly related to other modeling approaches presented in literature. Our methodology is closest in spirit to the compositional modeling (CM) paradigm (Falkenheiner & Forbus, 1991). In our methodology, models of individual processes correspond to model fragments in CM and combining schemes to combining rules in CM. Both CM and our methodology views model building as a search for an appropriate combination of model fragments. The PRET reasoning system for automated modeling also follows the CM paradigm to modeling, but it employs a slightly different kind of modeling knowledge (Stolle & Bradley, this volume; Easley & Bradley, this volume). The first kind of knowledge used in PRET is domain-specific knowledge in the form of "conservation rules". An example of such a rule in the spring mechanics domain specifies that "the sum of forces at any observed coordinate of the mechanical system is zero". These rules are more general than the domain knowledge about model fragments and their composition used in compositional modeling approaches. Therefore, PRET rules constrain the space of possible models much less. PRET compensates for this lack of constraints by using a second kind of domain-independent knowledge about models of dynamic systems based on ordinary differential equations. An example of such a rule specifies that "a model with oscillatory behavior has to be second-order". This kind of ODE rules allows PRET to efficiently rule out inappropriate models by high-level abstract (qualitative) reasoning. As we have illustrated in (Todorovski, 2003), both kinds of modeling knowledge, used in PRET, can be easily encoded within our formalism. Note, however, that LAGRAMGE is not capable of ruling out inappropriate candidate models based on qualitative reasoning, but rather tries to perform quantitative simulation of the candidate models and find out that they can not fit the measured data well.

Another related study is presented by Garrett et al. (this volume). They apply the compositional modeling approach to the task of inducing models of chemical reaction pathways from noisy measurement data. However, the models they induce are qualitative. Although the concepts introduced within the area of compositional modeling are also relevant for automated building of quantitative models of real-world systems, this idea has not been widely explored.

Our approach is similar to the ECOLOGIC approach (Robertson et al., 1991) in the sense that it allows for representing modeling knowledge and domain-specific knowledge. However, in ECOLOGIC, the user has to select among the alternative models, whereas in our approach observational data is used to select among the alternatives. It is also related to process-based approaches to qualitative physics (Forbus, 1984). We can think of the food-chain or domain-specific part of the knowledge as describing processes qualitatively, whereas the modeling part together with the data introduces the quantitative component. However, the ECOLOGIC approach is limited to modeling systems in the environmental domain, whereas our approach is applicable in a variety of domains. Salles and Bredeweg (2003) presents a framework similar to ECOLOGIC that can be used for building models of population and community dynamics. In contrast to the work presented here, they focus on building qualitative conceptual models that do not require numeric data nor provide precise simulation of system behavior.

The work on revising models presented here is related to two other lines of work. In the first, Saito et al. (this volume) address the same task of revising models based on equations. Their approach transforms a part of the model into a neural network, retrains the neural network on available data, and transforms the trained network back into an equation-based model. They obtained revised models with a considerably smaller error rate than the original one, but gained a slightly lower accuracy improvement than our method did. A limitation of their approach is that it requires some hand-crafting to encode the equations as a neural network. The authors state that "the need to translate the existing CASA model into a declarative form that our discovery system can manipulate" is a challenge to their approach.

The approach of transforming equation-based models to neural networks and using these for refinement is similar in spirit to the KBANN approach proposed in (Towell & Shavlik, 1994). There, an initial theory based on classification rules is first encoded as neural network. Then, the topology of the network is refined and the network is re-trained with the newly observed data. Finally, the network is transformed back into rules. However, the application of KBANN is limited to theories and models expressed as classification rules. In other related work, Whigham and Recknagel (2001) consider the task of revising an existing model for predicting chlorophyll-a by using measured data. They use a genetic algorithm to calibrate the equation parameters. They also use a grammar-based genetic programming approach to revise the structure of two subparts of the initial model, one at a time. A most general grammar that can derive an arbitrary expression using the specified arithmetic operators and functions was used for each subpart. Unlike the work presented here, Whigham and Recknagel (2001) do not present a general framework for the revision of equation-based models, although their approach is similar to ours in that they use grammars to specify possible revisions.

Different aspects of the work presented in this chapter has been already published in other articles and papers. First, Todorovski and Džeroski (1997) introduce the grammar-based equation discovery method LAGRAMGE. The paper illustrates the use of grammars for integrating different aspects of domain knowledge in the process of equation discovery — measurement units and process-based knowledge being among others. The successful application of LAGRAMGE to modeling phytoplankton growth in Lake Glumsø is the topic of the paper by Todorovski et al. (1998), which also identifies the difficulty of encoding knowledge into grammars as a main drawback of LAGRAMGE. The work presented by Todorovski and Džeroski (2001b) introduces a new higher-level formalism for encoding process-based knowledge about population dynamics. These ideas lead also to a separate line of work starting with (Langley et al., 2002). Finally, Todorovski and Džeroski (2001a) present a grammar-based methodology for revising equation-based models.

7 Conclusion

In the chapter, we presented a methodology for integrating various aspects of knowledge specific to the domain of interest in the process of inducing equation-based models from data. The methodology is based on idea of transforming the domain-specific knowledge to a language bias for induction that specifies the set of candidate hypotheses. Context free grammars are used as a formalism for specifying the language bias. We show how three different types of domain-specific knowledge can be easily transformed to grammars. We demonstrated the utility of our approach by performing experiments using LAGRAMGE, an induction method that can take into account declarative language bias encoded in a form of a grammar. The results of the experiments show that models induced using knowledge make sense to domain scientists and more importantly, scientists can easily understand and interpret the induced models. Furthermore, the models reveal important non-trivial relations between observed entities that are difficult to infer using black-box models. A comparison of models induced with and without using domain-specific knowledge in the last experiment shows that using knowledge can considerably reduce overfitting and increase model prediction performance on test cases unseen in the induction phase.

The immediate direction of further work is establishing libraries of encoded knowledge in different domains. These libraries will be built in cooperation with domain experts that have expertise in modeling real-world systems from measured data. Establishing such libraries will make the developed methods usable by domain experts that collect data about real-world systems, but are not experienced mathematical modelers. First steps toward establishing a library for modeling of aquatic ecosystems, based on recent developments in the domain, have been already made (Atanasova and Kompare 2003; personal communication). Furthermore, the same team of experts work on a library for establishing models of equipment used for waste water treatment. In both cases, the libraries will be used for automated modeling based on collections of measurements.

The automated modeling approach based on transformation to grammars is limited to modeling tasks where the domain expert is capable to provide processes that are expected to be important for modeling the observed system. However, there are many real-world tasks, where experts are not able to specify the list of processes. In these cases, a two level search procedure should be developed that is capable of discovering the processes that influence the behavior of the observed system. At the higher level, the search will look for the optimal set of processes. For each set of processes, the proposed modeling framework will be used at the lower level to find the model, based on the particular set of processes, that fits the measured data best.

The methodology presented in this chapter is still under development and different methods presented are developed and evaluated independently of the others. For example, the methods do not allow the revision of models based on partial differential equations, although in principle this should not be a problem. Furthermore, the formalism can easily encode domain knowledge about changes of the systems along a spatial dimension, but a method for discovering partial differential equations is not integrated within LAGRAMGE. There is a clear need for proper integration of the methods into a single modeling assistant that would allow establishing new and revising existing models based on algebraic, ordinary, and partial differential equations.

The integrated modeling assistant should be further integrated within standard data analysis and simulation environments[3] that are routinely used by mathematical modelers. Beside improved ease of use, the integration will enable standard techniques for parameter estimation and sensitivity analysis to be used in conjunction with the automated modeling framework to yield a proper scientific assistant.

References

Bendoricchio, G., Coffaro, G., DeMarchi, C.: A trophic model for Ulva Rigida in the Lagoon of Venice. Ecological Modelling 75/76, 485–496 (1994)

Bunch, D.S., Gay, D.M., Welsch, R.E.: Algorithm 717; subroutines for maximum likelihood and quasi-likelihood estimation of parameters in nonlinear regression models. ACM Transactions on Mathematical Software 19, 109–130 (1993)

Chomsky, N.: Three models for the description of language. IRE Transactions on Information Theory 2, 113–124 (1956)

Coffaro, G., Carrer, G., Bendoricchio, G.: Model for Ulva Rigida growth in the Lagoon of Venice (Technical Report). University of Padova, Padova, Italy. UNESCO MURST Project: Venice Lagoon Ecosystem (1993)

Džeroski, S., Todorovski, L.: Discovering dynamics: from inductive logic programming to machine discovery. Journal of Intelligent Information Systems 4, 89–108 (1995)

Falkenheiner, B., Forbus, K.D.: Compositional modeling: Finding the right model for the job. Artificial Intelligence 51, 95–143 (1991)

Forbus, K.D.: Qulitative process theory. Artificial Intelligence 24, 85–168 (1984)

Giordano, F.R., Weir, M.D., Fox, W.P.: A first course in mathematical modeling. 2nd edn. Brooks/Cole Publishing Company, Pacific Grove, CA (1997)

Hopcroft, J.E., Ullman, J.D.: Introduction to automata theory, languages and computation. Addison-Wesley, Reading, MA (1979)

Kokar, M.M.: Determining arguments of invariant functional descriptions. Machine Learning 4, 403–422 (1986)

Kompare, B., Džeroski, S.: Getting more out of data: automated modelling of algal growth with machine learning. In: Proceedings of the International symposium on coastal ocean space utilisation, Yokohama, Japan, pp. 209–220 (1995)

Križman, V.: Avtomatsko odkrivanje strukture modelov dinamičnih sistemov. Doctoral dissertation, Faculty of computer and information science, University of Ljubljana, Ljubljana, Slovenia. In Slovene (1998)

Langley, P., Sanchez, J., Todorovski, L., Džeroski, S.: Inducing process models from continuous data. In: Proceedings of the Fourteenth International Conference on Machine Learning, pp. 347–354. Morgan Kaufmann, Sidney, Australia (2002)

Langley, P., Simon, H., Bradshaw, G.: Heuristics for empirical discovery. In: Computational models of learning, pp. 21–54. Springer, Heidelberg (1987)

Langley, P., Żytkow, J.: Data-driven approaches to empirical discovery. Artificial Intelligence 40, 283–312 (1989)

Lavrač, N., Džeroski, S.: Inductive logic programming: Techniques and applications. Chichester: Ellis Horwood. (1994), Available for download at `http://www-ai.ijs.si/SasoDzeroski/ILPBook/`

[3] Examples of such systems are MatLab (`http://www.mathworks.com/`), SciLab (`http://www-rocq.inria.fr/scilab/`), and Octave (`http://www.octave.org/`).

Mitchell, T.M.: The need for biases in learning generalizations. In: Readings in machine learning, pp. 184–191. Morgan Kaufmann, San Mateo, CA (1991)

Murray, J.D.: Mathematical biology, 2nd corrected edn. Springer, Heidelberg (1993)

Nédellec, C., Rouveirol, C., Adé, H., Bergadano, F., Tausend, B.: Declarative bias in ILP. In: Raedt, L.D. (ed.) Advances in inductive logic programming, pp. 82–103. IOS Press, Amsterdam, The Netherlands (1996)

Ourston, D., Mooney, R.J.: Theory refinement combining analytical and empirical methods. Artificial Intelligence 66, 273–309 (1994)

Pazzani, M., Kibler, D.: The utility of background knowledge in inductive learning. Machine Learning 9, 57–94 (1992)

Pazzani, M.J., Mani, S., Shankle, W.R.: Acceptance of rules generated by machine learning among medical experts. Methods of Information in Medicine 40, 380–385 (2001)

Potter, C.S., Klooster, S.A.: Global model estimates of carbon and nitrogen storage in litter and soil pools: Response to change in vegetation quality and biomass allocation. Tellus 49B, 1–17 (1997)

Quinlan, J.R.: C4.5: Programs for machine learning. Morgan Kaufmann, San Mateo, CA (1993)

Robertson, D., Bundy, A., Muetzelfield, R., Haggith, M., Uschold, M.: Eco-logic: logic-based approaches to ecological modelling. MIT Press, Cambridge, MA (1991)

Salles, P., Bredeweg, B.: Qualitative reasoning about population and community ecology. AI Magazine 24, 77–90 (2003)

Todorovski, L.: Using domain knowledge for automated modeling of dynamic systems with equation discovery. Doctoral dissertation, Faculty of computer and information science, University of Ljubljana, Slovenia (2003)

Todorovski, L., Džeroski, S.: Declarative bias in equation discovery. In: Proceedings of the Fourteenth International Conference on Machine Learning, pp. 376–384. Morgan Kaufmann, San Mateo, CA (1997)

Todorovski, L., Džeroski, S.: Theory revision in equation discovery. In: Proceedings of the Fourth International Conference on Discovery Science, pp. 389–400. Springer, Heidelberg (2001a)

Todorovski, L., Džeroski, S.: Using domain knowledge on population dynamics modeling for equation discovery. In: Proceedings of the Twelfth European Conference on Machine Learning, pp. 478–490. Springer, Heidelberg (2001b)

Todorovski, L., Džeroski, S., Kompare, B.: Modelling and prediction of phytoplankton growth with equation discovery. Ecological Modelling 113, 71–81 (1998)

Towell, G.G., Shavlik, J.W.: Knowledge-based artificial neural networks. Artificial Intelligence 70, 119–165 (1994)

Utgoff, P.E.: Shift of bias for inductive concept learning. In: Machine learning: An artificial intelligence approach — vol. 2, pp. 107–148. Morgan Kaufmann, San Mateo, CA (1986)

Voit, E.O.: Computational analysis of biochemical systems. Cambridge University Press, Cambridge, UK (2000)

Washio, T., Motoda, H.: Discovering admissible models of complex systems based on scale-types and identity constraints. In: Proceedings of the Fifteenth International Joint Conference on Artificial Intelligence, pp. 810–817. Morgan Kaufmann, San Mateo, CA (1997)

Whigham, P.A., Recknagel, F.: Predicting chlorophyll-a in freshwater lakes by hybridising process-based models and genetic algorithms. Ecological Modelling 146, 243–251 (2001)

Williams, R.J., Zipser, D.: A learning algorithm for continually running fully recurrent neural networks. Neural Computation 1, 270–280 (1989)

Wrobel, S.: First order theory refinement. In: Raedt, L.D. (ed.) Advances in inductive logic programming, pp. 14–33. IOS Press, Amsterdam, The Netherlands (1996)

Zembowicz, R., Żytkow, J.M.: Discovery of equations: Experimental evaluation of convergence. In: Proceedings of the Tenth National Conference on Artificial Intelligence, pp. 70–75. Morgan Kaufmann, San Mateo, CA (1992)

Communicability Criteria of Law Equations Discovery

Takashi Washio and Hiroshi Motoda

Institute of Scientific and Industrial Research
Osaka University, Osaka, Japan
washio@sanken.osaka-u.ac.jp

Abstract. The "laws" in science are not the relations established by only the objective features of the nature. They have to be consistent with the assumptions and the operations commonly used in the study of scientists identifying these relations. Upon this consistency, they become communicable among the scientists. The objectives of this literature are to discuss a mathematical foundation of the communicability of the "scientific law equation" and to demonstrate "Smart Discovery System (SDS)" to discover the law equations based on the foundation. First, the studies of the scientific law equation discovery are briefly reviewed, and the need to introduce an important communicability criterion called "Mathematical Admissibility" is pointed out. Second, the axiomatic foundation of the mathematical admissibility in terms of measurement processes and quantity scale-types are discussed. Third, the strong constraints on the admissible formulae of the law equations are shown based on the criterion. Forth, the SDS is demonstrated to discover law equations by successively composing the relations that are derived from the criterion and the experimental data. Fifth, the generic criteria to discover communicable law equations for scientists are discussed in wider view, and the consideration of these criteria in the SDS is reviewed.

1 Introduction

Various relations among objects, events and/or quantity values are observed in natural and social behaviors. Especially, scientists call the relation as a "law", if it is commonly observed over the wide range of the behaviors in a domain. When the relation of the law can be represented in form of mathematical formulae constraining the values of some quantities characterizing the behaviors, the relation is called "law equations". In popular understanding, the relations of the laws and the law equations are considered to be objective in the sense that they are embedded in the behaviors independent of our processes of observation, experiment and interpretation. However, the definition of the laws and the law equations as communicable knowledge shared by scientists must be more carefully investigated. Indeed, they are not the relations established by only the objective features of the nature as discussed in this chapter. They have to be consistent with the assumptions and the operations commonly used

S. Džeroski and L. Todorovski (Eds.): Computational Discovery, LNAI 4660, pp. 98–119, 2007.

in the study of scientists identifying these relations. Upon this consistency, they become communicable among the scientists.

On the other hand, the studies to develop automated or semi-automated systems to discover scientific law equations have been performed in the last two decades. As the main goal of the studies is to discover law equations representing meaningful relations among quantities for scientists, i.e., communicable with scientists, the systems must take into account the communicability criteria to some extent. The objectives of this chapter are to discuss a mathematical foundation of the communicability of the "scientific law equation" and to demonstrate "Smart Discovery System (SDS)" to discover the law equations based on the foundation. Through the demonstration of the scientific law equation discovery and the subsequent discussion, the communicability criteria of law equation discovery are clarified.

2 Study of Law Equation Discovery

First, we briefly review the past studies of the scientific law equation discovery from the view point of the equation formulae having the communicability in science. The most well known pioneering system to discover scientific law equations under the condition where some quantities are actively controlled in a laboratory experiment is BACON (Langley et al., 1985). FAHRENHEIT (Koehn & Żytkow, 1986) and ABACUS (Falkenhainer & Michalski, 1986) are successors that basically use similar algorithms to BACON to discover law equations. LAGRANGE (Džeroski & Todorovski, 1995) and LAGRAMGE (Todorovski & Džeroski, 1997) are another type of scientific law equation discovery systems based on the ILP-like generate and test reasoning to discover equations representing the dynamics of the observed phenomenon.

Many of these succeeding systems introduced the constraint of the unit dimension of physical quantities to prune the search space of the equation formulae. The constraint is called "Unit Dimensional Homogeneity" (Bridgman, 1922; Buckingham, 1914) that all additive terms in a law equation formula must have an identical unit dimension. For example, a term having a length unit $[m]$ is not additive to another term having a different unit $[kg]$ in a law equation, even if the formula including their addition well fits to given data. Though the main purpose of the use of this constraint in these systems was to reduce the ambiguity in their results under noisy measurements and the high computational cost of their algorithms, the introduction also had an effect to increase the communicability of the discovered equations with scientists because the discovered equations are limited to more meaningful formulae. A law equation discovery system COPER (Kokar, 1986) more intensively applied the constraints deduced from the unit dimensional analysis. The limitation of the constraints is so strong that some parts of the equation formulae are almost predetermined without using the measurement data set, and the derived equations has high communicability with scientists. LAGRANGE and LAGRAMGE are also capable of introducing these constraints in principle. However, the main purpose of these works is to provide an elegant measure to implement the constraints in the scientific law equation discovery but not to propose the contents of the constraints to enhance the communicability.

A strong limitation of the use of the unit dimensional constraints is its narrow applicability only to the quantities whose units are clearly known. To overcome this drawback, a law equation discovery system named "Smart Discovery System (SDS)" has been proposed (Washio & Motoda, 1997; Washio et al., 2000). It discovers scientific law equations by limiting its search space to "Mathematically Admissible" equations in terms of the constraints of "scale-type" and "identity". They represent the important assumptions and operations commonly used in measurement and modeling processes identifying the relations among quantities by scientists. Since the use of scale-types and identity is not limited by the availability of the unit dimensions, SDS is applicable to non-physical domains including biology, sociology, and economics. In the following section, the axiomatic foundation of the mathematical admissibility in terms of measurement processes and quantity scale-types are discussed.

3 Scale-Types of Quantities

"Mathematical Admissibility" includes the constraints of some fundamental notions in mathematics such as arithmetic operations, but they are very weak to constrain the shape of the law equation formulae. Stronger constraints are deduced from the assumptions and operations used in measurement process. The value of a quantity is obtained through a measurement in most of the scientific domains, and some features of the quantity are characterized by the measurement process. Though the unit dimension is an example of such features, a more generic feature is called "scale-types". S.S. Stevens defined that a measurement is to assign a value to each element in a set of the objects and/or events under given rules, and claimed that the rule set defines the "scale type" of the measured quantity. He categorized the scale-types into "nominal", "ordinal", "interval" and "ratio" scales (Stevens, 1946). In the later study, another scale-type called absolute scale is added. Subsequently, D.H. Krantz et al. axiomatized the measurement processes and the associated scale-types (Krantz et al., 1971). In this section, their theory on the scale-types is reviewed.

Definition 1 (A Relation System). *The following series of finite length α is called "a relation system".*

$$\alpha =< A, R_1, R_2, \ldots, R_n >$$

where A is a non-empty set of elements and R_i: $R_i(a_1, a_2, \ldots, a_{m_i})$ is a relation among the elements $a_1, a_2, \ldots, a_{m_i} \in A$.

Definition 2 (Type and Similarity). *Given a relation system α, where each R_i is the relation among m_i elements in A, the series of positive integers $< m_1, m_2, \ldots, m_n >$ is called "type" of α. Two relation systems α and $\beta =< B, S_1, S_2, \ldots, S_n >$ are "similar", if they have identical types.*

Definition 3 (Isomorphism (Homomorphism)). *Given two relation systems α and β, if the following conditions are met, they are called "isomorphic (homomorphic)".*

(1) α and β are similar.

(2) A bijection (surjection) f from A to B exists where
$R_i(a_1, a_2, \ldots, a_{m_i}) \Leftrightarrow S_i(f(a_1), f(a_2), \ldots, f(a_{m_i}))$.

Definition 4 (Numerical and Empirical Relation Systems). *When a relation system α satisfies the following conditions, α is called a "numerical relation system".*

(1) The domain $A \subseteq \Re$.

(2) $R_i (i = 1, 2, \ldots, n)$ is a relation among values in $\Re$.

A relation system which is not a numerical relation system is called an "empirical system".

3.1 Extensional Measurement

Extensional measurement is the measurement done by using a facility to directly map each element in an empirical relation system to a numeral in a numerical relation system while preserving the relations in each system. For example, the measurement of length and the measurement of weight are the extensional measurements respectively since a ruler and a balance map the lengths and the weights of objects to numerals directly.

Definition 5 (Scale-type in Extensional Measurement). $< \alpha, \beta_f, f >$ *is called a "scale-type" where*

α: an empirical relation system,

β_f: a full numerical relation system,

f: isomorphic or homomorphic mapping from α to the subsystem of β_f.

Here, the "full numerical relation system" is the system which domain is the entire $\Re$, and the "subsystem" is the system where its domain is the sub-domain of the original system, and all relations of the subsystem have one-to-one correspondence to the relations of the original system.

When an empirical relation system $\alpha =< A, I >$ is the following classification system, the measurement by "nominal scale" is applicable.

Definition 6 (Classification System). *Given $\alpha =< A, I >$, if I is a binary relation on A, α is called a "binary system". Furthermore, if the following three axioms hold for I, I is called an "equivalence relation", α a "classification system" and the set of elements where I holds "I-equivalence class".*

Reflexive law: $\forall a \in A, I(a, a)$

Symmetric law: $\forall a, \forall b \in A, I(a, b) \Rightarrow I(b, a)$

Transitive law: $\forall a, \forall b, \forall c \in A, I(a, b) \wedge I(b, c) \Rightarrow I(a, c)$

An example of the classification system and the measurement in the nominal scale is explained through the empirical relation system α depicted in Fig. 1. The domain A is the power set of the set of 6 weights $\{a, b_1, b_2, c_1, c_2, c_3\}$. The

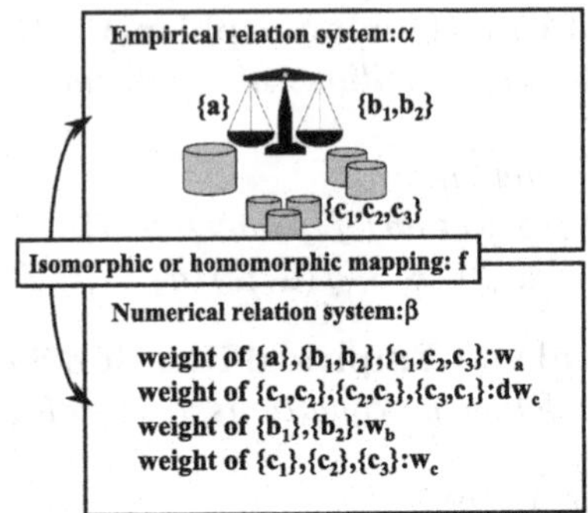

Fig. 1. Example of extensional measurement

equivalence relation I is that two sets of the weights are balanced. The reflexive law holds since two identical weight sets balance. The symmetric law also holds because the balance of the following pair wise sets is invariant for the exchange of their positions between the left dish and the right dish.

$$(\{b_1\},\{b_2\}),(\{c_1\},\{c_2\}),(\{c_2\},\{c_3\}),(\{c_3\},\{c_1\}),$$
$$(\{c_1,c_2\},\{c_2,c_3\}),(\{c_2,c_3\},\{c_3,c_1\}),(\{c_3,c_1\},\{c_1,c_2\}),$$
$$(\{a\},\{b_1,b_2\}),(\{b_1,b_2\},\{c_1,c_2,c_3\}),(\{c_1,c_2,c_3\},\{a\})$$

In addition, if a pair of sets in the following combinations balance, then the rest pairs also balance. Thus, the transitive law holds.

$$(\{c_1\},\{c_2\},\{c_3\}),(\{c_1,c_2\},\{c_2,c_3\},\{c_3,c_1\}),(\{a\},\{b_1,b_2\},\{c_1,c_2,c_3\})$$

Accordingly, this empirical relation system α is a classification system. For this α, given a numerical relation system β and its domain $B \subseteq \Re$, a surjection f which maps any weight sets in I-equivalence class to an identical number on B is introduced as follows.

$$\begin{aligned}
&w_a = f(\{a\}) = f(\{b_1,b_2\}) = f(\{c_1,c_2,c_3\}),\\
&w_b = f(\{b_1\}) = f(\{b_2\}),\\
&w_c = f(\{c_1\}) = f(\{c_2\}) = f(\{c_3\}),\\
&dw_c = f(\{c_1,c_2\}) = f(\{c_1,c_3\}) = f(\{c_2,c_3\})\\
&\text{where } w_a, w_b, w_c, dw_c \in B.
\end{aligned}$$

If a relation of β, which is the equality of two numbers, is considered, the reflexive law holds as the equality of identical numbers is trivial. Also, the symmetric law and the transitive law hold for the equality of w_a, w_b, w_c, dw_c. Hence, β is a classification system. Therefore, the homomorphic mapping f to assign an identical number to balanced weights is a measurement in nominal scale.

Editorial note. Following a nominal measurement scale, we assign qualitative labels to groups or categories of observed objects, where there is no greater-than or less-than relations between labels. Examples of nominal scale quantities include color of an object and gender of a person. Variables measured using nominal scale are referred to as categorical variables.

When an empirical relation system $\alpha =< A, P >$ is the following series, the measurement in "ordinal scale" is possible.

Definition 7 (Series). *Given a binary system $\alpha =< A, P >$, if the following three axioms hold for P, P is called "inequivalence relation" and α "series".*

Asymmetric law: $\forall a, \forall b \in A, P(a,b) \Rightarrow \neg P(b,a)$
Transitive law: $\forall a, \forall b, \forall c \in A, P(a,b) \wedge P(b,c) \Rightarrow P(a,c)$
Law of the excluded middle: $\forall a, \forall b \in A,$ *one of* $P(a,b)$ *and* $P(b,a)$ *holds.*

In case that the relation I holds on some of the elements in A, *i.e.*, $\alpha =< A, I, P >$, the elements in each I-equivalence class are grouped, and A is replaced by the "quotient set" A/I. Then, $\alpha/I =< A/I, P >$ is a series, and it can be measured in ordinal scale. In the example of Fig. 1, the domain of $\alpha/I =< A/I, P >$ is A/I where the sets of the weights which mutually balance are grouped. Then, given two elements r and s in A/I, the binary relation $P(r,s)$ is defined that the dish on which s is put comes down. This $P(r,s)$ satisfies the conditions of the aforementioned series. On the other hand, given the binary relation $P(w_r, w_s)$ which is inequality $w_r < w_s$ between two numbers in the domain B of a numerical relation system β. This also satisfies the conditions of the series. Then we define a surjection f which assigns real numbers w_a, w_b, w_c, dw_c to the elements in A/I respectively where $w_r < w_s$ holds for r and s under $P(r,s)$. This definition of f which holds $w_c < w_b < dw_c < w_a$ is the measurement in ordinal scale.

Editorial note. While the labels in the nominal scale are unordered, ordinal scale introduces ordering between the labels. The labels of the ordered scale, thus let us rank the observed objects. An example of a variable measured using ordered scale is educational experience of a person. Variables measured using ordinal scale are called ordinal or rank variables.

Furthermore, when an empirical relation system $\alpha =< A, D >$ is the following "difference system", the measurement in "interval scale" is possible.

Definition 8 (Difference System). *Given $\alpha =< A, D >$, if the relation D is a quadruple relation on A, α is called a "quadruple system". Moreover, α is called a "difference system" if the following axioms holds for $\{a,b,c,d,e,f\} \subseteq A$.*

$P(a,b) \not\Leftrightarrow D(a,b,a,a),$
$I(a,b) \Leftrightarrow D(a,b,b,a) \wedge D(b,a,a,b),$
$D(a,b,c,d) \wedge D(c,d,e,f) \Rightarrow D(a,b,e,f),$
One of $D(a,b,c,d)$ *and* $D(c,d,a,b)$ *holds,*
$D(a,b,c,d) \Rightarrow D(a,c,b,d),$
$D(a,b,c,d) \Rightarrow D(d,c,b,a),$
$\exists c \in A, D(a,c,c,b) \wedge D(c,b,a,c),$
$P(a,b) \wedge \neg D(a,b,c,d) \Rightarrow \exists e \in A, P(a,e) \wedge P(e,b) \wedge D(c,d,a,e),$
$\exists e, \exists f \in A, \exists$ *an integer* $n, P(a,b) \wedge D(a,b,c,d) \Rightarrow M_n(c,e,f,d).$

Here, M_n is the relation to locate e and f between c and d in A where the distance between c and e and that between f and d are identical and one n-th of the distance between c and f. Even if some elements in A satisfy the equivalence relation I, $\alpha/I =< A/I, D >$ is a difference system, and α/I can be measured by an interval scale quantity. In the example of Fig. 1, let the relation $D(r,s,t,u)$ on A/I be that the left dish comes down when two sets of weights r and u are put on the left dish and s and t on the right dish. Then α/I is a difference system. Let $D(w_r, w_s, w_t, w_u)$ on the domain B of a numerical relation system β be $(w_s - w_r) \leq (w_u - w_t)$, and let a surjection f be the assignment of numerals w_a, w_b, w_c, dw_c to the sets of weights in such a way that $D(w_r, w_s, w_t, w_u)$ holds in β when $D(r,s,t,u)$ holds in α/I. In this example, $w_a = w_c + 4(w_b - w_c)$ and $dw_c = w_c + 2(w_b - w_c)$ are obtained, and the numerals mapped by the surjection f are interval scale. The f which satisfies this relation is not unique. The different mappings f_1 and f_2 for two numerical relation systems β_1 and β_2 which are homomorphic with α/I respectively have a linear relation $f_2 = k \cdot f_1 + c$ where k and c are constants, and this is the admissible unit conversion. The interval scale quantities follow the axioms of the classification system, the series and the difference system, but do not have any absolute origins. The examples are position, time and musical sound pitch since the origins of their coordinate systems are arbitrarily introduced.

Editorial note. Although the labels of the ordinal scale are ordered, the difference between two label values does not have a uniform meaning over the whole scale. On the other hand, the distance between labels of an interval scale does have a meaning and the meaning remains the same over all the measurement scale. For example, the distance between 10° and 20° Centigrade is the same as the distance between 40° and 50° and its value is 10 measurement units. The interval scale is the first scale in the hierarchy which introduces the concept of a measurement unit.

The quantities of ratio scale are derived by the extension of the difference system. Given two difference systems α/I and β, define a surjection f from $A/I \times A/I$ to $B \times B$ satisfying $f(r,s) = w_s - w_r$ and $f(\phi, r) = w_r$. Under this mapping, α/I is measured by a ratio scale quantity. In the example of Fig. 1, the two weights c_1 on the left dish and the weights c_2 and c_3 on the right dish balance in α/I, and this is homomorphic with the following relation in β.

$$f(\phi, \{c_1\}) = f(\{c_1\}, \{c_2, c_3\}),$$

where $f(\phi, \{c_1\}) = w_c$ and $f(\{c_1\}, \{c_2, c_3\}) = dw_c - w_c$. This deduces the relation $dw_c = 2w_c$. By substituting this relation to the aforementioned $w_a = w_c + 4(w_b - w_c)$ and $dw_c = w_c + 2(w_b - w_c)$, $2w_b = 3w_c$ and $w_a = 3w_c$ are deduced, and the ratio scale of weight is derived. f satisfying these relations are not unique. Given two numerical relation systems β_1 and β_2 which are homomorphic with α/I, the corresponding f_1 and f_2 have a similarity relation $f_2 = k \cdot f_1$, *i.e.*, the admissible unit conversion. The ratio scale quantities have absolute origins. The examples are distance, elapsed time and physical mass.

Editorial note. Finally, ratio measurement scale always has an absolute zero label that is meaningful. Examples of ratio scale variables include distance, time, mass, or count. The ratio scale let us construct a meaningful fraction (i.e., ratio) between two measured values and say, we need twice as much time, or this object weights twice as much as this one.

3.2 Intentional Measurement

Besides the quantities defined in the extensional measurement, another sort of quantities which can not be directly measured by any facilities but indirectly measured by functions of the other quantities exit. The process of this indirect measurement is called "intentional measurement", and the quantities measured through this measurement are "derivative quantities" obtained from the other quantities. The nature of each scale-type in the intentional measurement is very similar to that of the extensional measurement, and the admissible unit conversion is identical for each scale-type. The descriptions on the rigorous definitions of this measurement and its scale-types are omitted due to the space limitation. An example of the derivative quantity obtained through the intentional measurement is the temperature. It can be measured only through some other measured quantities such as the expansion length of mercury. The other representative derivative quantities are density, energy and entropy.

An important scale-type which is defined by the intentional measurement is absolute scale-type. Given a quantity g defined by the other ratio scale quantities $f_1, f_2, \ldots, f_n$ through $g = \prod_{i=1}^{n} f_i^{\gamma_i}$, when the relation $\prod_{i=1}^{n} k_i^{\gamma_i} = 1$ holds under any unit conversions of $f_1, f_2, \ldots, f_n$, the scale-type of g is called "absolute scale". Because the value of g is invariant for any unit conversions, it is uniquely defined and called "dimensionless number". Its admissible unit conversion follows the identity group $g'_2 = g_1$. The examples are the ratio of two masses and angle in radian.

4 Admissible Formulae of Law Equations

In this section, we review some important theorems on the relations among observed quantities, and show their extension for the discovery of law equations as communicable knowledge.

R.D. Luce claimed that the group structure of each scale-type is conserved through the unit transformation, and this fact strongly limits the mathematically admissible relations among quantities having interval and ratio scale-types (Luce, 1959). For example, when x and y are ratio scale, the admissible unit conversions are $x' = kx$ and $y' = Ky$ respectively. When we assume the relation between x and y to be $y = \log x$, and apply a unit conversion on x, then the unit of y should be also converted as $y' = \log x' = \log kx = \log x + \log k$. This consequence that the origin of y is changed is contradictory to the above admissible unit conversions of y. Thus, the logarithmic relation between two ratio scale quantities is not admissible. R.D. Luce further proceeded this discussion, and derived the

Table 1. Admissible relations between two quantities

No.	scale-types independent quantity x	dependent quantity $y(x)$	admissible relation
1	ratio	ratio	$y(x) = \alpha x^{\beta}$
2.1	ratio	interval	$y(x) = \alpha x^{\beta} + \delta$
2.2			$y(x) = \alpha \log x + \beta$
3	interval	ratio	impossible
4	interval	interval	$y(x) = \alpha x + \beta$

admissible binary relations between ratio and interval scale quantities depicted in Table 1.

On the other hand, an important theorem called "Product Theorem" on the relation formula among multiple measured quantities had been presented in the unit dimensional analysis which was independently studied by old scientists (Bridgman, 1922). However, this theorem addresses on the relation among ratio scale quantities only. We derived the following "Extended Product Theorem" (Washio & Motoda, 1997) to the case where the quantities of ratio, interval and absolute scales are included in the formula by introducing the consequences of R.D. Luce.

Theorem 1 (Extended Product Theorem). *Given a set of ratio scale quantities R and a set of interval scale quantities I, a derivative quantity Π is related with each $x_i \in R \cup I$ through one of the following formulae.*

$$\Pi = (\prod_{x_i \in R} |x_i|^{a_i})(\prod_{I_k \in C} (\sum_{x_j \in I_k} b_{kj}|x_j| + c_k)^{a_k}),$$

$$\Pi = \sum_{x_i \in R} a_i \log |x_i| + \sum_{I_k \in C_{\bar{g}}} a_k \log(\sum_{x_j \in I_k} b_{kj}|x_j| + c_k) + \sum_{x_\ell \in I_g} b_{g\ell}|x_\ell| + c_g,$$

where R or I can be empty, and C is a covering of I, $C_{\bar{g}}$ a covering of $I - I_g$ ($I_g \subseteq I$). Π can be any of interval, ratio and absolute scale, and each coefficient is constant.

Here, a "covering" C of a set I is a set of finite subsets Is_is of I where $I = \cup_i Is_i$. The same definition applies to $C_{\bar{g}}$ for $I - I_g$. When the argument quantities appearing in a law equation are ratio scale and/or interval scale, the relation among the quantities sharing arbitrary unit dimensions has one of the above formulae.

Another major theorem called "Buckingham Π-theorem" on the structure of a law equation consisting of ratio scale quantities only had also been presented in the old work in the unit dimensional analysis (Buckingham, 1914). We further extended this theorem to include the interval, ratio and absolute scale quantities in the argument (Washio & Motoda, 1997).

Theorem 2 (Extended Buckingham Π-theorem). *Given a complete equation $\phi(x, y, z, \ldots) = 0$, if every argument of this equation is either of interval, ratio and absolute scales, then the equation can be rewritten in the following form.*

$$F(\Pi_1, \Pi_2, ..., \Pi_{n-r-s}) = 0$$

where n is the number of the arguments of ϕ, r and s are the numbers of the basic unit and the basic origin contained in $x, y, \ldots$, and Π_i is absolute scale for all i and represented by the formulae of the regime defined by Extended Product Theorem.

Here, the basic unit is the unit dimension which defines the scaling independent of the other unit in ϕ as length $[L]$, mass $[M]$ and time $[T]$, and the basic origin is the origin which is artificially chosen in the measurement of an interval scale quantity, for example, the origin of temperature in Celsius defined as the melting point of water under the standard atmosphere pressure. Each $\Pi_i = \rho_i(x, y, \ldots)$ defining Π_i is called a "*regime*" and $F(\Pi_1, \Pi_2, ..., \Pi_{n-r}) = 0$ an "*ensemble*". Because all arguments of $F = 0$ are absolute scale, *i.e.*, dimensionless, the shape of the formula does not constrained by the theorem 1, and the arbitrary formula is admissible for $F = 0$ in terms of the scale-type.

The following example of the nuclear decay of a radioactive element is an example of the theorem 1 and the theorem 2.

$$N = N_0 \exp[-\lambda(t - t_0)] \tag{1}$$

where $t[s]$: time, $t_0[s]$: time origin, $\lambda[s^{-1}]$: decay speed constant,
$N[kg]$: current element mass, $N_0[kg]$: $t_0$original element mass

t and t_0 are interval scale, and λ, N and N_0 are ratio scale. By introducing dimensionless Π_1 and Π_2, the equation can be rewritten as

$$\Pi_1 = \exp(-\Pi_2), \tag{2}$$
$$\Pi_1 = N/N_0, \tag{3}$$
$$\Pi_2 = \lambda(t - t_0), \tag{4}$$

which are an *ensemble* and two *regimes*. The *regimes* (3) and (4) follow the first formula in the theorem 1. The number of the original arguments n is 5. r is equal to 2 because t, t_0 and λ share a basic unit of time $[s]$ and N and N_0 share the basic unit of mass $[kg]$. s is equal to 1 since t and t_0 share a basic origin of time. Thus $n - r - s = 2$ holds, and this satisfies the theorem 2. As indicated in the above example, the scale-type of measurement quantities strongly constrains the formulae of the law equations which are communicable among scientists. Empirical equations which relate the measurement quantities in arbitrary formulae do not provide excellent knowledge representation for the understanding and the communication among domain experts.

5 Algorithm of Smart Discovery System (SDS)

In this section, an algorithm of our "Smart Discovery System (SDS)" to discover a law equation based on the mathematical admissibility and the experiments on the objective behaviors is explained. An important point to perform these procedures is to establish a method to check if an equation holds for all behaviors which can be occurred in the experiments on the objective behaviors. A natural approach is to collect all possible combinations of the values of the controllable quantities in experiments and to fit the various candidate equations to the collected data. However, this generate and test approach faces the combinatorial explosion in the data collection and the candidate equation generation. To avoid this difficulty, we introduce the following assumptions.

(a) The objective behaviors are represented by a complete equation, and all quantities except one dependent quantity are controllable at least.
(b) The objective behaviors are static, or the time derivatives of some quantities are directly observable if the behaviors are dynamic.
(c) Given a pair of any quantities observed in the objective behaviors, the bi-variate relation on the pair can be identified while fixing the values of the other quantities in experiments.

5.1 Discovery of Regime Equations

Bi-Variate Fitting: If the objective behaviors and the experimental conditions satisfy these assumptions, "*bi-variate fitting*" which searches a pair wise relation of two observed quantities can be applied to reduce the data for the search. In addition, the mathematical admissibility criterion on the scale-type is used to limit the equation formula to be fitted to the observed data. Initially, for a pair of interval scale quantities $\{x_i, x_j\}$, a linear relation

$$b_{ij}x_i + x_j = d_{ij}$$

is searched in the fitting based on the constraints in the table 1 where b_{ij} should be a constant coefficient. For a pair of ratio scale quantities, a power relation

$$x_i{}^{a_{ij}}x_j = d_{ij}$$

is searched. In case of a pair of an interval scale x_i and ratio scale x_j, the following two candidate relations are searched.

$$b_{ij}x_i + x_j^{a_{ij}} = d_{ij},\ b_{ij}x_i + \log x_j = d_{ij}.$$

The goodness of fitting is checked in every fitting by the statistical F-test (Beaumont et al., 1996). The same experiments are repeated $m = 10$ times. Then the bi-variate fittings to the data obtained in each experiment are conducted to check the reproducibility of the coefficient b_{ij}, *i.e.*, its constancy, through χ^2-test, and the effect of noise and error on their values are reduced by averaging the coefficients over $m = 10$ results. By applying these fitting to every pair of quantities in the data, all bi-variate relations satisfying the constraints in the table 1 are identified. The mathematical complexity of the bi-variate fitting is $O(mn^2)$ where n is the total number of the quantities in the given data.

Triplet Test: In the next step, the mutually consistent bi-variate relations are composed to multiple regime equations shown in the theorem 1. Each regime equation is composed in bottom up manner which searches the equation relating less number of quantities in the data. The consistent composition is made through the following "triplet test". The consistency among the values of the constant coefficients in a triplet of the bi-variate relations for three observed quantities is checked under the assumption of a linear relation among the interval scale quantities as indicated in the theorem 1.

For example, given a set of three interval scale quantities $\{x_i, x_j, x_k\}$, if the following three bi-variate relations among them are mutually consistent,

$$b_{ij}x_i + x_j = d_{ij},\ b_{jk}x_j + x_k = d_{jk},\ b_{ki}x_k + x_i = d_{ki}$$

the following relation holds among the coefficients.

$$1 = b_{ij}b_{jk}b_{ki}.$$

This condition can be tested by the normal distribution test considering the error bounds of the coefficients. The error bounds of b_{ij}, b_{jk} and b_{ki}, *i.e.*, $\Delta b_{ij}, \Delta b_{jk}$ and Δb_{ki}, can be statistically evaluated based on the errors of the m least square fittings of each relation. Then the total error bound Δb_{rhs} of the right hand side of the above relation is derived by the following formula of error propagation.

$$\Delta b_{rhs} = \sqrt{(b_{jk}b_{ki}\Delta b_{ij})^2 + (b_{ij}b_{ki}\Delta b_{jk})^2 + (b_{ij}b_{jk}\Delta b_{ki})^2}$$

This standard deviation error bound is used to judge if the value of the product of the three coefficients are sufficiently close to 1 under the normal distribution test.

The principle of this test can be applied to the other triplets containing of ratio and/or interval scale quantities. If the consistency is confirmed, they can be merged into a relation. In the above example, they are merged into

$$x_i + b_{jk}b_{ki}x_j + b_{ki}x_k = \pi_{ijk},$$

where π_{ijk} is an intermediate derivative quantities composed by b_{jk} and b_{ki} which are known to be dependent of the other quantities, i.e., constants.

This procedure is continued for another quantity x_ℓ and any two quantities in $\{x_i, x_j, x_k\}$. If every triplet among the bi-variate relations of $\{x_h, x_i, x_j, x_\ell\}$ is consistent, they can be merged to a relation among the four quantities since all constant coefficients in a linear formulae are mutually consistent. In this case, the following linear relation is obtained.

$$x_i + b_{j\ell}b_{\ell j}x_j + b_{k\ell}b_{\ell k}x_k + b_{\ell i}x_\ell = \pi_{ijk\ell},$$

This procedure further repeated until no larger sets of quantities having consistency are found. This is similar to the generalization of bi-variate relations to multi-variate relations in BACON (Langley et al., 1985). However, the computational complexity of the triplet test $O(n^3)$ is lower than the conventional approach. This is because of the use of the mathematical admissibility constraints and the systematic triplet consistency test. Through this procedure, the set of regime equations relating the many original quantities with less number of dimensionless quantities $\{\Pi_i | i = 1, \ldots, n-r-s\}$ can be discovered, and this efficiently reduces the computational cost for the discovery of a complete law equation.

5.2 Discovery of Ensemble Equation

Term Merge: Once all regimes to define Π_is are discovered, an ensemble equation among Π_is is searched. Because the ensemble equation does not follow the scale-type constraints, it can take any arbitrary formula. Accordingly, we introduce an assumption that the ensemble equation consists of only the arithmetic operators and elementary functions among Π_is to limit the search space of the formula. The most of the law equations follows this assumption, and it is widely used in the other equation discovery approaches (Falkenhainer & Michalski, 1986).

In our approach, a set CE of candidate binary relations such as addition, multiplication, linear, exponential and logarithmic relations is given. Then by the technique of the bi-variate fitting, each relation in CE is applied to the data of Π_is calculated by the regime equations. For example, the following bi-variate product form and linear form are applied.

$$\Pi_i^{a_{ij}} \Pi_j = b_{ij} \text{ (product form) and}, a_{ij}\Pi_i + \Pi_j = b_{ij}, \text{ (linear form)}.$$

First, the former product form is adopted to the least square fitting to every pair of Π_i and $\Pi_j (i, j = 1, \ldots, n-r-s)$. Then, the statistical F-tests mentioned earlier are applied.

This process is repeated over the $k = 10$ different data sets obtained in the random experiments. The bi-variate equations passed all these tests are stored, and the invariance of the exponent a_{ij} of each bi-variate relation against the value changes of any other quantities are checked by examining the $k = 10$ values of a_{ij} obtained in the experiments through χ^2-test. If a_{ij} is invariant, we observe a high possibility that a_{ij} is a constant characterizing the nature of the objective system within the scope of the experiment. The relations having the invariant a_{ij}s are marked, and every maximal convex set MCS of quantities is searched where all pairs of quantities in MCS are related by the bi-variate relations marked as having the invariant a_{ij}. Then the quantities in every MCS are merged into the following term.

$$\Theta_i = \prod_{x_j \in MCS_i} x_j{}^{a_j}.$$

Similar procedure is applied to the linear bi-variate form, in which case the merged term of an MCS is as follows.

$$\Theta_i = \sum_{x_j \in MCS_i} a_j x_j.$$

This procedure is recursively repeated for all bi-variate relations in CE among Π_is and Θ_is until no new term becomes available. A Θ_i is a unique derivative term in each relation which is dependent of the values of the other Π_is outside the relation.

Identity Constraint: If all terms are merged into one in the above term merge process, the relation is the ensemble equation. Otherwise the following procedure to merge the Θ_is further continues by applying an extra mathematical constraint based on the "identity" of the relations. The basic principle of the identity constraints comes by answering the question that "what is the relation among Θ_h, Θ_i and Θ_j, if $\Theta_i = f_{\Theta_j}(\Theta_h)$ and $\Theta_j = f_{\Theta_i}(\Theta_h)$ are known?" For example, if $a(\Theta_j)\Theta_h + \Theta_i = b(\Theta_j)$ and $a(\Theta_i)\Theta_h + \Theta_j = b(\Theta_i)$ are given, the following identity equation is obtained by solving each for Θ_h.

$$\Theta_h \equiv -\frac{\Theta_i}{a(\Theta_j)} + \frac{b(\Theta_j)}{a(\Theta_j)} \equiv -\frac{\Theta_j}{a(\Theta_i)} + \frac{b(\Theta_i)}{a(\Theta_i)}$$

Because the third expression is linear with Θ_j for any Θ_i, the second must be so. Accordingly, the following must hold.

$$1/a(\Theta_j) = \alpha_1\Theta_j + \beta_1,\ b(\Theta_j)/a(\Theta_j) = -\alpha_2\Theta_j - \beta_2.$$

By substituting these to the second expression,

$$\Theta_h + \alpha_1\Theta_i\Theta_j + \beta_1\Theta_i + \alpha_2\Theta_j + \beta_2 = 0$$

is obtained.

This principle is generalized to various relations among multiple terms. Table 2 shows such relations for multiple linear relations and multiple product relations. The relation is used to fit to the data and to merge Θ_is further into another new term Θ which is a coefficient of the relation dependent of the values of the other Θ_is outside the relation. Similarly to the bi-variate fitting, the goodness of fitting is checked by the statistical F-test. These merging operations are repeated until a complete ensemble equation among the terms is obtained where all coefficients are constant in a relation.

Table 2. Identity constraints

bi-variate relation	general relation
$ax + y = b$	$\sum_{(A_i \in 2^{LQ}) \& (p \not\subseteq A_i \forall p \in L)} a_i \prod_{x_j \in A_i} x_j = 0$
$x^a y = b$	$\prod_{(A_i \in 2^{PQ}) \& (p \not\subseteq A_i \forall p \in P)} \exp(a_i \prod_{x_j \in A_i} \log x_j) = 0$

L is a set of pair wise terms having a bi-variate linear relation and $LQ = \cup_{p \in L} p$. P is a set of pair wise terms having a bi-variate product relation and $PQ = \cup_{p \in P} p$.

6 Application to Law Equation Discovery

6.1 Discovery of Law-Based Models

The aforementioned principles have been implemented to "Smart Discovery System (SDS)" (Washio & Motoda, 1997). SDS receives the data and the scale-type information of the quantities observed in model simulations, and tries to discover a complete law equation governing the simulation without knowing the model.

First, the application of SDS to a circuit depicted in Fig. 2 is demonstrated. This is a circuit of photometer to measure the rate of increase of photo intensity within a certain time period. This is represented by the following complete equation containing 18 quantities.

$$(\frac{R_3 h_{fe_2}}{R_3 h_{fe_2} + h_{ie_2}} \frac{R_2 h_{fe_1}}{R_2 h_{fe_1} + h_{ie_1}} \frac{rL^2}{rL^2 + R_1})(V_1 - V_0) - \frac{Q}{C} - \frac{K h_{ie_3} X}{B h_{fe_3}} = 0 \quad (5)$$

Here, L and r are photo intensity and sensitivity of the Csd device which is one of popular optical sensors. X, K and B are the position of indicator, spring constant and the intensity of magnetic field of the current meter respectively. h_{ie_i} is the input impedance of the base of the i-th transistor. h_{fe_i} is the gain ratio of the currents at the base and the collector of the i-th transistor. The definitions of the other quantities follow the standard symbolic representations in the electric circuit domain.

The electric voltage levels V_1 and V_2 are interval scale and h_{fe_i}s absolute scale. Thus, the set of interval scale quantities is $IQ = \{V_1, V_2\}$, that of ratio scale quantities $RQ = \{L, r, R_1, R_2, R_3, h_{ie_1}, h_{ie_2}, h_{ie_3}, Q, C, X, K, B\}$ and that of absolute scale quantities $AQ = \{h_{fe_1}, h_{fe_2}, h_{fe_3}\}$. In the following equation fitting, the value of each coefficient is rounded into the nearest integer or the nearest inverse of integer, if the value is close enough to it within the error bound. This is due to the empirical observation that the coefficients are often the integers or their inverses in a law equation.

Initially, the bi-variate fitting was applied to IQ, and a binary relation $\Pi_1 = V_1 - V_0$ was obtained. Since IQ includes only the two quantities, the search for Πs in IQ was stopped. In the next step, the bi-variate fitting was applied to the quantities in RQ and Π_1. Because the basic origin of the voltage level has been cancel out between V_1 and V_0, Π_1 became a ratio scale quantity. The resultant binary relations were as follows.

$$L^2 r = b_1, L^{-2} R_1 = b_2, r^{-1} R_1 = b_3, R_2^{-1} h_{ie_1} = b_4, R_3^{-1} h_{ie_2} = b_5, Q^{-1} C = b_6,$$
$$h_{ie_3} X = b_7, h_{ie_3} K = b_8, h_{ie_3}^{-1} B = b_9, XK = b_{10}, X^{-1} B = b_{11}, K^{-1} B = b_{12}$$

Subsequently, the triplet tests were applied to these relations, and the following regime equations were obtained.

$$\Pi_1 = V_1 - V_0, \Pi_2 = R_1 r^{-1.0} L^{-2.0}, \Pi_3 = h_{ie_1} R_2^{-1.0},$$
$$\Pi_4 = h_{ie_2} R_3^{-1.0}, \Pi_5 = h_{ie_3} X K B^{-1.0}, \Pi_6 = Q C^{-1.0}$$

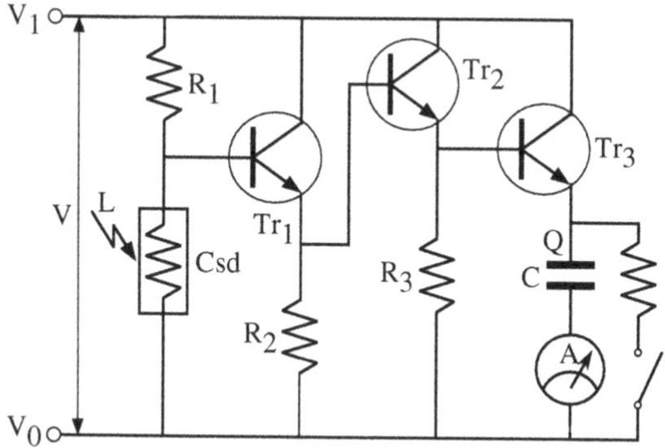

Fig. 2. A circuit of photometer

Then, the merge of these Πs and the quantities in AQ was performed by applying the binary relations in CE, and the following new terms were derived.

$$\begin{aligned}
\Theta_1 &= \Pi_2 h_{fe_1} = R_1 r^{-1.0} L^{-2.0} h_{fe_1}, \\
\Theta_2 &= \Pi_3 h_{fe_2} = h_{ie_1} R_2^{-1.0} h_{fe_2}, \\
\Theta_3 &= \Pi_4 h_{fe_3} = h_{ie_2} R_3^{-1.0} h_{fe_3}, \\
\Theta_4 &= \Pi_5 + \Pi_6 = h_{ie_3} XKB^{-1.0} + QC^{-1.0}, \\
\Theta_5 &= \Pi_1 \Theta_4^{-1.0} = (V_1 - V_0)(h_{ie_3} XKB^{-1.0} + QC^{-1.0})^{-1.0}
\end{aligned}$$

Thus, the quantities were merged into five terms $\{\Theta_1, \Theta_2, \Theta_3, \Theta_5\}$.

Furthermore, the identity constraint was applied to these terms since the binary linear relations were found in the combinations of $\{\Theta_1, \Theta_5\}, \{\Theta_2, \Theta_5\}$ and $\{\Theta_3, \Theta_5\}$. This derived the following multi-linear formula.

$$\Theta_1\Theta_2\Theta_3 + \Theta_1\Theta_2 + \Theta_2\Theta_3 + \Theta_1\Theta_3 + \Theta_1 + \Theta_2 + \Theta_3 + \Theta_5 + 1 = 0$$

Because every coefficient is independent of any terms, this is considered to be the ensemble equation. The equivalence of this result to Eq.(5) is easily checked by substituting the intermediate terms to this ensemble equation.

SDS has been also applied to non-physics domain. For example, given a sound frequency f and a musical sound pitch I where the former is ratio scale and the latter interval scale, the following two candidate relations have been derived by SDS.

$$I = \alpha f^{\beta} + \gamma, \text{ or } I = \alpha \log f + \beta$$

Because both equations show similar accuracy, and the latter contains less parameters, SDS prefers the latter by following the criterion of parsimony which will be discussed later. This equation has been called "*Fechner's Law*" in psychophysics. Another example is the law of spaciousness of a room in psychophysics (Kan et al., 1972).

$$S_p = c \sum_{i=1}^{n} RL_i^{0.3} W_i^{0.3},$$

where S_p, R, L_i and W_i are average spaciousness of a room, room capacity, light intensity and solid angle of window at the location i in the room. Though the unit dimension of S_p is unclear, its scale-type is known to be ratio scale since it was evaluated through the method of magnitude estimation which is a popular method to derive a ratio scale quantity in psychophysics. L and R are ratio scale, and W is absolute scale. SDS easily obtained the above expression.

6.2 Basic Performance of SDS

Table 3 shows the performance of SDS to discover various physical law equations. The relative CPU time of SDS normalized by the first case shows that its computational time is nearly proportional to n^2. For reference, the relative CPU time of ABACUS is indicated for the same cases except for the circuit examples of this paper (Falkenhainer & Michalski, 1986). Though ABACUS applies various heuristics including the information of unit dimension, its computational time is non-polynomial, and it could not derive the law equations for the complicated circuits within a tractable time.

The robustness of SDS against the noisy experimental environment has been also evaluated. The upper limitation of the noise level to obtain the correct result in the cases of more than 80% of 10 trials was investigated for each physical law, and they are indicated in the last column of Table 3. The noise levels shown here are the standard deviation of Gaussian noise relative to the real values of quantities, and were added to both controlled (input) quantities and measured (output) quantities at the same time. Thus actual noise level is higher than these levels. The results show the significant robustness of SDS. This is due to the bottom up approach of the bi-variate fitting where the fitting is generally robust because of its simplicity. SDS can provide appropriate results under any practical noise condition.

As shown in the above results, SDS can discover quite complex law-based models containing more than 10 quantities under practical conditions. As the modeling of the objective behaviors represented by many quantities is a difficult

Table 3. Performance of SDS and ABACUS in reconstructing physical laws

Example	n	TC(S)	TC(A)	NL(S)
Ideal Gas	4	1.00	1.00	±40%
Momentum	8	6.14	22.7	±35%
Coulomb	5	1.63	24.7	±35%
Stoke's	5	1.59	16.3	±35%
Kinetic Energy	8	6.19	285.	±30%
Circuit*1	17	21.6	-	±20%
Circuit*2	18	21.9	-	±20%

n: Number of Quantities, TC(S): Total CPU time of SDS, TC(A): Total CPU Time of ABACUS, NL(S): Limitation of Noise Level of SDS, *1: Case that electronic voltage is represented by a ratio scale V, *2: Case that electronic voltage is represented by two interval scale V_0 and V_1.

and time consuming task for scientists and engineers, the approach presented in this chapter provide a significant advantage.

7 Generic Criteria to Discover Communicable Law Equations

As we have seen in the previous sections, the "Mathematical Admissibility" plays an important role to discover the law equations as communicable knowledge shared by scientists, since it is based on the assumptions and the operations commonly used in the study of scientists. However, this is merely one of the criteria for the communicability. Many other important criteria must be considered in the process of the law equation discovery, and in fact the SDS takes these criteria into account under the environment where the data are experimentally obtained. In this section, the extra and important criteria are discussed. Probably, the complete axiomatization of the definitions and the conditions of law equations without any exception may be difficult since some relations might be named as "laws" in purely empirical manner. However, the clarification of its criteria is considered to be highly important to give a firm basis of the science.

Some of the important conditions on the scientific proposition are given by R. Descartes. They are clarity, distinctness, soundness and consistency in the deduction of the proposition (Descartes, 1637), and these conditions should be also take into account to clarify the scientific law criteria. I. Newton also proposed some conditions of the law equations (Newton, 1686). The first condition is the objectiveness where the relation reflects only the causal assumptions of the nature while excluding any human's mental effects, the second the parsimony of the causal assumptions supporting the relation, the third the generality where the relation holds over the various behaviors in a domain and the forth the soundness where the relation is not violated by any experimental result performed under the environment following the causal assumptions. H.A. Simon also claimed the importance of the parsimony of the law description (Simon, 1977). In the modern physics, the importance of the mathematical admissibility of the relation formulae under the nature of the time and the space also became to be stressed by some major physicists including R.P. Feynman (Feynman, 1965).

We introduce the following definitions and propositions associated with the criteria for the law equations discovery based on the above claims.

Definition 9 (A Scientific Region). *A scientific region T is represented by the following quadruplet.*

$$
\begin{aligned}
&T =< S, A, L, P > \\
&\textit{where} \\
&S = \{s_h | s_h \textit{ is a rule in syntax}, h = 1, \dots, p\}, \\
&A = \{a_i | a_i \textit{ is an axiom in semantics}, i = 1, \dots, q\}, \\
&L = \{\ell_j | \ell_j \textit{ is a postulate in semantics}, j = 1, \dots, r\}, \\
&P = \{o_k | o_k \textit{ is an objective behavior}, k = 1, \dots, s\}.
\end{aligned}
$$

S is the syntax of T, and for example its elements are the coordinate system, the definitions of quantities such as velocity and energy and the definitions of the algebraic operators in physics. The axioms in A are the set of the mathematical relations independent of objective behaviors, for example, the relations of distances among points in an Euclidean space. A postulate $\ell_j (\in L)$ is a law equation where its validity is empirically believed under some conditions which will be described later. An example is the following law of gravity in physics.

$$F = G\frac{M_1 M_2}{R^2}, \tag{6}$$

where $F[kg{\cdot}m/s^2]$ is the gravity force interacting between two mass points $M_1[kg]$ and $M_2[kg]$ when their interval distance is $R[m]$. $G[m^3/(kg{\cdot}s^2)]$ is the gravity constant. A and L give the semantics of T.

In addition, the definition of T involves a set of objective behaviors P which is analyzed in the scientific domain, since the scientific domain is established for the purpose to study some limited part of the universe. In other words, S, A and L must be valid within the analysis of P, and hence each ℓ_j is requested to satisfy the conditions of the law equations for P but not requested outside of PD

Moreover, an ℓ_j is used in the analysis of a part of P but not necessarily used for all of P. For example, the law of gravity is not necessarily used in the analysis of a spring behavior.

Definition 10 (Objective Behaviors of a Relation). *Given a mathematical relation e, if all quantities in e appear in the description of a behavior as mutually relevant quantities, the behavior is called an "objective behavior of e". A subset of P, in which the behaviors are the objective of e, is called "the set of the objective behaviors of e"* $P_e(\subseteq P)$.

For example, the gravity interaction between mass points characterized by the quantities of F, M_1, M_2 and R is an objective behavior of the aforementioned law of gravity.

Definition 11 (Satisfaction and Consistency of a Relation). *Given a mathematical relation e and its objective behavior, if the behavior is explicitly constrained by e, e is said to be "satisfactory" in the behavior. On the other hand, if the behavior does not explicitly violate e, e is said to be "consistent" with the behavior.*

When we consider the kinematic momentum conservation in the collision of two mass points, if the mass points are very heavy, this behavior is analyzed under the requirement that the law of gravity should be satisfactory. Otherwise, the law of gravity is ignored. But, it should be consistent in both cases.

Based on these definitions and the aforementioned claims of some major scientists, the criteria of a relation e to be a law equation are described as follows

(1) **Objectiveness:** All quantities appearing in e are observable directly and/or indirectly in the behaviors in P_e.

(2) **Generality:** The satisfaction of e is widely identified in the test on the behaviors included in P_e.

(3) **Reproducibility:** For every behavior in P_e, the identical result on the satisfaction and the consistency is identified in repeated tests.

(4) **Soundness:** The consistency of e is identified in the test on every behavior in P_e.

(5) **Parsimony:** e includes the least number of quantities to characterize the behaviors in P_e.

(6) **Mathematical: Admissibility** e follows the syntax S and the axioms of the semantics A.

Here, the "test" is an experiment or an observation, and the "identification" is to confirm a fact in the test while considering the uncertainty and/or the accuracy of the test. Though the objectiveness and the generality include the criteria of (3), (4) and (5) in wider sense, each criterion is more specifically defined in this literature to reduce their ambiguity.

Some widely known scientific relations are not identified as law equations among scientists. For example, given the enforced turbulence flow in a circular pipe, the heat transfer behavior from the flow liquid to the pipe wall is represented by the following Dittus-Boelter equation which is called as an "experimental equation" but not a "law equation" in thermo-hydraulics domain.

$$Nu = 0.023Re^{0.8}Pr^{0.4}, \tag{7}$$

where $Nu = hd/\lambda$, $Re = \rho ud/\eta$, and $Pr = \eta c_p/\lambda$. $h[W/(m^2 \cdot {}^\circ K)]$ is the coefficient of the heat transfer rate between the liquid and the wall, $d[m]$ the diameter of the circular pipe, $\lambda[W/(m \cdot {}^\circ K)]$, $\rho[kg/m^3]$, $u[m/s]$, $\eta[Pa \cdot s]$, $c_p[J/(kg \cdot {}^\circ K)]$ the heat conductance, density, velocity, viscosity and specific heat of the liquid under a constant pressure respectively (Kouzou, 1986).

This relation stands objectively independent of our interpretation. The set of the objective behaviors of the thermo-hydraulics P includes all behaviors over all value ranges of Nu, Re and Pr. Thus, according to the definition 10, P_e of the Dittus-Boelter equation is the set of all behaviors represented in some value ranges of Nu, Re and Pr in P. This equation meets the criterion of the objectiveness because Nu, Re and Pr are observable through some experiments. It is general over various enforced turbulence flows in circular pipes and reproducible for the repetition of the tests. It also has a parsimonious shape, and satisfies the unit dimensional constraint in terms of the mathematical admissibility. However, this equation is not sound in P_e, because it stands for only the value ranges of $10^4 \leq Re \leq 10^5$ and $1 \leq Pr \leq 10$, and is explicitly violated outside of these ranges. In this regard, this equation is not a law equation.

On the other hand, P of the classical mechanics includes the behaviors over all value ranges of mass, distance and force, and thus P_e of the law of gravity is the set of all behaviors represented in some value ranges of these quantities. This equation also meets the criteria of objectiveness, generality, reproducibility, parsimony and mathematical admissibility in P_e. Furthermore, as any behaviors in P_e do not violate this relation, it is sound.

Strictly speaking, the verifications of the generality and the soundness are very hard since they require the experimental knowledge on various behaviors. However, these can be checked if we relax the requirements to limit the verification within a given set of the objective behaviors. Under this premise, SDS seeks an equation having the generality to explain all behaviors shown by the combinations of the values of some quantities in the experiments on the objective behaviors. It also seeks the equation having the soundness not to contradict with all behaviors observed in the experiments. Eventually, the generality is subsumed by the soundness by limiting the behaviors for the verification. The objectiveness is ensured by seeking the relation among directly and indirectly observed quantities. The reproducibility is also ensured by checking if identical bi-variate relations are obtained multiple times in the repeated statistical tests. The parsimony is automatically induced in the algorithm to compose the equation in bottom up manner. The mathematical admissibility is well addressed as mentioned earlier.

8 Summary

In this chapter, the criteria on the relation among quantities observed in objective behaviors to be a the law equation as the communicable knowledge among domain experts were discussed through the demonstration of a law discovery system SDS. Especially, the criterion of the mathematical admissibility has been analyzed in detail on the axiomatic basis. The definitions of scale-types of quantities and the admissibility conditions on their relations based on the characteristics of the scale-types have been introduced, and the extension of the major theorems in the unit dimensional analysis was shown. Through these analyses, the communicability criteria of the law equation have been clarified.

Moreover, the superior performance of SDS was demonstrated through some simulation experiments. In the evaluation, the validity of the presented principles has been confirmed, and its power to systematically discover candidate law equations over various domains along the communicability criteria has been shown.

In the recent study, the function and ability of SDS have been further extended. It became to discover law-based models consisting of simultaneous equations (Washio & Motoda, 1998). Moreover, the most recent version of SDS can discover the law-based models from the data which are passively observed not in the artificial experiments but the natural environment (Washio et al., 1999). These developments extend the practical domains where communicable law equations are discovered for scientists.

References

Beaumont, A.P., Knowles, J.D., Beuamont, G.P.: Statistical tests: An introduction with minitab commentary. Prentice Hall, Upper Saddle River, NJ (1996)

Bridgman, P.W.: Dimensional analysis. Yale University Press, New Haven, CT (1922)

Buckingham, E.: On physically similar systems; Illustrations of the use of dimensional equations. Physical Review IV(4), 345–376 (1914)

Descartes, R.: Discours de la Methode/Discourse on the Method. Notre Dame. In: University of Notre Dame Press. Bilingual edition (1637/1994)

Džeroski, S., Todorovski, L.: Discovering dynamics: From inductive logic programming to machine discovery. Journal of Intelligent Information Systems 4, 89–108 (1995)

Falkenhainer, B.C., Michalski, R.S.: Integrating quantitative and qualitative discovery: The ABACUS system. Machine Learning 1, 367–401 (1986)

Feynman, R.P.: The character of physical law. MIT Press, Boston, MA (1965)

Kan, M., Miyata, N., Watanabe, K.: Research on spaciousness. Japanese Journal of Architecture 193, 51–57 (1972)

Koehn, B., Zytkow, J.M.: Experimeting and theorizing in theory formation. In: Proceedings of the International Symposium on Methodologies for Intelligent Systems, pp. 296–307. Knoxville, TN (1986)

Kokar, M.M.: Determining arguments of invariant functional descriptions. Machine Learning 1, 403–422 (1986)

Kouzou, T.: Material on heat transfer engineering. The Japan Society of Mechanical Engineers (JSME) 1, Ch. 2, 55–56 (1986)

Krantz, D.H., Luce, R.D., Suppes, P., Tversky, A.: Fundations of measurement. Academic Press, New York (1971)

Langley, P.W., Simon, H.A., Bradshaw, G., Zytkow, J.M.: Scientific discovery: Computational explorations of the creative process. MIT Press, Cambridge, MA (1985)

Luce, R.D.: On the possible psychological laws. The Psychological Review 66, 81–95 (1959)

Newton, I.: Principia, vol.II, The System of the World. Translated into English by Motte, A. (1729). University of California Press, Berkeley, CA, Copyright 1962 (1686)

Rissanen, J.: Modeling by shortest data description. Automatica 14, 465–471 (1978)

Simon, H.A.: Models of discovery. D. Reidel Publishing Company, Dordrecht, Holland (1977)

Stevens, S.S.: On the theory of scales of measurement. Science 103, 677–680 (1946)

Todorovski, L., Džeroski, S.: Declarative bias in equation discovery. In: Proceedings of the Fourteenth International Conference on Machine Learning, Nashville, TN, pp. 376–384 (1997)

Washio, T., Motoda, H.: Discovering admissible models of complex systems based on scale-types and identity constraints. In: Proceedings of Fifteenth International Joint Conference on Artificial Intelligence, Nagoya, Japan, pp. 810–817 (1997)

Washio, T., Motoda, H.: Discovering admissible simultaneous equations of large scale systems. In: Proceedings of Fifteenth National Conference on Artificial Intelligence, Madison, WI, pp. 189–196 (1998)

Washio, T., Motoda, H., Niwa, Y.: Discovering admissible model equations from observed data based on scale-types and identity constraints. In: Proceedings of Sixteenth International Joint Conference on Artificial Intelligence, Stockholm, Sweden, pp. 772–779

Washio, T., Motoda, H., Niwa, Y.: Enhancing the plausibility of law equation discovery. In: Proceedings of the Seventeenth International Conference on Machine Learning, Stanford, CA, pp. 1127–1134 (2000)

Quantitative Revision of Scientific Models

Kazumi Saito[1] and Pat Langley[2]

[1] NTT Communication Science Laboratories
NTT Corporation, Kyoto, Japan
saito@cslab.kecl.ntt.co.jp
[2] Institute for the Study of Learning and Expertise
Palo Alto, California, USA
langley@isle.org

Abstract. Research on the computational discovery of numeric equations has focused on constructing laws from scratch, whereas work on theory revision has emphasized qualitative knowledge. In this chapter, we describe an approach to improving scientific models that are cast as sets of equations. We review one such model for aspects of the Earth ecosystem, then recount its application to revising parameter values, intrinsic properties, and functional forms, in each case achieving reduction in error on Earth science data while retaining the communicability of the original model. After this, we consider earlier work on computational scientific discovery and theory revision, then close with suggestions for future research on this topic.

1 Research Goals and Motivation

Research on computational approaches to scientific knowledge discovery has a long history in artificial intelligence, dating back over two decades (e.g., Langley, 1979; Lenat, 1977; Lindsay et al., 1980). This body of work has led steadily to more powerful methods and, in recent years, to new discoveries deemed worth publication in the scientific literature, as reviewed by Langley (1998). However, despite this progress, mainstream work on the topic retains some important limitations.

One drawback is that few approaches to the intelligent analysis of scientific data can use available knowledge about the domain to constrain search for laws or explanations. Moreover, although early work on computational discovery cast discovered knowledge in notations familiar to scientists, more recent efforts have not. Rather, influenced by the success of machine learning and data mining, many researchers have adopted formalisms developed by these fields, such as decision trees and Bayesian networks. A return to methods that operate on established scientific notations seems necessary for scientists to understand their results.

Like earlier research on computational scientific discovery, our general approach involves defining a space of possible models stated in an established scientific formalism, specifically sets of numeric equations, and developing techniques to search that space. However, it differs from previous work in this area by starting from an existing scientific model and using heuristic search to revise the

S. Džeroski and L. Todorovski (Eds.): Computational Discovery, LNAI 4660, pp. 120–137, 2007.

model in ways that improve its fit to observations. Although there exists some research on theory refinement (e.g., Ourston & Mooney 1990; Towell, 1991), it has emphasized qualitative knowledge rather than quantitative models that relate continuous variables, which play a central role in many sciences.

In the pages that follow, we describe an approach to revising quantitative models of complex systems. We believe that our approach is general and is appropriate for many scientific domains, but we have focused our efforts on one area – certain aspects of the Earth ecosystem – for which we have a viable model, existing data, and domain expertise. We briefly review the domain and model before moving on to describe our approach to knowledge discovery and model revision. After this, we present some initial results that suggest our approach can improve substantially the model's fit to available data. We close with a discussion of related discovery work and directions for future research.

2 A Quantitative Model of the Earth Ecosystem

Data from the latest generation of satellites, combined with readings from ground sources, hold great promise for testing and improving existing scientific models of the Earth's biosphere. One such model, CASA, developed by Potter and Klooster (1997, 1998) at NASA Ames Research Center, accounts for the global production and absorption of biogenic trace gases in the Earth atmosphere, as well as predicting changes in the geographic patterns of major vegetation types (e.g., grasslands, forest, tundra, and desert) on the land.

CASA predicts, with reasonable accuracy, annual global fluxes in trace gas production as a function of surface temperature, moisture levels, and soil properties, together with global satellite observations of the land surface. The model incorporates difference equations that represent the terrestrial carbon cycle, as well as processes that mineralize nitrogen and control vegetation type. These equations describe relations among quantitative variables and lead to changes in the modeled outputs over time. Some processes are contingent on the values of discrete variables, such as soil type and vegetation, which take on different values at different locations. CASA operates on gridded input at different levels of resolution, but typical usage involves grid cells that are eight kilometers square, which matches the resolution for satellite observations of the land surface.

To run the CASA model, the difference equations are repeatedly applied to each grid cell independently to produce new variable values on a daily or monthly basis, leading to predictions about how each variable changes, at each location, over time. Although CASA has been quite successful at modeling Earth's ecosystem, there remain ways in which its predictions differ from observations, suggesting that we invoke computational discovery methods to improve its ability to fit the data. The result would be a revised model, cast in the same notation as the original one, that incorporates changes which are interesting to Earth scientists and which improve our understanding of the environment.

Because the overall CASA model is quite complex, involving many variables and equations, we decided to focus on one portion that lies on the model's

Table 1. Variables used in the NPPc portion of the CASA ecosystem model

NPPc is the net plant production of carbon at a site during the year.

E is the photosynthetic efficiency at a site after factoring various sources of stress.

T1 is a temperature stress factor ($0 < T1 < 1$) for cold weather.

T2 is a temperature stress factor ($0 < T2 < 1$), nearly Gaussian in form but falling off more quickly at higher temperatures.

W is a water stress factor ($0.5 < W < 1$) for dry regions.

Topt is the average temperature for the month at which MON-FAS-NDVI takes on its maximum value at a site.

Tempc is the average temperature at a site for a given month.

EET is the estimated evapotranspiration (water loss due to evaporation and transpiration) at a site.

PET is the potential evapotranspiration (water loss due to evaporation and transpiration given an unlimited water supply) at a site.

PET-TW-M is a component of potential evapotranspiration that takes into account the latitude, time of year, and days in the month.

A is a polynomial function of the annual heat index at a site.

AHI is the annual heat index for a given site.

MON-FAS-NDVI is the relative vegetation greenness for a given month as measured from space.

IPAR is the energy from the sun that is intercepted by vegetation after factoring in time of year and days in the month.

FPAR-FAS is the fraction of energy intercepted from the sun that is absorbed photosynthetically after factoring in vegetation type.

MONTHLY-SOLAR is the average solar irradiance for a given month at a site.

SOL-CONVER is 0.0864 times the number of days in each month.

UMD-VEG is the type of ground cover (vegetation) at a site.

'fringes' and that does not involve any difference equations. Table 1 describes the variables that occur in this submodel, in which the dependent variable, NPPc, represents the net production of carbon. As Table 2 indicates, the model predicts this quantity as the product of two unobservable variables, the photosynthetic efficiency, E, at a site and the solar energy intercepted, IPAR, at that site.

Photosynthetic efficiency is in turn calculated as the product of the maximum efficiency (0.56) and three stress factors that reduce this efficiency. One stress term, T2, takes into account the difference between the optimum temperature, Topt, and actual temperature, Tempc, for a site. A second factor, T1, involves the nearness of Topt to a global optimum for all sites, reflecting the intuition that plants which are better adapted to harsh temperatures are less efficient overall. The third term, W, represents stress that results from lack of moisture as

Table 2. Equations used in the NPPc portion of the CASA ecosystem model

$$
\begin{aligned}
&\text{NPPc} = \textstyle\sum_{month} \max\,(\text{E} \cdot \text{IPAR}, 0)\\
&\quad \text{E} = 0.56 \cdot \text{T1} \cdot \text{T2} \cdot \text{W}\\
&\qquad \text{T1} = 0.8 + 0.02 \cdot \text{Topt} - 0.0005 \cdot \text{Topt}^2\\
&\qquad \text{T2} = 1.18/[(1 + e^{0.2\cdot(\text{Topt}-\text{Tempc}-10)}) \cdot (1 + e^{0.3\cdot(\text{Tempc}-\text{Topt}-10)})]\\
&\qquad \text{W} = 0.5 + 0.5 \cdot \text{EET/PET}\\
&\qquad\quad \text{PET} = 1.6 \cdot (10 \cdot \text{Tempc} / \text{AHI})^A \cdot \text{PET-TW-M if Tempc} > 0\\
&\qquad\quad \text{PET} = 0 \text{ if Tempc} \leq 0\\
&\qquad\qquad A = 0.000000675 \cdot \text{AHI}^3 - 0.0000771 \cdot \text{AHI}^2 + 0.01792 \cdot \text{AHI} + 0.49239\\
&\quad \text{IPAR} = 0.5 \cdot \text{FPAR-FAS} \cdot \text{MONTHLY-SOLAR} \cdot \text{SOL-CONVER}\\
&\qquad \text{FPAR-FAS} = \min((\text{SR-FAS} - 1.08)/\text{SRDIFF(UMD-VEG)}, 0.95)\\
&\qquad\quad \text{SR-FAS} = -\ (\text{MON-FAS-NDVI} + 1000)\ /\ (\text{MON-FAS-NDVI} - 1000)
\end{aligned}
$$

reflected by EET, the estimated water loss due to evaporation and transpiration, and PET, the water loss due to these processes given an unlimited water supply. In turn, PET is defined in terms of the annual heat index, AHI, for a site, and PET-TW-M, another component of potential evapotranspiration.

The energy intercepted from the sun, IPAR, is computed as the product of FPAR-FAS, the fraction of energy absorbed photosynthetically for a given vegetation type, MONTHLY-SOLAR, the average radiation for a given month, and SOL-CONVER, the number of days in that month. FPAR-FAS is a function of MON-FAS-NDVI, which indicates relative greenness at a site as observed from space, and SRDIFF, an intrinsic property that takes on different numeric values for different vegetation types as specified by the discrete variable UMD-VEG.

Of the variables we have mentioned, NPPc, Tempc, MONTHLY-SOLAR, SOL-CONVER, MON-FAS-NDVI, and UMD-VEG are observable. Three additional terms – EET, PET-TW-M, and AHI – are defined elsewhere in the model, but we assume their definitions are correct and thus we can treat them as observables. The remaining variables are unobservable and must be computed from the others using their definitions. This portion of the model also contains some numeric parameters, as shown in the equations in Table 2.

3 An Approach to Quantitative Model Revision

As noted earlier, our approach to scientific discovery involves refining models like CASA that involve relations among quantitative variables. We adopt the traditional view of discovery as heuristic search through a space of models, with the search process directed by candidates' ability to fit the data. However, we assume this process starts not from scratch, but rather with an existing model, and the search operators involve making changes to this model, rather than constructing entirely new structures.

Our long-term goal is not to automate the revision process, but instead to provide an interactive tool that scientists can direct and use to aid their model development. As a result, the approach we describe in this section addresses the task of making local changes to a model rather than carrying out global optimization, as assumed by Chown and Dietterich (2000). Thus, our software takes as input not only observations about measurable variables and an existing model stated as equations, but also information about which portion of the model should be revised. The output is a revised model that fits the observed data better than the initial one.

Below we review two discovery algorithms that we utilize to improve the specified part of a model, then describe three distinct types of revision they support. We consider these in order of increasing complexity, starting with simple changes to parameter values, moving on to revisions in the values of intrinsic properties, and ending with changes in an equation's functional form.

3.1 The RF5 and RF6 Discovery Algorithms

Our approach relies on RF5 and RF6, two algorithms for discovering numeric equations described by Saito and Nakano (1997, 2000). Given data for some continuous variable y that is dependent on continuous predictive variables $x_1, \ldots, x_K$, the RF5 system searches for multivariate polynomial equations of the form

$$\begin{aligned} y &= w_0 + \sum_{j=1}^{J} w_j x_1^{w_{j1}} \cdots x_K^{w_{jK}} \\ &= w_0 + \sum_{j=1}^{J} w_j \exp\left(w_{j1} \ln(x_1) + \cdots + w_{jK} \ln(x_K)\right). \end{aligned} \quad (1)$$

For example, the equation $\mathrm{W} = 0.5 + 0.5 \cdot \mathrm{EET/PET}$ in this scheme becomes $\mathrm{W} = 0.5 + 0.5 \cdot \mathrm{EET}^{+1.0} \cdot \mathrm{PET}^{-1.0}$. When we want to obtain an annual quantity such as W_{annual} by summing up monthly ones like $\mathrm{W}_1, \ldots, \mathrm{W}_{12}$, the equation becomes $\mathrm{W}_{annual} = \sum_{j=1}^{12} \mathrm{W}_j = 6 + \sum_{j=1}^{12} 0.5 \cdot \mathrm{EET}_j^{+1.0} \cdot \mathrm{PET}_j^{-1.0}$. Such functional relations subsume many of the numeric laws found by previous computational discovery systems like BACON (Langley, 1979) and FAHRENHEIT (Żytkow, Zhu, & Hussam, 1990).

Given a functional form of this sort and observations for predictive variables $\mathbf{x}$ and dependent variable y, RF5 produces a polynomial equation with new parameter values by:

1. Transforming the functional form into a three-layer neural network;
2. Carrying out search through the weight space using the BPQ algorithm;
3. Transforming the revised network into a polynomial equation.

RF5's first step involves transforming a candidate functional form with J summed terms into a three-layer neural network based on the rightmost form of expression (1), in which the J hidden nodes in this network correspond to *product units*

(Durbin & Rumelhart, 1989). The system then carries out search through the weight space using the BPQ algorithm, a second-order learning technique that calculates both the descent direction and the step size automatically.

This process halts when it finds a set of weights that minimize the squared error on the dependent variable y. RF5 runs the BPQ method on networks with different numbers of hidden units, then selects the one that gives the best score on an MDL metric. Finally, the program transforms the resulting network into a polynomial equation, with weights on hidden units becoming exponents and other weights becoming coefficients.

The RF6 algorithm extends RF5 by adding the ability to find conditions on a numeric equation that involve nominal variables, which it encodes using one input variable for each nominal value. When Boolean predictive variables $q_1, \ldots, q_L$ are included, the RF6 system first searches for equations of the form

$$y = w_0 + \sum_{j=1}^{J} w_j f\left(v_{j1}q_1 + \cdots + v_{jL}q_L\right) x_1^{w_{j1}} \cdots x_K^{w_{jK}}, \qquad (2)$$

where $f(\cdot)$ denotes a transfer function. For example, consider a nominal variable with two possible values, say "Water" and "Air", and its associated intrinsic property P with positive values, say $P1$ and $P2$. By expressing "Water" as $q_1 = 1$ and $q_2 = 0$, and "Air" as $q_1 = 0$ and $q_2 = 1$, we can compute $P1$ and $P2$ by an expression $\exp(v_1 q_1 + v_2 q_2)$ as a function of the Boolean values q_1 and q_2, if the parameter values are set to $v_1 = \log P1$ and $v_2 = \log P2$. Note that in this example, the transfer function $f(\cdot)$ was exponential in order to guarantee that intrinsic values are always positive. More generally, since the set of the possible output values for $f(\cdot)$ is discrete due to its Boolean inputs, the system can discover conditional laws of the form

$$if\ \text{cond}^i(q_1, \ldots, q_L) = \text{true}, \quad y = w_0 + \sum_{j=1}^{J} w_j^i x_1^{w_{j1}} \cdots x_K^{w_{jK}}, \ i = 1, \ldots, I, \qquad (3)$$

where the polynomial coefficients are defined as $w_j^i = w_j f(\cdot)$ and where I denotes the number of rules.

In order to transform expression (2) to (3), the system first generates one such condition for each training case as a vector $(f(v_{11}q_1 + \cdots + v_{1L}q_L), \ldots, f(v_{J1}q_1 + \cdots + v_{JL}q_L))$ and then utilizes k-means clustering to generate a smaller set of more general conditions, with the number of clusters determined through cross validation. Finally, RF6 invokes decision-tree induction to construct a classifier that discriminates among these clusters, which it transforms into rules that form the nominal conditions on the polynomial equation that RF5 has generated.

As Table 2 reveals, the NPPc portion of the CASA ecosystem model includes some functions like max and min, whose derivatives are undefined at some points in their value range, and therefore we cannot apply the BPQ algorithm directly. One approach to overcoming this problem is to approximate such functions by using a smooth nonlinear transformation like soft-max. However, when both the max and min functions were eliminated, the root mean squared error (RMSE) for the original model on the available data degraded only slightly, from 465.213 to 467.910. Thus, we utilized this simplified NPP model in our experiments on model revision. Here we should emphasize that, given an existing model like in Table 2, the current system requires some handcrafting to encode the equations as a neural network.

3.2 Three Types of Model Refinement

There exist three natural types of refinement within the class of models, like CASA, that are stated as sets of equations that refer to unobservable variables. These include revising the parameter values in equations, altering the values for an intrinsic property, and changing the functional form of an existing equation.

Improving the parameters for an equation is the most straightforward process. The NPPc portion of CASA contains some parameterized equations that our Earth science team members believe are reliable, like that for computing the variable A from AHI, the annual heat index. However, it also includes equations with parameters about which there is less certainty, like the expression that predicts the temperature stress factor T2 from Tempc and Topt. Our approach to revising such parameters relies on creating a specialized neural network that encodes the equation's functional form using ideas from RF5, but also including a term for the unchanged portion of the model. We then run the BPQ algorithm to find revised parameter values, initializing weights based on those in the model.

We can utilize a similar scheme to improve the values for an intrinsic property like SRDIFF that the model associates with the discrete values for some nominal variable like UMD-VEG (vegetation type). We encode each nominal term as a set of dummy variables, one for each discrete value, making the dummy variable equal to one if the discrete value occurs and zero otherwise. We introduce one hidden unit for the intrinsic property, with links from each of the dummy variables and with weights that correspond to the intrinsic values associated with each discrete value. To revise these weights, we create a neural network that incorporates the intrinsic values but also includes a term for the unchanging parts of the model. We can then run BPQ to revise the weights that correspond to intrinsic values, again initializing them to those in the initial model.

Altering the form of an existing equation requires somewhat more effort, but maps more directly onto previous work in equation discovery. In this case, the details depend on the specific functional form that we provide, but because we have available the RF5 and RF6 algorithms, the approach supports any of the forms that they can discover or specializations of them. Again, having identified a particular equation that we want to improve, we create a neural network that encodes the desired form, then invoke the BPQ algorithm to determine its parametric values, in this case initializing the weights randomly.

In the next section, we provide examples of neural networks that result from different types of revisions. This approach to model refinement can modify more than one equation or intrinsic property at a time. However, we can reasonably assume that scientists may want to change a small portion, and this is consistent with the interactive process described earlier. We envision the scientist identifying a portion of the model that he thinks could be better, running one of the three revision methods to improve its fit to the data, and repeating this process until he is satisfied.

4 Initial Results on Ecosystem Data

In order to evaluate our approach to scientific model revision, we obtained data relevant to the NPPc model from the Earth science members of our team. These data consisted of observations from 303 distinct sites with known vegetation type and for which measurements of Tempc, MON-FAS-NDVI, MONTHLY-SOLAR, SOL-CONVER, and UMD-VEG were available for each month during the year. In addition, other portions of CASA were able to compute values for the variables AHI, EET, and PET-TW-M. The resulting 303 training cases seemed sufficient for initial tests of our revision methods, so we used them to drive three different changes to the handcrafted model of carbon production.

4.1 Results on Parameter Revision

Our Earth science team members identified the equation for T2, one of the temperature stress variables, as a likely candidate for revision. As noted earlier, the handcrafted expression for this term was

$$T2 = 1.8/[(1 + e^{0.2(Topt - Tempc - 10)})(1 + e^{0.3(Tempc - Topt - 10)})],$$

which produces a Gaussian-like curve that is slightly asymmetrical. This reflects the intuition that photosynthetic efficiency will decrease when temperature (Tempc) is either below or above the optimal (Topt).

To improve upon this equation, we defined $x = \text{Topt} - \text{Tempc}$ as an intermediate variable and recast the expression for T2 as the product of two sigmoidal functions of the form $\sigma(a) = 1/(1 + \exp(-a))$ and a parameter. We transformed these into a neural network and used BPQ to minimize the error function

$$\mathcal{F}_1 = \sum_{site} \left(\text{NPPc} - \sum_{month} w_0 \cdot \sigma(v_{10} + v_{11} \cdot x) \cdot \sigma(v_{20} - v_{21} \cdot x) \cdot \text{Rest}\right)^2,$$

over the parameters $\{w_0, v_{10}, v_{11}, v_{20}, v_{21}\}$, where $\text{Rest} = 0.56 \cdot \text{T1} \cdot \text{W} \cdot \text{IPAR}$. The resulting equation generated in this manner was

$$T2 = 1.80/[(1 + e^{0.05(Topt - Tempc - 10.8})(1 + e^{-0.03(Tempc - Topt - 90.33})],$$

which has reasonably similar values to the original ones for some parameters but quite different values for others.

The root mean squared error (RMSE) for the original model on the available data was 467.910. In contrast, the error for the revised model was 457.757 on the training data and 461.466 using leave-one-out cross validation. Thus, RF6's

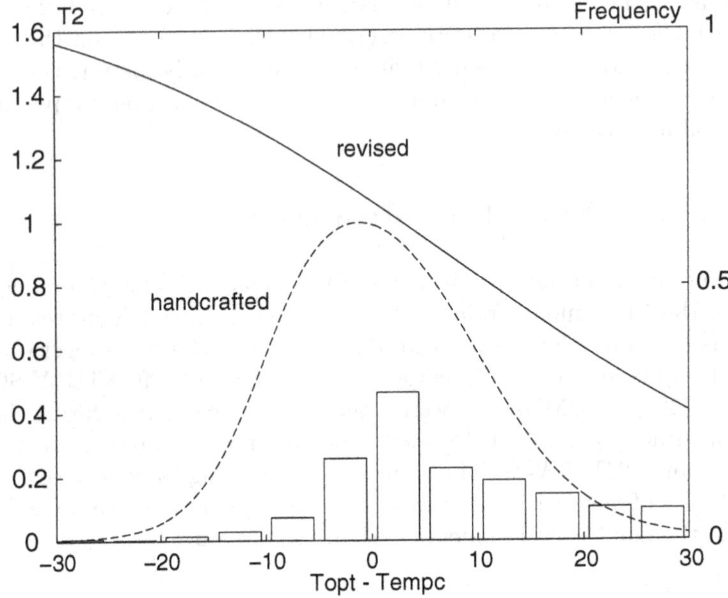

Fig. 1. Behavior of handcrafted and revised equations for the stress variable T2

modification of parameters in the T2 equation produced slightly more than one percent reduction in overall model error, which is somewhat disappointing.

However, inspection of the resulting curves reveals a more interesting picture. Figure 1 plots the values of the temperature stress factor T2, using the revised equations, as a function of the difference Topt − Tempc, where the histogram denotes the frequency of samples for various values of this difference. Although the new curve remains Gaussian-like, its values within the effective range (from −30 to 30 Celsius) decrease monotonically. This seems counterintuitive but interesting from an Earth science perspective, as it suggests this stress factor has little influence on NPPc. Moreover, the original equation for T2 was not well grounded in principles of plant physiology, making empirical improvements of this sort beneficial to the modeling enterprise.

4.2 Results on Intrinsic Value Revision

Another portion of the NPPc model that held potential for revision concerns the intrinsic property SRDIFF associated with the vegetation type UMD-VEG. For each site, the latter variable takes on one of 11 nominal values, such as grasslands, forest, tundra, and desert, each with an associated numeric value for SRDIFF that plays a role in the FPAR-FAS equation. This gives 11 parameters to revise, which seems manageable given the number of observations available.

Table 3. Original and revised values for the SRDIFF intrinsic property, along with the frequency for each vegetation type

vegetation type	A	B	C	D	E	F	G	H	I	J	K
original	3.06	4.35	4.35	4.05	5.09	3.06	4.05	4.05	4.05	5.09	4.05
revised	2.57	4.77	2.20	3.99	3.70	3.46	2.34	0.34	2.72	3.46	1.60
clustered	2.42	3.75	2.42	3.75	3.75	3.75	2.42	0.34	2.42	3.75	2.42
frequency	3.3	8.9	0.3	3.6	21.1	19.1	15.2	3.3	19.1	2.30	3.60

As outlined earlier, to revise these intrinsic values, we introduced one dummy variable, UMD-VEG$_k$, for each vegetation type such that UMD-VEG$_k$ $= 1$ if UMD-VEG $= k$ and 0 otherwise. We then defined SRDIFF(UMD-VEG) as $\exp(-\sum_k v_k \cdot \text{UMD-VEG}_k)$ and, since SRDIFF's value is independent of the month, we used BPQ to minimize, over the weights $\{v_k\}$, the error function

$$\mathcal{F}_2 = \sum_{site} (\text{NPPc} - \exp(\textstyle\sum_k v_k \cdot \text{UMD-VEG}_k) \cdot \text{Rest})^2 ,$$

where Rest $= \sum_{month} E \cdot 0.5 \cdot (\text{SR-FAS} - 1.08) \cdot \text{MONTHLY-SOLAR} \cdot \text{SOL-CONVER}$.

Table 3 shows the initial values for this intrinsic property, as set by the CASA developers, along with the revised values produced by the above approach when we fixed other parts of the NPPc model. The most striking result is that the revised intrinsic values are nearly always lower than the initial values. The RMSE for the original model was 467.910, whereas the error using the revised values was 432.410 on the training set and 448.376 using cross validation. The latter constitutes an error reduction of over four percent, which seems substantial.

However, since the original 11 intrinsic values were grouped into only four distinct values, we applied RF6's clustering procedure over the trained neural network to group the revised values in the same manner. Table 4 shows the effect on error rate as one varies the number of clusters from one to five and that of the basic neural network, which effectively has 11 clusters. As expected, the training RMSE decreased monotonically, but the cross-validation RMSE was minimized for a small number of clusters, specifically three. The estimated error for this revised model is about one percent better than for the one with 11 distinct values.

Again, the clustered values, which constrained the intrinsic values for certain groups of vegetation types to be equal, are nearly always lower than the initial ones, a result that is certainly interesting from an Earth science viewpoint. We suspect that measurements of NPPc and related variables from a wider range of sites would produce intrinsic values closer to those in the original model. However, such a test must await additional observations and, for now, empirical fit to the available data should outweigh the theoretical basis for the initial settings.

Table 4. Error rates for different numbers of distinct SRDIFF values

No. clusters	Training RMSE	Cross-validation RMSE
1	501.56	503.34
2	448.31	453.50
3	436.16	442.95
4	434.72	445.60
5	433.79	446.49
⋮	⋮	⋮
11	432.41	448.38

4.3 Results on Revising Equation Structure

We also wanted to demonstrate our approach's ability to improve the functional form of the NPPc model. For this purpose, we selected the equation for photosynthetic efficiency,

$$E = 0.56 \cdot T1 \cdot T2 \cdot W,$$

which states that this term is a product of the water stress term, W, and the two temperature stress terms, T1 and T2. Because each stress factor takes on values less than one, multiplication has the effect of reducing photosynthetic efficiency E below the maximum 0.56 possible (Potter & Klooster, 1998).

Since E is calculated as a simple product of the three variables, one natural extension was to consider an equation that included exponents on these terms. To this end, we borrowed techniques from the RF5 system to create a neural network for such an expression, then used BPQ to minimize the error function

$$\mathcal{F}_3 = \sum_{site} \left(\text{NPPc} - \sum_{month} u_0 \cdot \text{T1}^{u_1} \cdot \text{T2}^{u_2} \cdot \text{W}^{u_3} \cdot \text{IPAR}\right)^2,$$

over the parameters $\{u_0, u_1, u_2, u_3\}$, which assumes the equations that predict IPAR remain unchanged. We initialized u_0 to 0.56 and the other parameters to 1.0, as in the original model, and constrained the latter to be positive. The revised equation found in this manner,

$$E = 0.521 \cdot T1^{0.00} \cdot T2^{0.03} \cdot W^{0.00},$$

has a small exponent for T2 and zero exponents for T1 and W, suggesting the former influences photosynthetic efficiency in minor ways and the latter not at all. On the available data, the root mean squared error for the original model was 467.910. In contrast, the revised model has an RMSE of 443.307 on the training set and an RMSE of 446.270 using cross validation. Thus, the revised equation produces a substantially better fit to the observations than does the original model, in this case reducing error by almost five percent.

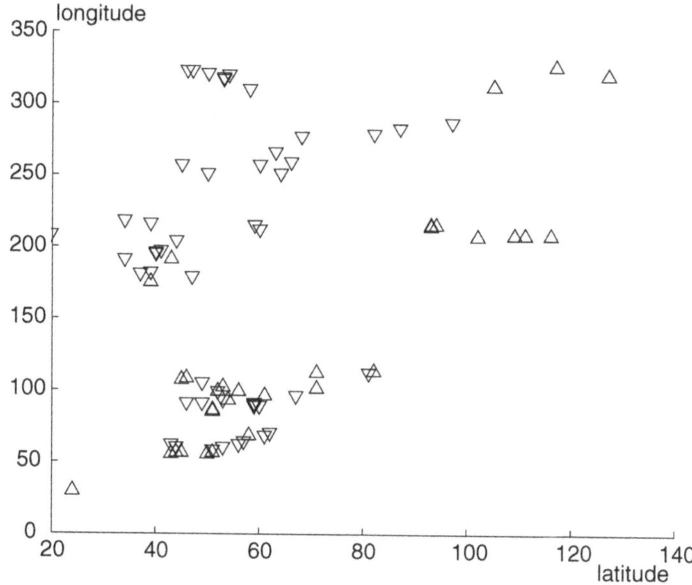

Fig. 2. Sites with large prediction errors as a function of latitude and longitude

With regards to Earth science, these results are plausible and the most interesting of all, as they suggest that the T1 and W stress terms are unnecessary for predicting NPPc. One explanation is that the influence of these factors is already being captured by the NDVI measure available from space, for which the signal-to-noise ratio has been steadily improving since CASA was first developed.

To further understand these results, we identified sites with large errors in predictive accuracy and plotted them by longitude and latitude, as shown in Figure 2. In this graph, upward-pointing triangles indicate situations in which the difference between predicted and observed NPP value was more than 400, whereas downward-pointing triangles depict sites in which this difference was less than −400. This error plot is consistent with the CASA team's previous analyses, which suggests the model overestimates the observed NPP at higher (temperate) latitudes and underestimates at the lower (tropical) latitudes.

4.4 Combining Multiple Revisions

In the previous sections, we were careful to let RF6 revise the NPP submodel one component at a time. However, a natural extension is to combine all three types of model refinement in a single revision step. For instance, we could let RF6 revise, in a single run, the parameter values in the T2 expression, the intrinsic values associated with vegetation type, and the functional form of photosynthetic efficiency E. We tried this approach using the same experimental setting as in our former studies, except for fixing the exponent of T2 at 1 because its parameters were being revised.

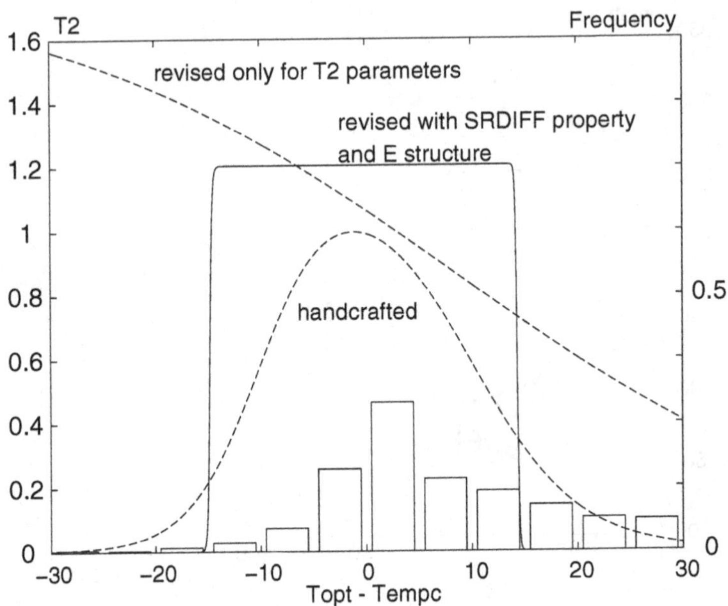

Fig. 3. Behavior of revised equation for the stress variable T2 by changing the other portions

This strategy produced some interesting results. For example, the system transformed the equation for the temperature stress factor T2 into

$$\mathrm{T2} = 1.2044/[(1 + e^{-8.06(Topt-Tempc+14.7)})(1 + e^{-12.7(Tempc-Topt+14.4)})],$$

which has quite different parameter values from those obtained when RF6 revised it in isolation. In this case, the two sigmoidal functions have nearly the same intercepts but opposite signs for their slopes. Figure 3 plots the predicted values of T2 as a function of the difference Topt − Tempc. Recall that the histogram plot denotes the frequency of samples for distinct values of the difference Topt − Tempc. This curve is relatively similar to the original, hand-crafted one, having higher values in some regions but a sharper drop off on each side.

Table 5 shows the revised values for the intrinsic property SRDIFF. The new values are uniformly larger than those shown in Table 3 from the isolated revision process. However, the correlation between the two sets of revised values was 0.867, which suggests that their apparent difference results from a shift in measurement scale. The relation of the new parameters to the originals is more complex, with half taking on higher values and the rest lower ones.

The one model change that was almost unaffected by the combined revision process involved the photosynthetic efficiency; this became

$$E = 0.56 \cdot T1^{+0.00} \cdot T2 \cdot W^{+0.00},$$

Table 5. Original and revised values for the SRDIFF intrinsic property, along with the frequency for each vegetation type

vegetation type	A	B	C	D	E	F	G	H	I	J	K
original	3.06	4.35	4.35	4.05	5.09	3.06	4.05	4.05	4.05	5.09	4.05
revised	3.46	7.67	5.20	7.58	5.43	4.81	3.97	0.79	3.98	4.64	3.10
clustered	3.84	7.64	5.10	7.64	5.10	5.10	3.84	0.79	3.84	5.10	3.84
frequency	3.3	8.9	0.3	3.6	21.1	19.1	15.2	3.3	19.1	2.3	3.6

which, except for the fixed exponent on T2, is nearly identical to the expression produced when this was altered in isolation. This provides even stronger evidence that the T1 and W terms provide little in the way of explanatory power to the overall model.

The RMSE for the original model was 467.910, whereas the error using the revised model was 412.350 on the training set and 429.369 using cross validation. Thus, the revised model reduced the error about eight percent over the initial version, which is nearly equal to the summed reduction from the three previous runs. We also examined the effect of applying RF6's clustering procedure to the SRDIFF values, which used cross-validated error to produce the four clusters shown in Table 6. As before, the estimated error for this revised model is about one percent better than for the model with 11 distinct values. In terms of error reduction, invoking the three types of model refinement appears to give similar results to running them individually.

5 Related Research on Computational Discovery

Our research on computational scientific discovery draws on two previous lines of work. One approach, which has an extended history within artificial intelligence, addresses the discovery of explicit quantitative laws. Early systems for numeric law discovery like BACON (Langley, 1979; Langley et al., 1987) carried out a heuristic search through a space of new terms and simple equations. Numerous successors like FAHRENHEIT (Żytkow et al., 1990) and RF5 (Saito & Nakano, 1997) incorporate more sophisticated and more extensive search through a larger space of numeric equations.

The most relevant equation discovery systems take into account domain knowledge to constrain the search for numeric laws. For example, Kokar's (1986) COPER utilized knowledge about the dimensions of variables to focus attention and, more recently, Washio and Motoda's (1998) SDS extends this idea to support different types of variables and sets of simultaneous equations. Todorovski and Džeroski's (1997) LAGRAMGE takes a quite different approach, using domain knowledge in the form of context-free grammars to constrain its search for differential equation models that describe temporal behavior.

Table 6. Error rates for different numbers of distinct SRDIFF values

No. clusters	Training RMSE	Cross-validation RMSE
1	448.70	451.02
2	428.35	434.84
3	417.75	430.61
4	414.34	425.62
5	413.57	427.19
⋮	⋮	⋮
11	412.35	429.37

Although research on computational discovery of numeric laws has emphasized communicable scientific notations, it has focused on constructing such laws rather than revising existing ones. In contrast, another line of research has addressed the refinement of existing models to improve their fit to observations. For example, Ourston and Mooney (1990) developed a method that used training data to revise models stated as sets of propositional Horn clauses. Towell (1991) reports another approach that transforms such models into multilayer neural networks, then uses backpropagation to improve their fit to observations, much as we have done for numeric equations. Work in this paradigm has emphasized classification rather than regression tasks, but one can view our work as adapting the basic approach to equation discovery, as Todorovski and Džeroski's (2001) have done in more recent work on LaGramge.

We should also mention related work on the automated improvement of ecosystem models. Most AI work on Earth science domains focuses on learning classifiers that predict vegetation from satellite measures like NDVI, as contrasted with our concern for numeric prediction. Chown and Dietterich (2000) describe an approach that improves an existing ecosystem model's fit to continuous data, but their method only alters parameter values and does not revise equation structure. On another front, Schwabacher et al. (this volume) use a rule-induction algorithm to discover piecewise linear models that predict NDVI from climate variables, but their method takes no advantage of existing models.

6 Directions for Future Research

Although we have been encouraged by our results to date, there remain a number of directions in which we must extend our approach before it can become a useful tool for scientists. As noted earlier, we envision an interactive discovery aide that lets the user focus the system's attention on those portions of the model it should attempt to improve. To this end, we need a graphical interface that supports marking of parameters, intrinsic properties, and equations that can be revised, as well as tools for displaying errors as a function of space, time, and predictive variables.

In addition, the current system requires some handcrafting to encode the equations as a neural network, as well as manual creation of the error function to be minimized for a particular set of revisions. Future versions should provide a library that maps functional forms to neural network encodings, so the system can transform the former into the latter automatically. They should also generate an appropriate error function from the set of revisions that the user indicates he desires.

Naturally, we also hope to evaluate our approach on its ability to improve other portions of the CASA model, as additional data becomes available. Another test of generality would be application of the same methods to other scientific domains in which there already exist formal models that can be revised. In the longer term, we should evaluate our interactive system not only in its ability to increase the predictive accuracy of an existing model, but in terms of the satisfaction to scientists who use the system to that end.

Another challenge that we have encountered in our research has been the need to translate the existing CASA model into a declarative form that our discovery system can manipulate. In response, another long-term goal involves developing a modeling language in which scientists can cast their initial models and carry out simulations, but that can also serve as the declarative representation for our discovery methods. The ability to automatically revise models places novel constraints on such a language, but we are confident that the result will prove a useful aid to the discovery process.

7 Concluding Remarks

In this paper, we addressed the computational task of improving an existing scientific model that is composed of numeric equations. We illustrated this problem with an example model from the Earth sciences that predicts carbon production as a function of temperature, sunlight, and other variables. We identified three activities that can improve a model – revising an equation's parameters, altering the values of an intrinsic property, and changing the functional form of an equation, then presented results for each type on an ecosystem modeling task that reduced the model's prediction error, sometimes substantially.

Our research on model revision builds on previous work in numeric law discovery and qualitative theory refinement, but it combines these two themes in novel ways to enable new capabilities. Clearly, we remain some distance from our goal of an interactive discovery tool that scientists can use to improve their models, but we have also taken some important steps along the path, and we are encouraged by our initial results on an important scientific problem.

Acknowledgements: This work was supported by Grant NAG 2-1335 from NASA Ames Research Center and by NTT Communication Science Laboratories, Nippon Telegraph and Telephone Corporation. We thank Christopher Potter, Steven Klooster, and Alicia Torregrosa for making available both the

CASA model and relevant data, as well as for their time in formulating the discovery task and analyzing our results. We also thank Trond Grenager, Stephanie Sage, Mark Schwabacher, and Jeff Shrager for helping with early stages of the project. An earlier version of this chapter appeared in the *Proceedings of the Fourth International Conference on Discovery Science.*

References

Chown, E., Dietterich, T.G.: A divide and conquer approach to learning from prior knowledge. In: Proceedings of the Seventeenth International Conference on Machine Learning, Stanford, CA, pp. 143–150 (2000)

Durbin, R., Rumelhart, D.E.: Product units: A computationally powerful and biologically plausible extension. Neural Computation 1, 133–142 (1989)

Kokar, M.M.: Determining arguments of invariant functional descriptions. Machine Learning 1, 403–422 (1986)

Langley, P.: Rediscovering physics with BACON.3. In: Proceedings of the Sixth International Joint Conference on Artificial Intelligence, Tokyo, Japan, pp. 505–507 (1979)

Langley, P.: The computer-aided discovery of scientific knowledge. In: Proceedings of the First International Conference on Discovery Science. Fukuoka, Japan (1998)

Langley, P., Simon, H.A., Bradshaw, G.L., Żytkow, J.M.: Scientific discovery: Computational explorations of the creative processes. MIT Press, Cambridge, MA (1987)

Lenat, D.B.: Automated theory formation in mathematics. In: Proceedings of the Fifth International Joint Conference on Artificial Intelligence, Cambridge, MA, pp. 833–842 (1977)

Lindsay, R.K., Buchanan, B.G., Feigenbaum, E.A., Lederberg, J.: Applications of artificial intelligence for organic chemistry: The DENDRAL project. McGraw-Hill, New York (1980)

Ourston, D., Mooney, R.: Changing the rules: A comprehensive approach to theory refinement. In: Proceedings of the Eighth National Conference on Artificial Intelligence, Boston, MA, pp. 815–820 (1990)

Potter, C.S., Klooster, S.A.: Global model estimates of carbon and nitrogen storage in litter and soil pools: Response to change in vegetation quality and biomass allocation. Tellus 49B, 1–17 (1997)

Potter, C.S., Klooster, S.A.: Interannual variability in soil trace gas (CO_2, N_2O, NO) fluxes and analysis of controllers on regional to global scales. Global Biogeochemical Cycles 12, 621–635 (1998)

Saito, K., Nakano, R.: Law discovery using neural networks. In: Proceedings of the Fifteenth International Joint Conference on Artificial Intelligence, Nagoya, Japan, pp. 1078–1083 (1997)

Saito, K., Nakano, R.: Discovery of nominally conditioned polynomials using neural networks, vector quantizers and decision trees. In: Proceedings of the Third International Conference on Discovery Science, Kyoto, Japan, pp. 325–329 (2000)

Schwabacher, M., Langley, P.: Discovering communicable scientific knowledge from spatio-temporal data. In: Proceedings of the Eighteenth International Conference on Machine Learning, Williamstown, MA, pp. 489–496 (2001)

Todorovski, L., Džeroski, S.: Declarative bias in equation discovery. In: Proceedings of the Fourteenth International Conference on Machine Learning, Nashville, TN, pp. 376–384 (1997)

Todorovski, L., Dzeroski, S.: Theory revision in equation discovery. In: Proceedings of the Fourth International Conference on Discovery Science, Washington DC, pp. 389–400 (2001)

Towell, G.: Symbolic knowledge and neural networks: Insertion, refinement, and extraction. Doctoral dissertation, Computer Sciences Department, University of Wisconsin, Madison (1991)

Washio, T., Motoda, H.: Discovering admissible simultaneous equations of large scale systems. In: Proceedings of the Fifteenth National Conference on Artificial Intelligence, Madison, WI, pp. 189–196 (1998)

Żytkow, J.M., Zhu, J., Hussam, A.: Automated discovery in a chemistry laboratory. In: Proceedings of the Eighth National Conference on Artificial Intelligence, Boston, MA, pp. 889–894 (1990)

Discovering Communicable Models from Earth Science Data

Mark Schwabacher[1], Pat Langley[2], Christopher Potter[3], Steven Klooster[3,4], and Alicia Torregrosa[3,4]

[1] Intelligent Systems Division
NASA Ames Research Center, Moffett Field, California, USA
Mark.A.Schwabacher@nasa.gov
[2] Institute for the Study of Learning and Expertise
Palo Alto, California, USA
langley@isle.org
[3] Earth Science Division
NASA Ames Research Center, Moffett Field, California, USA
Christopher.S.Potter@nasa.gov
[4] Earth System Science and Policy
California State University Monterey Bay, Seaside, California, USA

Abstract. This chapter describes how we used regression rules to improve upon results previously published in the Earth science literature. In such a scientific application of machine learning, it is crucially important for the learned models to be *understandable* and *communicable*. We recount how we selected a learning algorithm to maximize communicability, and then describe two visualization techniques that we developed to aid in understanding the model by exploiting the spatial nature of the data. We also report how evaluating the learned models across time let us discover an error in the data.

1 Introduction and Motivation

Many recent applications of machine learning have focused on commercial data, often driven by corporate desires to better predict consumer behavior. Yet scientific applications of machine learning remain equally important, and they can provide technological challenges not present in commercial domains. In particular, scientists must be able to *communicate* their results to others in the same field, which leads them to agree on some common formalism for representing knowledge in that field. This need places constraints on the representations and learning algorithms that we can utilize in aiding scientists' understanding of data.

Moreover, some scientific domains have characteristics that introduce both challenges and opportunities for researchers in machine learning. For example, data from the Earth sciences typically involve variation over both space and time, in addition to more standard predictive variables. The spatial character of these data suggests the use of visualization in both understanding the discovered

S. Džeroski and L. Todorovski (Eds.): Computational Discovery, LNAI 4660, pp. 138–157, 2007.

knowledge and identifying where it falls short. The observations' temporal nature holds opportunities for detecting developmental trends, but it also raises the specter of calibration errors, which can occur gradually or when new instruments are introduced.

In this chapter, we explore these general issues by presenting the lessons we learned while applying machine learning to a specific Earth science problem: the prediction of Normalized Difference Vegetation Index (NDVI) from predictive variables like precipitation and temperature. This chapter describes the results of a collaboration among two computer scientists (Schwabacher and Langley) and three Earth scientists (Potter, Klooster, and Torregrosa). It describes how we combined the computer scientists' knowledge of machine learning with the Earth scientists' domain knowledge to improve upon a result that Potter had previously published in the Earth science literature (Potter & Brooks, 1998).

We begin by reviewing the scientific problem, including the variables and data, and proposing regression learning as a natural formulation. After this, we discuss our selection of piecewise linear models to represent learned knowledge as consistent with existing NDVI models, along with our selection of Quinlan's Cubist (RuleQuest, 2002) to generate them. Next we compare the results we obtained in this manner with models from the Earth science literature, showing that Cubist produces significantly more accurate models with little increase in complexity.

Although this improved predictive accuracy is good news from an Earth science perspective, we found that the first Cubist models we created were not sufficiently understandable or communicable. In our efforts to make the discovered knowledge understandable to the Earth scientists on our team, we developed two novel approaches to visualizing this knowledge spatially, which we report in some detail. Moreover, evaluation across different years revealed an error in the data, which we have since corrected.

Having demonstrated the value of Cubist in Earth science by improving upon a previously published result, we set out to use Cubist to fit models to data to which models had not previously been fit. Doing so produced models that we believe to be very significant.

We discuss some broader issues that these experiences raise and propose some general approaches for dealing with them in other spatial and temporal domains. In closing, we also review related work on scientific data analysis in this setting and propose directions for future research.

2 Monitoring and Analysis of Earth Ecosystem Data

The latest generation of Earth-observing satellites is producing unprecedented amounts and types of data about the Earth's biosphere. Combined with readings from ground sources, these data hold promise for testing existing scientific models of the Earth's biosphere and for improving them. Such enhanced models would let us make more accurate predictions about the effect of human activities on our planet's surface and atmosphere.

One such satellite is the NOAA (National Oceanic and Atmospheric Administration) Advanced Very High Resolution Radiometer (AVHRR). This satellite has two channels which measure different parts of the electromagnetic spectrum. The first channel is in a part of the spectrum where chlorophyll absorbs most of the incoming radiation. The second channel is in a part of the spectrum where spongy mesophyll leaf structure reflects most of the light. The difference between the two channels is used to form the Normalized Difference Vegetation Index (NDVI), which is correlated with various global vegetation parameters. Earth scientists have found that NDVI is useful for various kinds of modeling, including estimating net ecosystem carbon flux. A limitation of using NDVI in such models is that they can only be used for the limited set of years during which NDVI values are available from the AVHRR satellite. Climate-based prediction of NDVI is therefore important for studies of past and future biosphere states.

Potter and Brooks (1998) used multiple linear regression analysis to model maximum annual NDVI[1] as a function of four climate variables and their logarithms:[2]

- Annual Moisture Index (AMI): a unitless measure, ranging from -1 to +1, with negative values for relatively dry, and positive values for relatively wet. Defined by Willmott & Feddema (1992);
- Chilling Degree Days (CDD): the sum of the number of days times mean monthly temperature, for months when the mean temperature is less than 0° C;
- Growing Degree Days (GDD): the sum of the number of days times mean monthly temperature, for months when the mean temperature is greater than 0° C; and
- Total Annual Precipitation (PPT).

These climate indexes were calculated from various ground-based sources, including the World Surface Station Climatology at the National Center for Atmospheric Research. Potter and Brooks interpolated the data, as necessary, to put all of the NDVI and climate data into one-degree grids. That is, they formed a 360 × 180 grid for each variable, where each grid cell represents one degree of latitude and one degree of longitude, so that each grid covers the entire Earth. They used data from 1984 to calibrate their model. Potter and Brooks decided, based on their knowledge of Earth science, to fit NDVI to these climate variables by using a piecewise linear model with two pieces. They split the data into two sets of points: the warmer locations (those with $GDD \geq 3000$), and the cooler locations (those with $GDD < 3000$). They then used multiple linear regression to fit a different linear model to each set, resulting in the piecewise linear model shown in Table 1. They obtained correlation coefficients (r values)

[1] They obtained similar results when modeling minimum annual NDVI. We chose to use maximum annual NDVI as a starting point for our research, and all of the results in this chapter refer to this variable.

[2] They did not use the logarithm of AMI, since AMI can be negative.

Table 1. The piecewise linear model from Potter and Brooks (1998)

```
Rule 1:
  if
    GDD<3000
  then
    ln(NDVI) = 0.715 ln(GDD) + 0.377 ln(PPT) - 0.448

Rule 2:
  if
    GDD>= 3000
  then
    NDVI = 189.89 AMI + 44.02 ln(PPT) + 227.99
```

of 0.87 on the first set and 0.85 on the second set, which formed the basis of a publication in the Earth science literature (Potter & Brooks, 1998).

3 Problem Formulation and Learning Algorithm Selection

When we began our collaboration, we decided that one of the first things we would do would be to try to use machine learning to improve upon their NDVI results. The research team had already formulated this problem as a regression task, and in order to preserve communicability, we chose to keep this formulation, rather than discretizing the data so that we could use a more conventional machine learning algorithm. We therefore needed to select a *regression learning* algorithm — that is, one in which the outputs are continuous values, rather than discrete classes.

In selecting a learning algorithm, we were interested not only in improving the correlation coefficient, but also in ensuring that the learned models would be both understandable by the scientists and communicable to other scientists in the field. Since Potter and Brooks' previously published results involved a piecewise linear model that used an inequality constraint on a variable to separate the pieces, we felt it would be beneficial to select a learning algorithm that produces models of the same form. Fortunately, Potter and Brooks' model falls within the class of models used by Ross Quinlan's M5 and Cubist machine learning systems. M5 (Quinlan, 1992) learns a decision tree, similar to a C4.5 decision tree (Quinlan, 1993), but with a linear model at each leaf; the tree thus represents a piecewise linear model. Cubist (RuleQuest, 2002) learns a set of rules, similar to the rules learned by C4.5rules (Quinlan, 1993), but with a linear model on the right-hand side of each rule; the set of rules thus also represents a piecewise linear model. Cubist is a commercial product; we selected it over M5 because it is a newer system than M5, which, according to Quinlan (personal communication, 2001), has much better performance than M5.

Table 2. The effect of Cubist's minimum rule cover parameter on the number of rules in the model and the model's correlation coefficient

MINIMUM RULE COVER	NUMBER OF RULES	r
1%	41	0.91
5%	12	0.90
10%	7	0.89
15%	4	0.88
20%	3	0.86
25%	2	0.85
100%	1	0.84

4 First Results

We ran Cubist (version 1.09) using the same data sets that Potter and Brooks had used to build their model, but instead of making the cuts in the piecewise linear model based on knowledge of Earth science, we let Cubist decide where to make the cuts based on the data. The results exceeded our expectations. Cubist produced a correlation coefficient of 0.91 (using ten-fold cross-validation), which was a substantial improvement over the 0.86 correlation coefficient obtained in Potter and Brooks' earlier work. The Earth scientists on our team were pleased

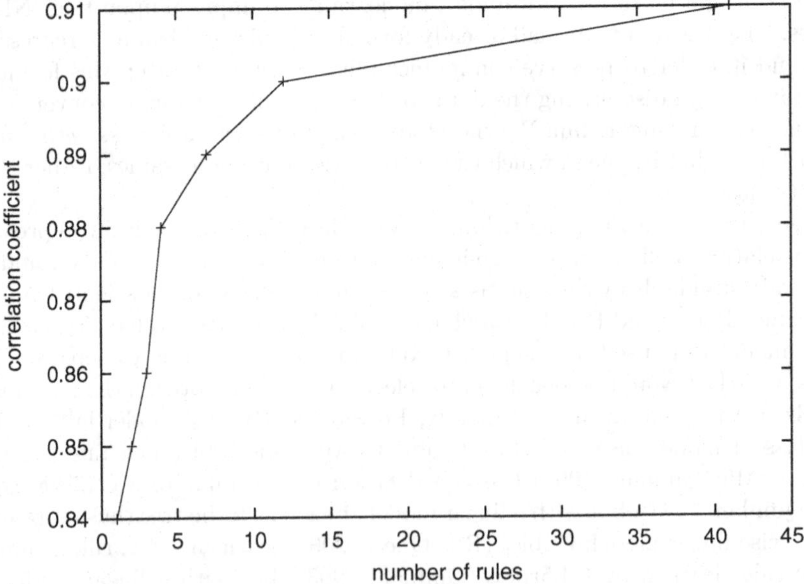

Fig. 1. The number of rules in the Cubist model and the correlation coefficient for several different values of the minimum rule cover parameter

Table 3. The two rules produced by Cubist when the minimum rule cover parameter is set to 25%

```
Rule 1:
 if
   PPT <= 25.457
 then
   NDVI = -3.22 + 7.07 PPT + 0.0521 CDD - 84 AMI + 0.4 ln(PPT) + 0.0001 GDD

Rule 2:
 if
   PPT > 25.457
 then
   NDVI = 386.327 + 316 AMI + 0.0294 GDD - 0.99 PPT + 0.2 ln(PPT)
```

with the 0.91 correlation coefficient, but when presented with the 41 rules produced by Cubist, they had difficulty interpreting them. Some of the rules clearly did not make sense, and were probably a result of Cubist overfitting the data. More importantly, the large number of rules — some 41 as compared with two in the earlier work — was simply overwhelming.

The first step we took in response to this understandability problem was to change the parameters to Cubist so that it would produce fewer rules. One of these parameters specifies the minimum percentage of the training data that must be covered by each rule. The default value of 1% produced 41 rules. We experimented with different values of this parameter between 1% and 100%; the results appear in Table 2 and Figure 1. Using a model with only one rule — that is, using conventional multiple linear regression analysis — results in a correlation coefficient of 0.84, whereas adding rules gradually improves accuracy. Interestingly, when using two rules, Cubist split the data on a different variable than the one the Earth scientists selected. Potter and Brooks split the data on GDD (essentially temperature), while Cubist instead chose precipitation, which produced a very similar correlation coefficient (0.85 versus 0.86). The two-rule model produced by Cubist is shown in Table 3. A comparison between Table 1 and Table 3 reveals that Potter and Brooks modeled ln(NDVI) in one rule, and NDVI in the other rule, while Cubist modeled NDVI in both rules. Cubist does not have the ability to model the logarithm of the class variable in some rules while modeling the original class variable in other rules (there can only be one class variable), so the space of rules searched by Cubist did not include Potter and Brooks' model. Interestingly, Cubist produced similar accuracy even though it searched a more limited rule space.

In machine learning there is frequently a tradeoff between accuracy and understandability. In this case, we are able to move along the tradeoff curve by adjusting Cubists' minimum rule cover parameter. Figure 1 illustrates this tradeoff by plotting the number of rules and the correlation coefficient produced by Cubist for each value of the minimum rule cover parameter in Table 2. We believe that generally a model with fewer rules is easier to understand, so the figure essentially plots accuracy against understandability. We used trial and error to select values for the minimum rule cover parameter that produced the

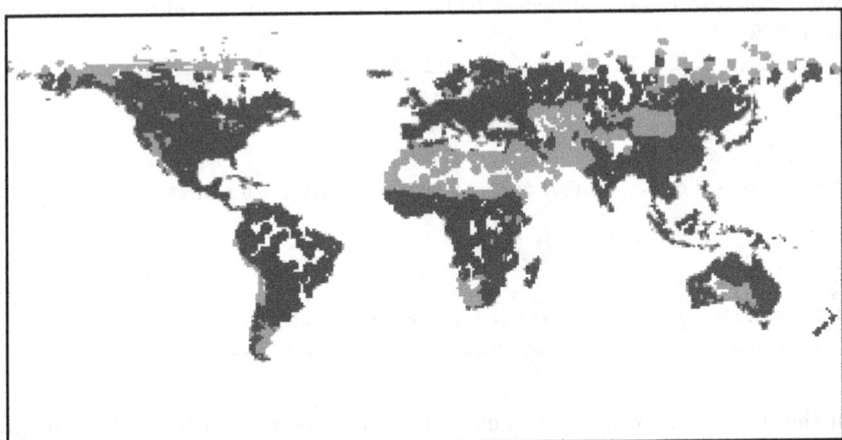

Fig. 2. Map showing which of the two Cubist rules from Table 3 are active across the globe

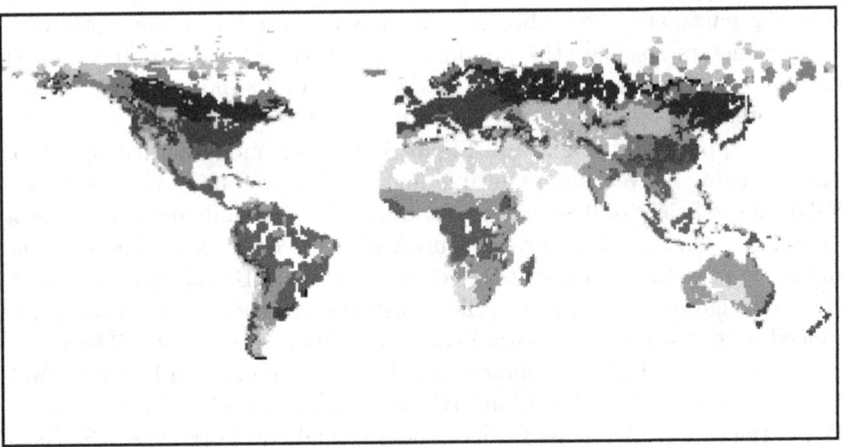

Fig. 3. Map showing which of the seven Cubist rules from Table 4 are active across the globe

number of rules we wanted for understandability reasons. Based on this experience, We concluded that a useful feature for future machine learning algorithms would be the ability to directly specify the maximum number of rules in the model as a parameter to the learning algorithm. After reviewing a draft of a conference paper on our NDVI work (Schwabacher and Langley, 2001), Ross Quinlan decided to implement this feature in the next version of Cubist - see Section 7.2.

5 Visualization of Spatial Models

Reducing the number of rules in the model by modifying Cubists' parameters made the model more understandable, but to further understand the rules, we decided to plot which ones were active where. We developed special-purpose C code, which produced the map in Figure 2. In this figure, the white areas represent portions of the globe that were excluded from the model because they are covered with water or ice, or because there was insufficient ground-based data available. After excluding these areas, we were left with 13,498 points that were covered by the model. The light gray areas are the areas in which Rule 1 from Table 3 applies (the drier areas), and the dark gray areas are the areas in which Rule 2 from Table 3 applies (the wetter areas).

Figure 3 shows where the various rules in a seven-rule model are active. In this figure, the white regions were excluded from the model, as before. The gray areas represent regions in which only one rule applies; the seven shades of gray correspond to the seven rules. (We normally use different colors for the different rules, but resorted to different shades of gray for this book.) The black areas are regions in which more than one rule in the model applied. (In these cases, Cubist uses the average of all applicable rules.) The seven rules used to produce this map are shown in Table 4.

The Earth scientists on our team found these maps very interesting, because one can see many of the Earth's major topographical and climatic features. The maps provide valuable clues as to the scientific significance of each rule. With the aid of this visualization, the scientists were better able to understand the seven-rule model. Before seeing the map, the scientists had difficulty interpreting Rule 7, since its conditions specified that CDD and GDD were both high, which appears to specify that the region is both warm and cold. After seeing the map showing where Rule 7 is active, they determined that Rule 7 applies in the northern boreal forests, which are cold in the winter and fairly warm in the summer. The seven-rule model, which is made understandable by this visualization, is almost as accurate as the incomprehensible 41-rule model (see Table 2). This type of visualization could be used whenever the learning task involves spatial data and the learned model is easily broken up into discrete pieces that are applicable in different places, such as rules in Cubist or leaves in a decision tree.

A second visualization tool that we developed (also as special-purpose C code) shows the error of the Cubist predictions across the globe. In Figure 4, white represents either zero error or insufficient data, black represents the largest error, and shades of gray represent intermediate error levels. From this map, it is possible to see that the Cubist model has large errors in Alaska and Siberia, which is consistent with the belief of the Earth scientists on our team that the quality of the data in the polar regions is poor. Such a map can be used to better understand the types of places in which the model works well and those in which it works poorly. This understanding in turn may suggest ways to improve the model, such as including additional attributes in the training data or using a different learning algorithm. Such a visualization can be used for any learning task that uses spatial data and regression learning.

Table 4. The seven rules for NDVI produced by Cubist when the minimum rule cover parameter is set to 10%

```
Rule 1:
  if
    CDD <= 16.52
    PPT <= 25.457
  then
    NDVI = 3.48 + 7.17 PPT - 161 AMI - 0.0082 GDD - 9.9 ln(PPT) + 0.0003 CDD

Rule 2:
  if
    CDD > 16.52
    PPT <= 25.457
  then
    NDVI = -69.99 + 16.08 PPT - 0.0449 GDD - 263 AMI + 0.0352 CDD + 0.4 ln(PPT)

Rule 3:
  if
    AMI <= -0.09032081
    PPT > 25.457
  then
    NDVI = 375.9 + 367 AMI + 0.0257 GDD - 0.01 PPT + 0.2 ln(PPT)

Rule 4:
  if
    GDD <= 1395.62
    PPT > 25.457
  then
    NDVI = 267.3 + 0.12 GDD + 0.0036 CDD + 3 AMI - 0.01 PPT + 0.2 ln(PPT)

Rule 5:
  if
    AMI > -0.09032081
    GDD > 5919.36
  then
    NDVI = 601.1 - 0.0063 GDD - 0.11 PPT + 3 AMI + 0.2 ln(PPT) + 0.0001 CDD

Rule 6:
  if
    AMI > -0.09032081
    CDD <= 908.73
    GDD > 1395.62
    GDD <= 5919.36
  then
    NDVI = 359.8 + 317 AMI + 0.037 GDD + 0.0425 CDD - 1 PPT + 0.2 ln(PPT)

Rule 7:
  if
    AMI > -0.09032081
    CDD > 908.73
    GDD > 1395.62
  then
    NDVI = 373.13 + 0.0645 GDD + 249 AMI - 1.32 PPT + 0.0134 CDD + 0.2 ln(PPT)
```

6 Discovery of Quantitative Errors in the Data

Having successfully trained Cubist using data for one year, we set out to see how well an NDVI model trained on one year's data would predict NDVI for another year. We thought this exercise would serve two purposes. If we generally found transfer across years, that would be good news for Earth scientists, because it

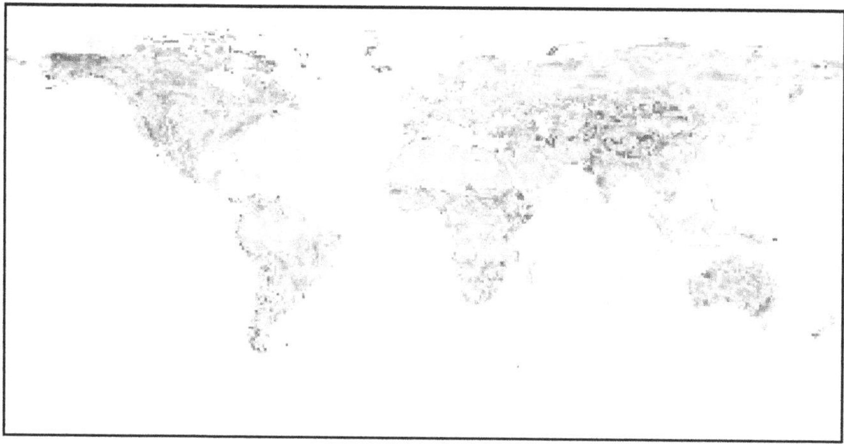

Fig. 4. Map showing the errors of the prediction of the Cubist model from Table 4 for NDVI across the globe

would let them use the model to obtain reasonably accurate NDVI values for years in which satellite-based measurements of NDVI are not available. On the other hand, if the model learned from one year's data transferred well to some years but not others, that would indicate some change in the world's ecosystem across those years. Such a finding could lead to clues about temporal phenomena in Earth science such as El Niños or global warming.

What we found, to our surprise, is that the model trained on 1983 data worked very well when tested on the 1984 data, and that the model trained on 1985 data worked very well on data from 1986, 1987, and 1988, but that the model trained on 1984 data performed poorly when tested on 1985 data. The second column of

Table 5. Correlation coefficients obtained when cross-validating using one year's data and when training on one year's data and testing on the next year's data, using the original data set and using the corrected data set

Data Set	r, original	r, corrected
cross-validate 1983	0.97	0.91
cross-validate 1984	0.97	0.91
cross-validate 1985	0.92	0.92
cross-validate 1986	0.92	0.92
cross-validate 1987	0.91	0.91
cross-validate 1988	0.91	0.91
train 1983, test 1984	0.97	0.91
train 1984, test 1985	0.80	0.91
train 1985, test 1986	0.91	0.91
train 1986, test 1987	0.91	0.91
train 1987, test 1988	0.90	0.90

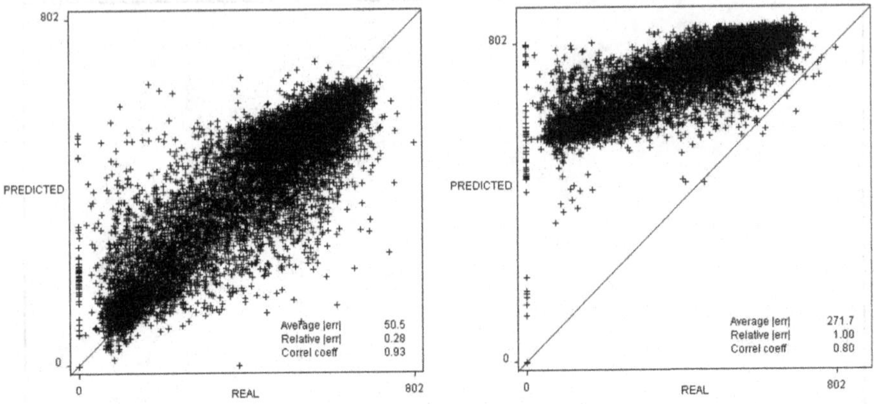

Fig. 5. Predicted NDVI against actual NDVI for *(left)* cross-validated 1985 data and *(right)* training on 1984 data and testing on 1985 data

Table 5 shows the tenfold cross-validated correlation coefficients for each year, as well as the correlation coefficients obtained when testing each year's model on the next year's data. Clearly, something changed between 1984 and 1985. At first we thought this change might have been caused by the El Niño that occurred during that period.

Further light was cast on the nature of the change by examining the scatter plots that Cubist produces. In Figure 5, the graph on the left plots predicted NDVI against actual NDVI for the 1985 cross-validation run. The points are clustered around the $x = y$ line, indicating a good fit. The graph on the right plots predicted against actual NDVI when using 1985 data to test the model learned from 1984 data. In this graph, the points are again clearly clustered around a line, but one that has been shifted away from the $x = y$ equation. This shift is so sudden and dramatic that the Earth scientists on our team believed that it could not have been caused by a natural phenomenon, but rather that it must be due to problems with the data.

Further investigation revealed that there was in fact an error in the data. In the data set given to to us, a recalibration that should have been applied to the 1983 and 1984 data had not been done. We obtained a corrected data set and repeated each of the Cubist runs from Table 5, obtaining the results in the third column.[3] With the corrected data set, the model from any one year transfers very well to the other years, so these models should be useful to Earth scientists in order to provide NDVI values for years in which no satellite-based measurements of NDVI are available.

Our experience in finding this error in the data suggests a general method of searching for calibration errors in time-series data, even when no model of the data is available. This method involves learning a model from the data for

[3] All of the results presented in the previous sections are based on the corrected data set.

each time step and then testing this model on data from successive time steps. If there exist situations in which the model fits the data unusually poorly, then those are good places to look for calibration errors in the data. Of course, when such situations are found, the human experts must examine the relevant data to determine, based on their domain knowledge, whether the sudden change in the model results from an error in the data, from a known discontinuity in the natural system being modeled, or from a genuinely new scientific discovery. This idea can be extended beyond time-series problems to any data set that can be naturally divided into distinct sets, including spatial data.

7 New Data Sets

7.1 Using Other Variables to Predict NDVI

Having demonstrated the value of Cubist to Earth science by improving upon a previously published result, we set out to use Cubist to fit models to data to which models had not previously been fit. First, we tried using additional variables to predict NDVI, beyond the four variables that were used in Potter and Brooks (1998). The additional variables we tried were:

- Potential Evapotranspiration (PET): potential loss of water from the soil both by evaporation and by transpiration from the plants growing thereon, as defined by Thornthwaite (1948).
- Elevation (DEM)
- Percentage wetland (WETLND)
- HET2SOLU: a two-dimensional measure of heterogeneity that counts the number of different combinations of soil and landuse polygons within each grid cell.
- HET3SOLU: a three-dimensional measure of heterogeneity that takes elevation into account.
- Vegetation type according to the University of Maryland (UMDVEG)
- Vegetation type according to the CASA model (CASAVEG)

We found that the variable that produced the largest improvement in accuracy when used together with the original four variables was UMDVEG. Including UMDVEG together with the original four variables increased the cross-validated correlation coefficient (with a minimum rule cover of 1%) from 0.91 to 0.94. Further investigation of this variable, however, revealed that it was derived from NDVI, so that using it to predict NDVI would not be useful.

We found that including PET, DEM, WETLND, and HET2SOLU (along with the original four variables) increased the cross-validated correlation coefficient (using a minimum rule cover of 1%) from 0.91 to 0.93. This model has 40 rules, and is very difficult to understand. Increasing the minimum rule cover parameter to 10% produced a model with seven rules and a cross-validated correlation coefficient 0.90. This model is slightly more accurate than the model produced from the original four variables (which had a cross-validated correlation coefficient of 0.89) and is somewhat harder to understand.

We concluded that the four variables chosen by Potter and Brooks (1998) appear to be a good choice of variables for building a model that is both accurate and understandable. In applications for which accuracy is more important than understandability, it may be better to use the model with eight variables and 40 rules.

7.2 Predicting NPP

We decided to try using Cubist to predict another measure of vegetation: Net photosynthetic accumulation of carbon by plants, also known as net primary production (NPP). While NDVI is used as an indicator of the type of vegetation at different places, NPP is a measure of the rate of vegetation growth. It is usually reported in grams of carbon per square meter per year.

NPP provides the energy that drives most biotic processes on Earth. The controls over NPP are an issue of central relevance to human society, mainly because of concerns about the extent to which NPP in managed ecosystems can provide adequate food and fiber for an exponentially growing population. In addition, accounting of the long-term storage potential in ecosystems of atmospheric carbon dioxide (CO_2) from industrial pollution sources begins with an understanding of major climate controls on NPP.

NPP is measured in two ways. The first method, known as "destructive sampling," involves harvesting and weighing all of the vegetation in a defined area, and estimating the age of the vegetation using techniques such as counting the number of rings in the cross-sections of trees. The second method uses towers that sample the atmosphere above the vegetation, and estimating NPP from the net CO_2 uptake. Both methods are expensive and provide values for only one point at at time, so until recently NPP values were only available for a small number of points on the globe.

Previous ecological research has shown that surface temperature and precipitation are the strongest controllers of yearly terrestrial NPP at the global scale (eg., Potter et al., 1999). Lieth (1975) used single linear regression to predict NPP from either temperature or precipitation, using a data set containing NPP values from only a handful of sites.

We recently obtained a new, much larger NPP data set from the Ecosystem Model-Data Intercomparison (EMDI) project, sponsored by the National Center for Ecological Analysis and Synthesis (NCEAS) in the U.S. and the International Geosphere Biosphere Program (IGBP). This data set contains NPP values from 3,855 points across the globe. We decided to try using Cubist to predict NPP from the following three variables:

- annual total precipitation in millimeters, 1961–1990 (PPT)
- average mean air temperature in degrees centigrade, 1961–1990 (AVGT)
- biome type, a discreet variable with 12 possible values (BIOME)

After Ross Quinlan reviewed a draft of a conference paper on our NDVI work (Schwabacher & Langley, 2001), he implemented a new feature in Cubist that

Table 6. The four rules produced by Cubist for predicting NPP

```
Rule 1:
  if
    PPT <= 653
  then
    NPP = 63.8 + 0.49 PPT - 6.5 AVGT

Rule 2:
  if
    BIOME in {Grassland, Wooded-grassland, Shrubland, ENL-forest-boreal}
  then
    NPP = 94.3 + 0.418 PPT - 7.3 AVGT

Rule 3:
  if
    BIOME in {Forest-temperate, Forest-boreal, Forest-xeric}
  then
    NPP = 215.1 + 0.377 PPT - 2.4 AVGT

Rule 4:
  if
    BIOME in {Savanna, EBL-forest-tropical, Forest-tropical,
              DBL-forest-tropical}
  then
    NPP = 115.4 + 29.1 AVGT + 0.056 PPT
```

allows the user to directly specify the maximum number of rules, rather than having to use trial and error to pick a value of the minimum rule cover parameter that will produce the desired number of rules. For the NPP prediction, we used a new version of Cubist (version 1.10) that includes this new feature. We specified a maximum of five rules. Cubist produced the four rules shown in Table 6, and a cross-validated correlation coefficient of 0.98.

The Earth scientists on our team were very happy with the 0.98 correlation coefficient, and felt that the rules generally made sense. They liked the idea of having different linear models for different groups of biome types. Initially, however, they were surprised that the coefficient on AVGT was negative in three of the four rules. After giving it more thought, they came up with a plausible explanation of why this coefficient is negative. AVGT is acting mainly as a predictor of relatively higher (or lower) heat fluxes that tend to severely dry out (or leave moist) the soils and plants, given a similar PPT. This explanation still requires further investigation.

To help understand these four rules, we produced a map showing where the rules are active. Initially we produced a map with four colors representing the four rules, and black representing multiple rules being active or no rules being active (as in Figure 3). The result was a map in which almost all of the land area was

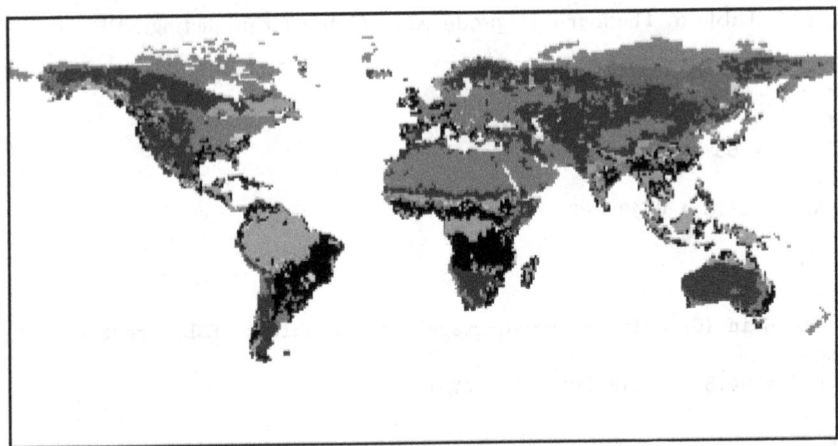

Fig. 6. Map showing which combinations of the four Cubist rules for NPP are active across the globe

black, which of course was not useful. It turns out that with this set of rules, for much of the land area, two rules are active. Since there are only four rules, and the last three are mutually exclusive, we were able to assign a different color to each of the eight possible combinations of rules. Also, the 3,855 points in our NPP data are from 12 biome types, while approximately 48% of the world's land area has biome types other than these 12, resulting in no rule being active. Most of these points are tundra, desert, or cultivated land, which are biome types for which NPP has not been measured. We assigned a ninth color to represent these areas. In addition, the global data set and the NPP data set use two different sets of discrete values for biome type. Some of the biome types in the global data set map into more than one biome type in the NPP data set, and in some cases these multiple biome types appear in multiple Cubist rules, making it unclear which Cubist rules are active. These ambiguous points account for approximately 17% of the world's land area; we assigned a tenth color to these points. The resulting map (translated into shades of gray for this book) is shown in Figure 6. The black areas in this map are the ambiguous points.

The Earth scientists on our team felt that this map was useful in understanding the rules, and in understanding the coverage of the model. It showed them that the current EMDI data set of measured NPP values allows for a somewhat limited extrapolation of the Cubist model (with no deserts, tundra, or cultivated areas), but that the extrapolation still covers a substantial portion of the global land surface, and that it covers most of the naturally "green" areas.

8 Related Work

Robust algorithms for flexible regression have been available for some time. Breiman, Friedman, Olshen, and Stone's (1984) CART first introduced the notion of inducing regression trees to predict numeric attributes. CART trees have a numeric constant at each leaf, yielding a piecewise constant model. Weiss and Indurkhya (1993) extended the idea to rule induction, inducing a set of rules where the right-hand side of each rule has a numeric constant. Quinlan extended the idea to piecewise linear models, by putting a linear model at each leaf of a decision tree in M5 (Quinlan, 1992) or on the right-hand side of each of a set of rules in Cubist (RuleQuest, 2002). Each approach has proved successful in many domains, and both CART and Cubist have achieved commercial success. However, neither approach has yet seen much application to Earth science data, despite the considerable work on classification learning for tasks like assigning ground cover types to pixels (e.g., Brodley & Friedl, 1999) and clustering adjacent pixels into groups (e.g., Ester, Kriegel, Sander, & Xu, 1996).

The work on communicability and understandability described in this chapter builds on previous work in comprehensibility. Our requirement for communicability is similar to Michalski's (1983) "comprehensibility postulate" which states that the results of computer induction should be in a form that is syntactically and semantically similar to that used by humans experts. A collection of papers on comprehensibility can be found in Kodratoff and Nédellec (1995).

Researchers have also carried out extensive work on techniques for visualizing data and learned knowledge. Tufte (1983) did early influential work on the former topic, whereas Keim and Kriegel (1996) review many of the existing approaches. Rheingans and desJardins (2000) describe a technique for using self-organizing maps to display high-dimensional data, predictions, and errors in two dimensions. Within the data-mining community, researchers have developed a variety of methods for the graphical display of learned knowledge (e.g., Brunk, Kelly, & Kohavi, 1996). However, although much of this work employs a spatial metaphor, little has focused on learned spatial knowledge itself.

Applications of machine learning to Earth science data, as in methods for ground cover prediction (e.g., Brodley & Friedl, 1999), regularly display classes on maps. Smyth, Ghil, and Ide (1999) plot predictions of a learned mixture model on the globe, but our approach to visualizing areas in which regression rules match, as well as anomalous regions, appears novel.

The European project SPIN! (2002) is seeking to develop a spatial data mining system by combining data mining tools like C4.5 (Quinlan, 1993) with tools for visualizing spatial data like Descartes (Andrienko & Andrienko, 1999). The planned system will let its users visualize geographically-referenced data on maps, and mine the data using the data-mining tools, from a unified user interface. The researchers plan to test the SPIN! system on applications involving seismic and volcano data. The visualization component of the project seems focused on letting users visualize the data, rather than visualizing the knowledge learned through data mining.

There has also been considerable research on using machine-learned knowledge to detect and either ignore or correct errors in training data. Much of this work has focused on removing cases with faulty class labels (e.g., John, 1995; Brodley & Friedl, 1999), but some has addressed detecting errors in the values of predictive variables. GritBot, a product of Quinlan's RuleQuest Research (2002), detects both errors in the class labels and errors in the predictive values by finding what it calls anomalies: items in the training data that are outliers. We ran GritBot on both the NDVI and the NPP data sets, and it found a number of anomalies. For example, it found a point that had the unusual combination of a high maximum NDVI and a low minimum NDVI. All of the anomalies that GritBot finds are single-point anomalies — each anomaly is one item in the training data, which in the applications described in this chapter means that it is a single point on the globe at a single point in time — so GritBot is not capable of finding the type of systematic error that we describe in Section 6. Naturally, there are established methods for detecting and correcting calibration problems in remote-sensing systems (e.g., Chen, 1997), but these rely on predefined models. Thus, our use of regression rules to detect systematic errors appears novel to both the machine learning and calibration communities.

9 Future Work

Our collaboration is in its early stages, and we still have many research avenues to explore. Our next step in modeling NDVI will incorporate time explicitly by adding the year to the continuous variables used in regression equations, rather than building a separate model for each year. We hope that by examining the resulting multi-year models, we can learn something about climate change over time.

In this chapter, we have assumed that models with fewer rules are more understandable. In future work, we plan to test this assumption by having the Earth scientists on our team examine various sets of rules that Cubist produces for different parameter values and telling us which sets they think are easier to understand. Naturally, we will also ask them to judge the rules' plausibility and interestingness from the perspective of Earth science.

Another direction for future work is to develop an extension to the Cubist algorithm that would allow it to take advantage of background knowledge. One possible form of background knowledge would be knowledge of the sign of the coefficients on some of the variables within the linear models. For example, we believe that the coefficient on PPT in the NPP model should always be positive. Pazzani and Bay (1999) describe an algorithm that uses knowledge of the signs of the coefficients to constrain the construction of regression equations. Their algorithm accepts input about the sign of each term, then use an optimization method to find the best weights given the constraints. The resulting equations were just as accurate as the unconstrained linear models on separate test sets, and domain experts found them more comprehensible. It would be interesting to combine Pazzani and Bay's algorithm with the Cubist algorithm to produce decision rules with linear models that obey sign constraints.

The NDVI predictive model is only one piece of a larger framework, known as CASA (Potter & Klooster, 1998), that Potter's team has developed to model the Earth's ecosystem. CASA takes the form of a process model, stated in terms of differential equations, for the production and absorption of biogenic trace gases in the Earth's atmosphere. CASA's output is NPP. We have achieved very good accuracy by using Cubist to predict NPP, but for the reasons of understandability and communicability described earlier, we would like our learned models to take the same form as the CASA model, which means we cannot rely on Cubist alone in our future efforts.

There has been some research on discovering laws that take the form of differential equations (Todorovski & Dzeroski, 1997), but this work has not used an existing set of equations as the starting point. We plan to develop an algorithm that will begin with the current CASA model and search through the space of possible equations to find an improved model. We will consider developing a Cubist-like algorithm that learns a model with a set of rules to select among different sets of differential equations (instead of different linear models). We hope that this effort will improve the accuracy of the CASA model to the point where it is as accurate as the Cubist model of NPP, while retaining CASA's communicability and its scientific plausibility. We also hope that the changes our system makes to the model will suggest new insights about Earth science.

10 Lessons Learned

In their editorial on applied research in machine learning, Provost and Kohavi (1998) claimed that a good application paper will "focus research on important unsolved problems that currently restrict the practical applicability of machine learning methods." In this chapter, we have identified, and provided initial solutions for, three such problems that arise in scientific applications:

Communicability. In scientific domains, it is important for the form of the learned models to match the form that is customarily used in the relevant literature, so that the learned models can be communicated to other scientists.

Understandability. In domains that involve spatial data, understanding of the models can be increased by visualizing the spatial distribution of the model's errors and visualizing the locations in which the model's components (e.g., rules) are active. Adjusting the parameters to the learning algorithm in order to produce a smaller model can also aid understandability.

Quantitative errors. In applications that involve time-series numerical data, machine learning methods can be used to identify quantitative errors by testing a learned model for one time period against data from other time periods.

Although we have developed these ideas in the context of a specific scientific application – the prediction of NDVI and NPP from climate variables – we believe they have general applicability to any domain that involves scientific understanding of spatio-temporal data. As we continue utilizing machine learning to improve the CASA model, we expect that the challenging nature of the task will reveal other methods and principles that contribute to both Earth science and the science of machine learning.

Acknowledgments: We would like to thank Vanessa Brooks for her help in creating the data files. We would also like to thank Jeff Shrager for his help in formulating the problem, and for numerous discussions in which he has participated. Finally, we would like to thank Kazumi Saito and Ross Quinlan for reviewing drafts of this paper. This research was funded by the NASA Intelligent Systems Program.

References

Andrienko, G.L., Andrienko, N.V.: Interactive maps for visual data exploration. International Journal Geographic Information Science 13, 355–374 (1999)

Breiman, L., Friedman, J.H., Olshen, R.A., Stone, C.J.: Classification and regression trees. Wadsworth, Belmont, CA (1984)

Brodley, C.E., Friedl, M.A.: Identifying mislabeled training data. Journal of Artificial Intelligence Research 11, 131–167 (1999)

Brunk, C., Kelly, J., Kohavi, R.: MineSet: An integrated system for data mining. In: Proceedings of the Second International Conference of Knowledge Discovery and Data Mining, Portland, OR, pp. 135–138 (1996)

Chen, H.S.: Remote sensing calibration systems: An introduction. A. Deepak Publishing, Hampton, VA (1997)

Ester, M., Kriegel, H.-P., Sander, J., Xu, X.: A density-based algorithm for discovering clusters in large spatial databases with noise. In: Proceedings of the Second International Conference of Knowledge Discovery and Data Mining, Portland, OR, pp. 226–231 (1996)

John, G.A.: Robust decision trees: Removing outliers from data. In: Proceedings of the First International Conference of Knowledge Discovery and Data Mining, Montreal, Canada, pp. 174–179 (1995)

Keim, D.A., Kriegel, H.-P.: Visualization techniques for mining large databases: A comparison. Transactions on Knowledge and Data Engineering 8, 923–938 (1996)

Kodratoff, Y., Nédellec, C. (eds.): Working Notes of the IJCAI-95 Workshop on Machine Learning and Comprehensibility, Montreal, Canada (1995)

Lieth, H.: Modeling the primary productivity of the world. In: Lieth, H., Whittaker, R.H. (eds.) Primary Productivity of the Biosphere, Springer, Heidelberg, pp. 237–263 (1975)

Michalski, R.S.: A theory and methodology of inductive learning. Artificial Intelligence 20, 111–161 (1983)

Pazzani, M.J., Bay, S.D.: The independent sign bias: gaining insight from multiple linear regression. In: Proceeding of the Twenty-First Annual Meeting of the Cognitive Science Society, Vancouver, Canada (1999)

Potter, C.S., Brooks, V.: Global analysis of empirical relations between annual climate and seasonality of NDVI. International Journal of Remote Sensing 19, 2921–2948 (1998)

Potter, C.S., Klooster, S.A.: Interannual variability in soil trace gas (CO_2, N_2O, NO) fluxes and analysis of controllers on regional to global scales. Global Biochemical Cycles 12, 621–635 (1998)

Potter, C.S., Klooster, S.A., Brooks, V.: Interannual variability in terrestrial net primary production: Exploration of trends and controls on regional to global scales. Ecosystems 2(1), 36–48 (1999)

Provost, F., Kohavi, R.: On applied research in machine learning. Machine Learning 30, 127–132 (1998)

Quinlan, J.R.: Learning with continuous classes. In: Proceedings of the Australian Joint Conference on Artificial Intelligence, Hobart, Australia, pp. 343–348 (1992)

Quinlan, J.R.: C4.5: Programs for Machine Learning. Morgan Kaufmann, San Mateo, CA (1993)

Rheingans, P., desJardins, M.: Visualizing high-dimensional predictive model quality. In: Proceedings of the Eleventh IEEE Visualization Conference, Salt Lake City, UT, pp. 493–496 (2000)

RuleQuest. RuleQuest Research data mining tools (2002), `http://www.rulequest.com`

Schwabacher, M., Langley, P.: Discovering communicable scientific knowledge from spatio-temporal data. In: Proceedings of the Eighteenth International Conference on Machine Learning, Stanford, CA, pp. 489–496 (2001)

Smyth, P., Ghil, M., Ide, K.: Multiple regimes in Northern hemisphere height fields via mixture model clustering. Journal of the Atmospheric Sciences 56 (1999)

SPIN!, Spatial mining for data of public interest (2002), `http://www.ais.fraunhofer.de/KD/SPIN/`

Thornthwaite, C.W.: An approach toward rational classification of climate. Geographical Review 38, 55–94 (1948)

Todorovski, L., Dzeroski, S.: Declarative bias in equation discovery. In: Proceedings of the Fourteenth International Conference on Machine Learning, Nashville, TN, pp. 376–384 (1997)

Tufte, E.R.: The visual display of quantitative information. Graphics Press, Cheshire, CT (1983)

Weiss, S., Indurkhya, N.: Rule-based regression. In: Proceedings of the Thirteenth International Joint Conference on Artificial Intelligence, Chambéry, France, pp. 1072–1078 (1993)

Willmott, C.J., Feddema, J.J.: A more rational climate moisture index. Professional Geographer 44, 84–87 (1992)

Structure Discovery from Massive Spatial Data Sets Using Intelligent Simulation Tools

Feng Zhao[1], Chris Bailey-Kellogg[2], Xingang Huang[3], and Iván Ordóñez[4]

[1] Palo Alto Research Center (PARC)
Palo Alto, California, USA
zhao@parc.com

[2] Department of Computer Sciences
Purdue University, West Lafayette, Indiana, USA
cbk@cs.purdue.edu

[3] The Ohio State University
Columbus, Ohio, USA
huang@cis.ohio-state.edu

[4] Bios Group
Santa Fe, New Mexico, USA
ivan.ordonez@biosgroup.com

Abstract. Extracting structures as communicable knowledge is a central problem in spatio-temporal data analysis. Spatial Aggregation is an effective way for discovering structures. To address the computational challenges posed by applications such as weather data analysis or engineering optimization, Spatial Aggregation recursively aggregates local data into higher-level descriptions, exploiting the fact that these physical phenomena can be described as spatio-temporally coherent "objects" due to continuity and locality in the underlying physics. This paper uses several problem domains — weather data interpretation, distributed control optimization, and spatio-temporal diffusion-reaction pattern analysis — to demonstrate that intelligent simulation tools built upon the principles of Spatial Aggregation are indispensable for scientific discovery and engineering analysis.

1 Introduction

Information technology has fundamentally changed the way we conduct scientific experiments and synthesize engineering artifacts. For instance, powerful computers have routinely been used to extract interesting features in satellite images, diagnose abnormalities in nuclear reactors, allocate resources in air traffic control, and discern subtle trends in stock markets, to name just a few applications.

This paper describes a coherent body of theories and techniques, collectively known as *intelligent simulation*, that have recently been developed by researchers in Artificial Intelligence and Computational Science and Engineering to process massive spatio-temporal data sets arising from many scientific and engineering

S. Džeroski and L. Todorovski (Eds.): Computational Discovery, LNAI 4660, pp. 158–174, 2007.

applications. Simply put, intelligent simulation is about automatic computer interpretation of numerical simulation or measurement data to produce high-level structures understandable by human experts working in the domain. Intelligent simulation combines the reasoning and representational power of computational intelligence with an arsenal of numeric, symbolic, and geometric tools to solve problems not amenable to traditional numeric or symbolic methods alone (Abelson et al., 1989). Suitable task domains for intelligent simulation include mining scientific data and designing and controlling engineering systems. Instead of attempting to survey the huge body of existing work in the field, we will introduce and focus on concrete instances from our own work to demonstrate that intelligent simulation tools are essential for the rapid prototyping and execution of application programs in many challenging scientific and engineering domains. We will describe future research directions to develop and realize the full potential of intelligent simulation.

2 Challenges in Analyzing Physical Fields

A major class of physical problems for which intelligent simulation techniques are particularly well-suited are the so-called *field* problems. Continuous, distributed parameter fields are common physical phenomena: consider the temperature field in a building, the air flow around an airplane wing, or the noise from a copy machine. Figure 1 gives examples of fields — a fluid field and a weather field (map). Other examples of fields are phase spaces for describing behaviors of dynamical systems (Arnold, 1987) and configuration spaces for motion of mechanical systems (Lozano-Perez, 1983). Yet another example is the force field synthesized from a distributed micro-electro-mechanical system (MEMS) actuator array; Böhringer and Donald developed force vector fields suitable for transporting and orienting small parts such as semiconductor wafers (Bohringer & Donald, 1998).

Many practical applications rely on the ability to reason about and control these processes and systems. For instance, the drag on an airplane can be reduced by analyzing and controlling the air flow around the wings. Temperature in a "smart" building can be regulated to maximize occupant comfort while minimizing energy consumption. Because of the rapid advances in micro-fabrication technology that can integrate and produce MEMS devices on a massive scale, we are becoming increasingly reliant on large networks of sensors, actuators, and computational elements to augment our ability to interact with and control the physical environment (Williams & Nayak, 1996).

However, the challenges one faces in interpreting data from physical fields and controlling the behaviors of the fields are enormous. The difficulties arise from three sources. First, a distributed parameter field is conceptually harder to reason about and model than a lumped parameter system such as a circuit. In addition to the combinatorial structures, spatial topology, metric, material properties and physical laws all come into play. The underlying physical processes might be nonlinear and defy analytic, closed-form solution. Second, numerical

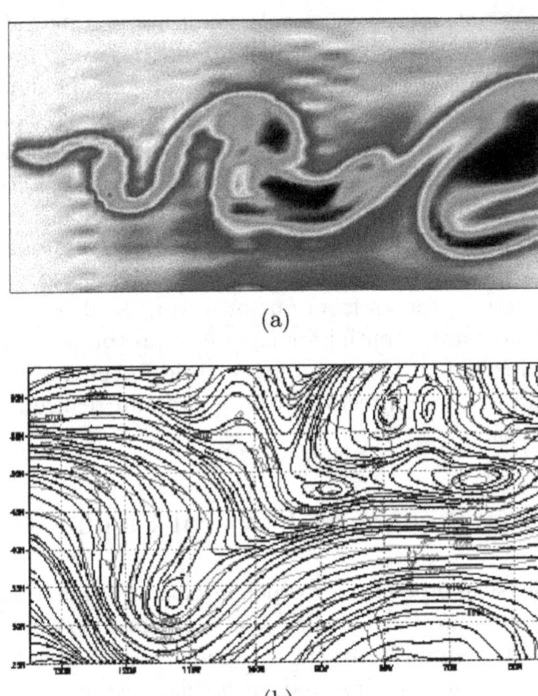

Fig. 1. Examples of fields: (a) A fluid flow. The fluid field describes how objects such as high density regions and large vortex structures are spatially distributed (shown here) and temporally evolving (not shown). (b) A 300mb weather map over North America. The data in a typical meteorological map includes pressure, temperature, and wind velocities on a spatial grid. An experienced meteorologist could identify qualitatively important weather features such as the location of a cold front and the direction of its movement, by extracting and correlating geometric features such as pressure troughs and thermal packing.

methods developed for designing and controlling physical fields require solving large systems of equations and hence are prohibitively expensive for large, irregular geometric domains and highly non-uniform, nonlinear phenomena. Third, many control applications rely on networks of sensors and controllers to interact with a physical field. The physical laws constrain the ability of spatially distributed, local agents to sense and affect the environment. Local sensors and control elements measure and interact with small neighborhoods around them. Macroscopic consequences are aggregated from local actions. Consequently, the design, programming, and coordination of distributed computational agents immersed in physical media require abstraction mechanisms, inference methods, and programming languages different from those for reasoning about and controlling centralized, lumped parameter models.

3 Interpreting Fields: An Example of Intelligent Simulation

Much of the field information amassed by sensors is in an analogue, data-rich form such as images, videos, or spatially distributed and continuous measurements of physical processes. For many reasoning and control tasks, we need to extract large-scale structures and behaviors from small-scale data descriptions.

As an example of how intelligent simulation provides organizational principles and building blocks to facilitate the development of programs for engineering problems, consider an interpretation task in dynamical system analysis, simplified from Yip's KAM program for analysis of dynamical systems (Yip, 1991). The input is a field of sampled states as points in phase space — a vector field — shown in Figure 2(a). The objective is to group states into trajectories and then trajectories into trajectory bundles that share the same qualitative behaviors, as shown in Figure 2(d) and Figure 2(f) respectively. We use the vocabulary of Spatial Aggregation (SA) (Yip & Zhao, 1996) to describe the computational patterns in this example. SA provides a unified description for a number of problem solvers for interpreting dynamical systems (Yip, 1991; Sacks, 1991; Nishida et al., 1991; Zhao, 1994), synthesizing controllers (Bradley & Zhao, 1993), and analyzing mechanical mechanisms (Joskowicz & Sacks, 1991). SA introduces the concepts of field ontology, multi-layer spatial aggregates mediated by neighborhood structures, and a uniform vocabulary for constructing and transforming the spatial aggregates. Even though this problem involves a 2D space, the framework applies to higher-dimensional spaces as well.

1. *Points* to *trajectory curves*
   (a) Given an input point field (Figure 2(a)).
   (b) *Aggregate* the points into a minimal spanning tree neighborhood graph (Figure 2(b)). Aggregation explicates a task-specific neighborhood relation (e.g. MST, Voronoi diagram, or nearness criteria) on a set of spatial objects.
   (c) *Classify* connected points into the same equivalence class if the edge connecting them isn't too long relative to nearby edges (Figure 2(c)). Classification identifies equivalence classes of objects in a neighborhood graph, according to a task-specific clustering mechanism and equivalence relation.
   (d) *Redescribe* equivalence classes of points as trajectory curves (Figure 2(d)). Redescription shifts the level of abstraction so that the aggregation process can repeat at a higher level.
2. *Trajectory curves* to *trajectory bundles*
   (a) *Aggregate* trajectory curves such that curves are adjacent if any of their constituent points are neighbors in the underlying minimal spanning tree (Figure 2(e)).

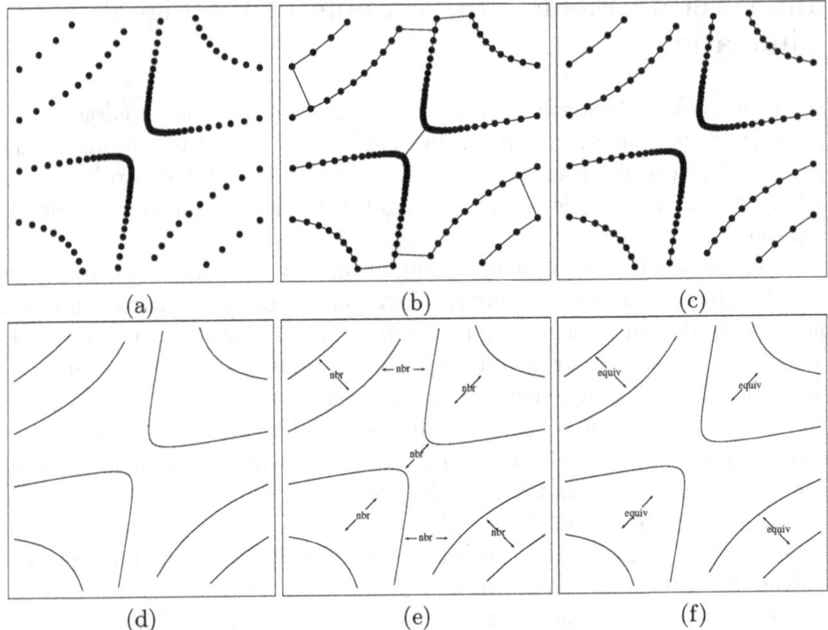

Fig. 2. Example steps in trajectory bundling application. (a) Input points. (b) Points aggregated into a minimal spanning tree. (c) Equivalence classes of points joined by short-enough edges. (d) Equivalence classes redescribed as trajectories. (e) Trajectories aggregated based on adjacencies of constituent points. (f) Equivalence classes of trajectories with similar-enough limit behavior.

(b) *Classify* connected curves into the same equivalence class if their shape is relatively similar (Figure 2(f)).
(c) *Redescribe* equivalence classes of trajectory curves as trajectory bundles.

More generally, SA takes as input measured data from a physical field or numerical data from simulating an ODE or PDE that models a field. The structure discovery process generates high-level descriptions of the physical field in terms of more abstract spatial objects and their configurations, e.g., the one-dimensional trajectory curves and the two dimensional trajectory bundles as in the above example. Other examples of spatial objects include temperature contour objects or "cold/warm" regions on a weather map. Section 5 will overview Spatial Aggregation operators that build such abstractions hierarchically.

As the example illustrates, SA programs are modular, using a common data structure (neighborhood graph) and an identical set of generic operators reusable for multiple levels of problem abstraction. They are concise and make explicit the important characteristics of the problem: neighborhood and equivalence relations.

4 Why Intelligent Simulation?

A simulation is a computational experiment performed on a model of a system in order to answer a set of questions about the system. As the above applications demonstrate, several characteristics distinguish techniques of intelligent simulation from those of conventional simulation.

- *Rich models*
  Numerical simulation typically generates overwhelming amounts of data that require human interpretation, while conventional symbolic methods are often data poor and do not scale up well with the size of problems. In contrast, models in intelligent simulation are structured and explicitly encode geometries, spatio-temporal scales, causes and effects, and even simplifying assumptions that are introduced during model construction in order to make the analysis tractable (Falkenhainer & Forbus, 1991; Bobrow et al., 1996). At the bottom level, the models are grounded on the rich, continuous physical sensory data. At the higher levels, the models describe class generic properties of physical entities such as the C-space free region diagram for mechanical mechanisms, or the iso-thermal structure of a temperature field. It is useful to note that intelligent simulation moves beyond qualitative simulation (Kuipers, 1986) of lumped parameter dynamical systems to analyze the more challenging spatially distributed parameter systems, or systems modeled by PDEs. Because of the richness in the representation, the models can be used for a variety of data-rich applications while simultaneously supporting high-level reasoning tasks such as explanation generation, fault diagnosis, or computer-aided tutoring. For example, the output of equivalence classes of trajectories in the dynamical system analysis example can potentially be used by system identification tools, for example (Bradley et al., 2001), to constrain the space of possible models.
- *Software modularity and reusability*
  In the past, numeric simulation software was monolithic and often difficult to maintain. MATLAB and other recent numerical computation environments have changed the image of simulation. Tools of intelligent simulation share many of the same goals as MATLAB: to provide a set of data types and operators to express commonly occurring computational patterns and to facilitate rapid prototyping of programs. Just as MATLAB uses matrices to organize computation for linear systems, intelligent simulation tools supply data types such as the neighborhood graph for problems involving distributed physical fields. These tools isolate and localize domain-specific knowledge and cross-cut concerns such as metric space, equivalence relation, and consistency maintenance. Programs are built by choosing and instantiating commonly used component implementations in libraries. Additionally, intelligent simulation provides tools for filtering data, maintaining consistency, and managing models. We view intelligent simulation as complementary to conventional simulation. Intelligent simulation provides building blocks for rationalizing existing programs and for constructing new problem solvers.

In the rest of the paper, we describe an implemented language, SAL, supporting a subset of the desired capabilities for intelligent simulation. We then overview three challenging applications developed using SAL: weather data interpretation, distributed control optimization, and diffusion-reaction spatio-temporal pattern analysis.

5 The Spatial Aggregation Language (SAL): A Rapid Prototyping Environment

We have developed the SAL programming environment (Bailey-Kellogg et al., 1996) to provide commonly used data types and generic Spatial Aggregation operators for application development. Application programs are written by mixing and matching components from prefabricated libraries. The SAL components explicate and localize important domain specific knowledge so that the resulting programs are more modular and easier to maintain than traditional numerical simulation software. They allow a programmer to focus on the knowledge relevant to an application, rather than on implementation details.

Spatial Aggregation uncovers structures at multiple levels of abstraction, with the structures uncovered at one level becoming the input to the structure-discovery process at the next level. For example, in a weather data analysis application (Huang & Zhao, 2000), Spatial Aggregation could extract from pressure data the isobars, pressure cells, and pressure troughs. Such multi-layer structures arise from continuities in fields at multiple scales. Due to the continuity, fields exhibit regions of uniformity, and these regions of uniformity can be abstracted as higher-level structures which in turn exhibit their own continuities. Task-specific domain knowledge specifies metrics and defines similarity and closeness of both field objects and their features. For example, isothermal contours are connected curves of equal (or similar enough) temperature.

SAL provides a small set of uniform data types and generic operators for constructing the spatial aggregate hierarchy. The central data type of SAL, the *neighborhood graph*, is an explicit representation of an object adjacency relation. The definition of adjacency is domain-specific and depends on the metric properties of the input field. Common adjacency relations include Delaunay triangulations, minimal spanning trees, and uniform grids. The neighborhood graph serves as computational glue, localizing interactions between neighboring objects. The main SA operators *aggregate* objects into neighborhood graphs satisfying an adjacency predicate, *classify* neighboring nodes into equivalences classes with respect to an equivalence predicate, *redescribe* equivalence classes into higher-level objects, and *localize* higher-level objects back into their constituent equivalence classes. Additional operators search through neighborhood graphs, check consistency of objects, extract geometric properties, and so forth. The earlier dynamical system analysis example illustrates how these operators can be applied iteratively. By instantiating these operators with proper

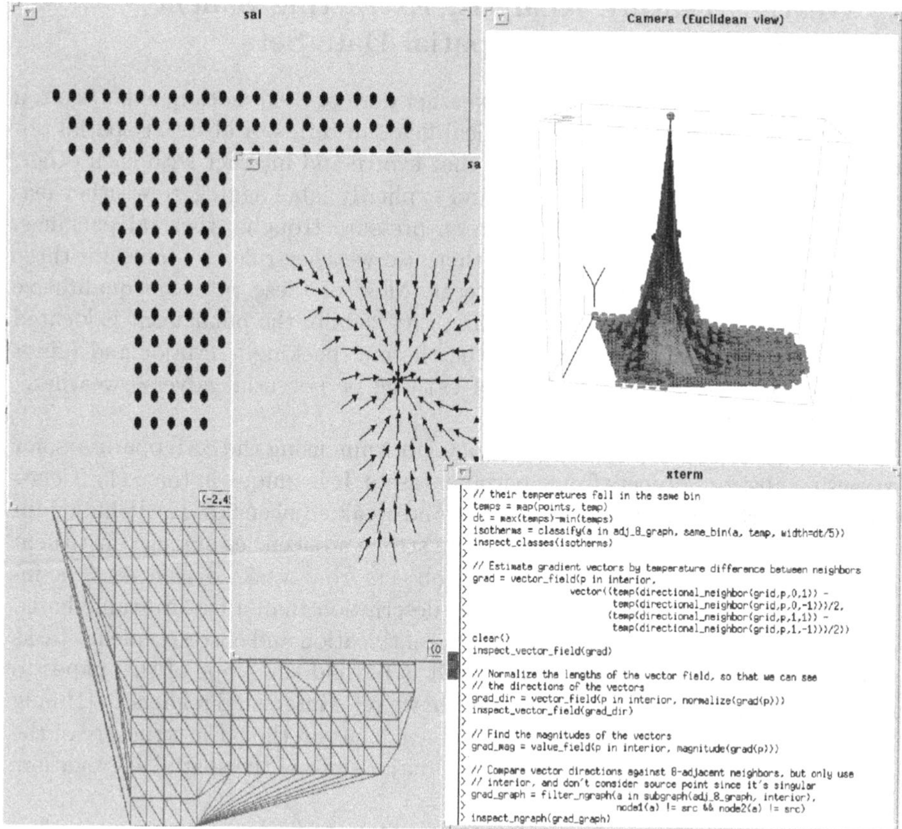

Fig. 3. The SAL interpreter in action: interactive code evaluation and graphical inspection of results

knowledge at different levels of abstraction, Spatial Aggregation allows specification of a variety of application programs.

The SAL implementation comprises a C++ library and an interpreted, interactive environment layered over the library. The library supports construction of efficient C++ programs with access to a large set of data type implementations and operations supporting the Spatial Aggregation programming style. The interpreter (Figure 3) supports rapid prototyping of application programs by providing a high-level interface to the main data type implementations and operators of the SAL library. Programmers can conveniently explore trade-offs in the specification of domain knowledge such as neighborhood relations and equivalence predicates, interactively examining and modifying the results without having to recompile a program. Graphical inspection tools support manipulation and exploration of the structures in physical data.

6 Weather Feature Analysis: Extracting Salient Structures from Large Spatial Data Sets

In analyzing spatial data sets such as weather data or fluid motion, experts often perceive and reason about these physical fields in terms of abstract spatial objects, also called features or patterns, that evolve and interact with each other. For example, meteorologists identify and explicitly label aggregate weather features such as high/low pressure centers, pressure troughs, thermal packings, fronts, and jet streams. The experts then use weather rules to correlate these features and establish prediction patterns. Most of these rules are qualitative and have a rich geometric interpretation: "At 850mb, the polar front is located parallel to and on the warm side of the thermal packing." "Major and minor 500mb troughs are good indicators of existing or potential adverse weather." (Air Water Service, 1975)

We have developed a structure finding algorithm, using the SAL operators, for extracting abstract objects from spatial datasets. It is unique in that: (1) It classifies neighborhood relations into strong and weak adjacencies. (2) It builds internal structures of aggregate objects from strong adjacencies among constituent objects, and relations between aggregate objects from weak adjacencies. The internal structures of objects form a richer description than a feature-value based representation and are essential for many identification and correspondence tasks in interpreting large scientific data sets. It is beyond the scope of this paper to present algorithmic details of the approach; we refer interested readers to (Huang & Zhao, 2000) for additional details. Here, we focus on the basic structure of the interpretation task: the input/output, the major steps of the spatial aggregation, and the quality of the interpretation.

We illustrate the algorithm with an application to finding troughs and ridges in weather data. Troughs and ridges are important features in weather analysis: high-altitude troughs correlate with bendings of jet streams and are useful for extended weather forecast; surface troughs are usually closely related to fronts and are important for locating the fronts. What are troughs and ridges? Visually, troughs and ridges are sequences of iso-bar segments bending consistently to one direction, with troughs bending from low iso-bars to high iso-bars and ridges from high iso-bars to low iso-bars. Though the extraction of troughs seems effortless and immediate to human eyes, it is only qualitatively defined.

Our algorithm extracts trough features from weather data using a multi-level approach. We use the raw data (pressure, temperature, wind streamlines, air density, etc.) from the National Weather Services website as input. The algorithm identifies salient structures from the data and outputs a geometric description of the troughs and ridges (as a curve graph). The algorithm goes through the following major steps: (1) It extracts salient iso-bar segments using an iterative thresholding technique. (2) It builds a neighborhood structure for the segments from a Delaunay triangulation of iso-points. (3) It then classifies the neighborhood relations and uses the strong adjacencies to extract the linear structures among the segments to obtain troughs. Figure 4 compares the high altitude

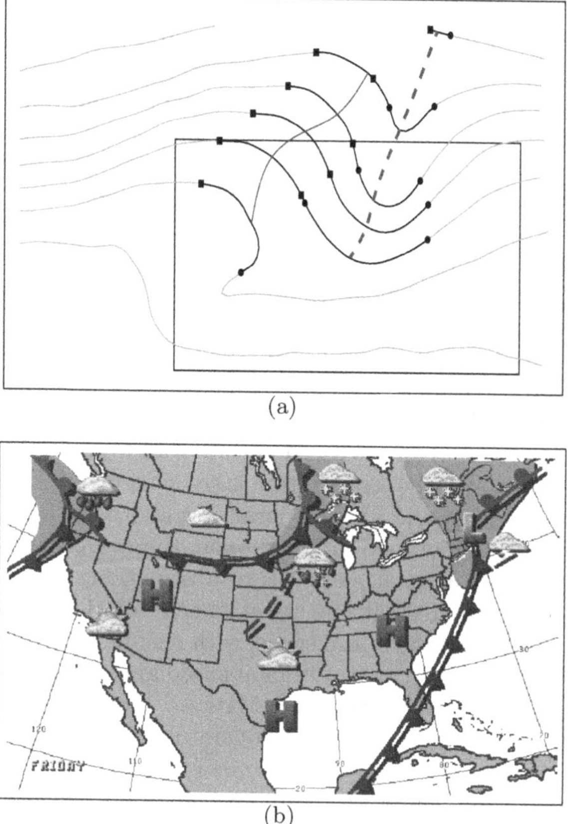

Fig. 4. Labeling weather chart: (a) The high-altitude trough (dashed line) detected by the spatial aggregation algorithm. (b) The corresponding trough (dashed line) drawn by meteorologists for the national weather forecast map for roughly the same area as the box in (a).

trough detected by our algorithm with that drawn by the meteorologists for the national weather forecast map, using the data for Friday, Jan. 15, 1999.

The rich structures extracted by the algorithm can be used to identify causal relations among weather events and to generate high-level explanations for predictions.

7 Distributed Control Optimization: Divide-and-Conquer by Exploiting Structural Information in Fields

Many sensor-rich control applications require *decentralized control* to attain adaptivity, robustness, and scalability. For example, a controller for a smart building regulates building temperature with networks of sensors and actuators; decentralized control allows the network to tolerate failures in individual elements and to scale up more gracefully with the number of nodes. As another example, rapid thermal processing in semiconductor curing employs separately controlled lamp zones to maintain a uniform temperature profile over semiconductor wafers in order to avoid defects (Kailath et al., 1996). Designing and optimizing decentralized controls for a distributed system requires achieving global control objectives through appropriate combinations of local control actions. To determine the placement of control nodes and their parameters, one must search a large design space subject to structural, behavioral, and performance constraints.

We have employed SAL to model a distributed thermal regulation problem and to design controls (Bailey-Kellogg & Zhao, 2001). The first task is to model the heat flow in a problem domain. Figure 5(a) overviews major computational patterns in this process. A geometric description of a problem domain (e.g. a piece of material for which the temperature is to be regulated) is discretized into a space of spatial objects and an associated neighborhood graph. Local interaction rules in the neighborhood graph support the computation of heat flow. This process can occur at multiple levels of abstraction, with approximations at coarse resolutions driving refinements at finer resolutions. It can also occur separately in subregions of the initial space, with results iteratively combined. In this manner, SAL supports many traditional paradigms of engineering computation (e.g. domain decomposition (Chan & Mathew, 1994) and multigrid (Briggs, 1987)), but with a set of concise, generic operators in a vocabulary natural for the domain.

The second task, overviewed in Figure 5(b), is to design both control placements and control parameters to achieve a desired control objective. SAL operators perform control design based on high-level characteristics of heat flow in a field. These characteristics are extracted by classifying the field's response to a set of control probes. In the example, the geometric constraint imposed by the narrow channel in the dumbbell-shaped piece of material results in similar field responses to the two probes in the left half of the dumbbell and similar responses to the two probes in the right half of the dumbbell. Based on the resulting classes, the field is decomposed into regions to be separately controlled. In this case, the left half of the dumbbell is decomposed from the right half. Controls are placed in the regions and optimized by adjusting their outputs in response to their effects on the field. The structural information also supports parametric optimization by indicating the influence of control nodes on field nodes. This allows efficient computation of changes in the field due to actions of a control, and the ability to focus computation and communication on the part of the field most strongly affected by a control's actions.

The resulting control placement designs are comparable to those achieved by traditional engineering optimization methods (e.g. simulated annealing (Metropolis et al., 1953)), but use a small, fixed number of function evaluations to

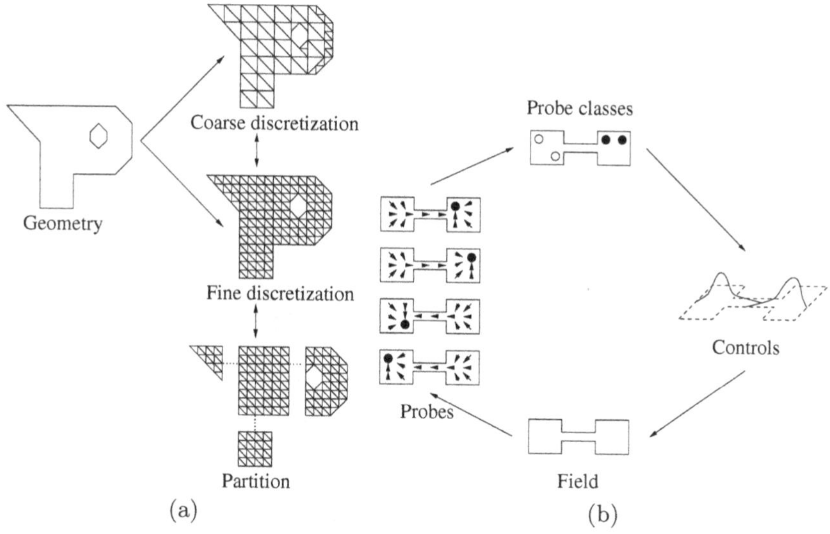

Fig. 5. Overview of SAL-based computational patterns for decentralized thermal regulation. (a) Modeling heat flow. (b) Designing controllers.

extract and exploit structural descriptions of the field. Basing the algorithms on structural descriptions allows explanation of and higher-level reasoning about the resulting designs. Use of structural information in parametric optimization results in drastic (several orders of magnitude) speed-ups in field evaluation, and methods for explicitly trading off between computation, communication, and control quality (Bailey-Kellogg & Zhao, 2001).

8 Diffusion-Reaction Patterns Analysis: Modeling Spatio-temporal Processes

We have also developed an algorithm for identifying and tracking coherent structures in spatio-temporal fields produced by a class of diffusion-reaction systems. Such systems describe a number of physical, chemical, and biological phenomena, and their study may shed light on how nature constructs and evolves structures that exhibit a high degree of regularity. For instance, it is important to describe the qualitative features of behaviors for the Gray-Scott diffusion-reaction model of glycolysis, shown in Figure 6.

The algorithm adaptively samples a spatial field using a particle system, aggregates the sampling particles into a neighborhood graph, classifies the structure into coherent regions, and tracks the regions over time to produce a qualitative description of the temporal evolution of the field. Because the adaptive sampling grid varies smoothly with the temporal evolution of the underlying field, the algorithm is able to efficiently track the corresponding objects over successive time frames by minimally updating the grid.

Fig. 6. A snapshot of a pattern in the Gray-Scott model, generated for a particular set of parameters. The pattern, denoting the concentration of a substance that diffuses and reacts with others, evolves over time.

The details of the algorithm can be found in (Ordonez & Zhao, 2000). Here, we sketch the major steps of the algorithm, and illustrate what a qualitative output of the interpretation looks like. The simulator for the algorithm takes as input a field model, its parameters, and a set of initial conditions, and numerically integrates the model to generate a time varying field. After initialization, the particle system uses the field gradient information to form a sampling mesh. The triangle elements in the particle mesh are aggregated to form polygons describing regions of uniformity. The mesh is updated locally as the sampled field changes, and the updates are propagated to the polygonal aggregates. Tracking these polygonal objects over time, the simulator generates a high-level description of the history of the objects — births and deaths, fissions, separations, assimilations, and fusions. At each time instant, the description consists of identified coherent regions in the field. Higher-level spatio-temporal aggregations spatially collect the regions into related shape clusters and temporally collect event sequences among the shape clusters. Figure 7 shows an example of a typical run.

The histories produced by the algorithm are important for further analysis of causal structures in the spatio-temporal processes. Because the histories are based on multi-level aggregates of an adaptive particle system, maintaining object identity over time is much more straightforward than for traditional techniques that analyze frames independently and then establish correspondence.

```
Events
  ...
  At step 108 body 3 (born 45) fused into body 2
  At step 108 body 1 (born 45) fused into body 0
  At step 112 body 2 (born 45) fused into body 0
  ...
  At step 176 body 4 arose from the fission of body 0 (born 45)
  At step 178 body 5 arose from the fission of body 0 (born 45)
  At step 180 body 6 arose from the fission of body 0 (born 45)
  ...

Detected Shape Clusters
  Shape cluster 2:
    body 0 @ [45,108], body 1 @ [45,-], body 2 @ [45,-],
    body 3 @ [45,-]
  Shape cluster 3:
    body 0 @ [176,-], body 4 @ [176,-], body 5 @ [178,-],
    body 6 @ [180,-]
  Shape cluster 5:
    body 0 @ [112,176]
  ...

Detected Event Sequences
  Sequence (shape cluster 2 -> shape cluster 5) observed @ 112
  Sequence (shape cluster 5 -> shape cluster 3) observed @ 176
  ...
```

Fig. 7. An example of a history of spatio-temporal objects identified by the Spatial Aggregation analysis: time-stamped object events, aggregated shape clusters, and temporal sequences of events. Cluster 2 contains four small L-shaped objects, cluster 3 contains four medium-sized L-shaped objects, and cluster 5 is a single large, squared-shaped object with a hole inside. The event sequences describe fusion of the small Ls into the square, and fission of the square into the medium-sized Ls.

9 Related Work

Intelligent simulation builds on techniques from many research areas that investigate the problem of extracting and exploiting structures in distributed physical data. Computer vision researchers have long tried to find structures in image data. In the spirit of Ullman's visual routines (Ullman, 1984), Spatial Aggregation attempts to identify a set of primitive, generic operators out of which more complex applications can be composed. Spatial data mining algorithms identify correlations in spatial data; for example, finding general climate patterns across regions by mining correspondences in geographic and weather data (Lu et al., 1993). Scientific data analysis applications find patterns in large data sets; for example, automatically cataloging sky images, and identifying volcanos in images of the surface of Venus (Fayyad et al., 1996). Scientific visualization displays physical data in a manner that helps scientists use perceptual processes to understand it (Rosenblum et al., 1994). An important component of scientific visualization is the detection of features in data, such as vortices in fluid data. Advanced data interpretation techniques, such as the visiometrics algorithms (Samtaney et al., 1994) and the Fluid Dynamicist's Workbench (Yip, 1995) use these feature detectors in order to track structures over time. Finally, qualitative spatial reasoning systems (e.g. (Forbus et al., 1991)) and diagrammatic reasoning applications (e.g. (Glasgow et al., 1995)) manipulate abstract descriptions of spatial phenomena.

10 Conclusion

The rapid advances in sensor and information processing technology have unleashed an unprecedented amount of data about the dynamic physical world. To interpret and utilize the information from the massive data, we will have to rely on and collaborate with computational discovery tools. It is important that these tools uncover spatial-temporal structures in the data and communicate them in high-level terms, in the same way as scientists would describe these structures as coherent "objects".

Intelligent simulation is an exciting area in which to do research because of the wealth of problem domains and the practical impact of the tools on real-world problems. Many research issues are wide open: Intelligent simulation leverages techniques of computer vision and computational geometry to model and process complex data. Are there techniques from other areas (e.g. statistical/learning techniques — see below) that are suitable and can be incorporated into intelligent simulation tools? What additional operators are required? For example, what operators are useful for merging bottom-up and top-down processing or correlating features from different data sets? Can we build up large domain libraries and automatically search them to build programs? What are the classes of problems for which intelligent simulation techniques are most natural?

In summary, spatial data interpretation is a rich source of problems in computational discovery of communicable knowledge. Techniques of Spatial Aggregation and intelligent simulation tools will become indispensable for scientists and engineers in these discovery processes.

Acknowledgments: An earlier version of this paper appeared under the title "Intelligent Simulation Tools for Mining Large Scientific Data Sets", New Generation Computing, 17:333-347, Ohmsha, Ltd. and Springer-Verlag, 1999.

This paper describes research conducted at Ohio State University and Xerox Palo Alto Research Center, supported in part by FZ's ONR YI grant N00014-97-1-0599, NSF NYI grant CCR-9457802, NSF grant CCR-9308639, and Alfred P. Sloan Foundation Research Fellowship.

References

Abelson, H., Eisenberg, M., Halfant, M., Katzenelson, J., Sussman, G.J., Yip, K.: Intelligence in scientific computing. Communications of the ACM 32, 546–562 (1989)

Air Weather Service, Back to Basics. AWS FOT Seminar, STT-Q9-0004. Air Weather Service, Scott Air Force Base, IL (1975)

Arnold, V.: Ordinary differential equations. MIT Press, Cambridge, MA (1987)

Bailey-Kellogg, C., Zhao, F.: Influence-Based Model Decomposition For Reasoning About Spatially Distributed Physical Systems. Artificial Intelligence 130, 125–166 (2001)

Bailey-Kellogg, C., Zhao, F., Yip, K.: Spatial aggregation: language and applications. In: Proceedings of the National Conference on Artificial Intelligence (1996)

Bobrow, D., Falkenhainer, B., Farquhar, A., Fikes, R., Forbus, K., Gruber, T., Iwasaki, Y., Kuipers, B.: A compositional modeling language. In: Proceedings of the Tenth International Workshop on Qualitative Reasoning, Stanford, CA, pp. 12–21 (1996)

Böhringer, K.-F., Donald, B.: Algorithmic MEMS. In: Agarwal, P.K., Kavraki, L.E., Mason, M.T. (eds.) Robotics: The Algorithmic Perspective, A.K. Peters, Natick, MA, pp. 1–20 (1998)

Bradley, E., Easley, M., Stolle, R.: Reasoning about nonlinear system identification. Artificial Intelligence 133, 139–188 (2001)

Bradley, E., Zhao, F.: Phase-space control system design. IEEE Control Systems 13, 39–47 (1993)

Briggs, W.L.: A multigrid tutorial. Lancaster, Richmond, VA (1987)

Chan, T., Mathew, T.: Domain decomposition algorithms. In: Acta Numerica 1994, vol. 3, pp. 61–143. Cambridge University Press, Cambridge, UK (1994)

Falkenhainer, B., Forbus, K.: Compositional modeling: finding the right model for the job. Artificial Intelligence 51, 95–143

Fayyad, U., Haussler, D., Stolorz, P.: KDD for science data analysis: issues and examples. In: Proceedings of Second International Conference on Knowledge Discovery and Data Mining, Portland, OR, pp. 50–56 (1996)

Forbus, K., Nielsen, P., Faltings, B.: Qualitative spatial reasoning: the CLOCK project. Artificial Intelligence 51, 417–471 (1991)

Glasgow, J., Narayanan, N., Chandrasekaran, B.: Diagrammatic reasoning: cognitive and computational perspectives. AAAI Press, Menlo Park, CA (1995)

Huang, X., Zhao, F.: Relation based aggregation: Finding objects in large spatial datasets. Intelligent Data Analysis 4, 129–147 (2000)

Joskowicz, L., Sacks, E.: Computational kinematics. Artificial Intelligence 51, 381–416 (1991)

Kailath, T., Schaper, C., Cho, Y., Gyugyi, P., Norman, S., Park, P., Boyd, S., Franklin, G., Sarasunt, K., Maslehi, M., Davis, C.: Control for advanced semiconductor device manufacturing: A case history. In: Levine, W. (ed.) The Control Handbook, CRC Press, Boca Raton, FL, pp. 1243–1259 (1996)

Kuipers, B.J.: Qualitative simulation. Artificial Intelligence 29, 289–338 (1986)

Lozano-Perez, T.: Spatial planning: a configuration-space approach. IEEE Transactions on Computers 32, 108–120 (1983)

Lu, W., Han, J., Ooi, B.: Discovery of general knowledge in large spatial databases. In: Proceedings of Far East Workshop on Geographic Information Systems, Singapore, pp. 275–289 (1993)

Metropolis, N., Rosenbluth, A., Rosenbluth, M., Teller, M., Teller, E.: Equation of state calculations by fast computing machines. Journal of Chemical Physics 21, 1087–1092 (1953)

Nishida, T., Mizutani, K., Kubota, A., Doshita, S.: Automated phase portrait analysis by integrating qualitative and quantitative analysis. In: Proceedings of the Ninth National Conference on Artificial Intelligence, Anaheim, CA, pp. 811–816 (1991)

Ordonez, I., Zhao, F.: STA: Spatio-Temporal Aggregation with Applications to Analysis of Diffusion-Reaction Phenomena. In: Proceedings of the Seventeenth National Conference on Artificial Intelligence. Austin, TX (2000)

Rosenblum, L., Earnshaw, R.A., Encarnacao, J., Hagen, H.: Scientific visualization: Advances and challenges. Academic Press, San Diego, CA (1994)

Sacks, E.: Automatic analysis of one-parameter planar ordinary differential equations by intelligent numerical simulation. Artificial Intelligence 51, 27–56 (1991)

Samtaney, R., Silver, D., Zabusky, N., Cao, J.: Visualizing features and tracking their evolution. IEEE Computer Magazine 27, 20–27 (1994)

Ullman, S.: Visual routines. Cognition 18, 97–159 (1984)

Williams, B., Nayak, P.: Immobile robots: AI in the new millenium. AI Magazine 17, 17–35 (1996)

Yip, K.: KAM: A system for intelligently guiding numerical experimentation by computer. MIT Press, Cambridge, MA (1991)

Yip, K.: Reasoning about fluid motion I: Finding structures. In: Proceedings of the Fourteenth International Joint Conference on Artificial Intelligence, Montreal, Canada, pp. 1782–1788 (1995)

Yip, K., Zhao, F.: Spatial aggregation: theory and applications. Journal of Artificial Intelligence Research 5, 1–26

Zhao, F.: Extracting and representing qualitative behaviors of complex systems in phase spaces. Artificial Intelligence 69, 51–92 (1994)

Computational Discovery in Pure Mathematics

Simon Colton

Department of Computing
Imperial College, London, United Kingdom
sgc@doc.ic.ac.uk

Abstract. We discuss what constitutes knowledge in pure mathematics and how new advances are made and communicated. We describe the impact of computer algebra systems, automated theorem provers, programs designed to generate examples, mathematical databases, and theory formation programs on the body of knowledge in pure mathematics. We discuss to what extent the output from certain programs can be considered a discovery in pure mathematics. This enables us to assess the state of the art with respect to Newell and Simon's prediction that a computer would discover and prove an important mathematical theorem.

1 Introduction

In a seminal paper predicting future successes of artificial intelligence and operational research, Alan Newell and Herbert Simon suggested that:

> 'Within ten years a digital computer will discover and prove an important mathematical theorem.' (Simon & Newell, 1958)

As theorem proving involves the discovery of a proof, their predictions are about automated discovery in mathematics. In this chapter, we explore what constitutes knowledge in pure mathematics and therefore what constitutes a discovery. We look at how automated techniques fit into this picture: which computational processes have led to new knowledge and to what extent the computer can be said to have discovered that knowledge.

To address the state of the art in automated mathematical discovery, we first look at what constitutes mathematical knowledge, so that we can determine the ways in which a discovery can add to this knowledge. We discuss these issues in Section 2. In Sections 3 to 7, we look at five broad areas where computational techniques have been used to facilitate mathematical discovery. In particular, we assess the contributions to mathematical knowledge from computer algebra systems, automated theorem provers, programs written to generate examples, mathematical databases, and programs designed to form mathematical theories.

We then return to Newell and Simon's prediction and consider whether important discoveries in mathematics have been made by computer yet. We conclude that no theorem accepted as important by the mathematical community has been both discovered and proved by a computer, but that there have been discoveries of important conjectures and proofs of well known results by computer.

S. Džeroski and L. Todorovski (Eds.): Computational Discovery, LNAI 4660, pp. 175–201, 2007.

Furthermore, some systems have both discovered and proved results which could potentially be important. We further conclude that, with the ubiquitous use of computers by mathematicians and an increasing dependence on computer algebra systems, many important results in pure mathematics are facilitated – if not autonomously discovered – by computer. We then discuss possibilities for the use of other software packages in the mathematical discovery process.

1.1 Scope of the Survey

We restrict our investigation to pure mathematics, to avoid discussion of any application of the mathematics discovered. For instance, a program might invent an algorithm, which when implemented, leads to a better design for an aircraft wing. From our point of view, the algorithm is the discovery of importance, not the wing design. By restricting our investigation to pure mathematics, we hope to make this explicit.

(Valdés-Pérez, 1995) makes a distinction between (a) programs which have been designed to model discovery tasks in a human-like manner and (b) programs which act as scientific collaborators. While these two classes are certainly not mutually exclusive, there have been many mathematics programs which have not been used for discovery tasks. These include the AM and Eurisko programs (Lenat, 1982) (Lenat, 1983), the DC program (Morales, 1985), the GT program (Epstein, 1987), the ARE program (Shen, 1987), the Cyrano program (Haase, 1986), the IL program (Sims, 1990), and more recently the SCOT program (Pistori & Wainer, 1999) and the MCS program (Zhang, 1999). These systems are surveyed in (Colton, 2002b), but we restrict ourselves here to a survey of programs which have actually added to mathematics.

The Graffiti program, as discussed in Section 7.1, has been applied to chemistry (Fajtlowicz, 2001), and the HR program, discussed in Section 7.3, is currently being used in biology (Colton, 2002a). However, our final restriction in this survey is to look only at the mathematical applications of the discovery programs. For comprehensive surveys of automated discovery in science, see (Langley, 1998) and (Valdés-Pérez, 1999).

2 Mathematical Knowledge and Discoveries

We can classify knowledge in pure mathematics into ground knowledge about mathematical objects such as groups, graphs and integers, as well as meta-level knowledge about how mathematical explorations are undertaken. To a large extent, only the ground knowledge is communicated via the conferences, journals, and textbooks, with the meta-level knowledge discussed mainly between individuals and in a few books as listed in Section 2.2.

2.1 Ground Mathematical Knowledge

Many journal papers and textbooks in pure mathematics proceed with quartets of background information, concept definitions, theorem and proof. The background information usually states some results in the literature that provide a context for the new results presented in the paper.

Under the 'definition' heading, new concepts are defined and sometimes obvious properties of the concept and/or some examples satisfying the definition are provided. Concept definitions include, but are not limited to: classes of object, (e.g., prime numbers), functions acting on a set of objects to produce an output, (e.g., the calculation of the chromatic number of a connected graph (Gould, 1988)), and maps taking one set of objects to another (e.g., isomorphisms in group theory (Humphreys, 1996)).

Following the concept definitions, a theorem is proposed as a statement relating known concepts and possibly some new concepts. Theorems include statements that one class of objects has a logically equivalent definition as another class, i.e., if-and-only-if theorems. For example: an even number is perfect – defined as being equal to twice the sum of its divisors – if and only if it is of the form $2^n(2^{n+1} - 1)$ where $2^{n+1} - 1$ is prime (Hardy & Wright, 1938). Theorems also include statements that one class of objects subsumes another, i.e., implication theorems. For example: all cyclic groups are Abelian (Humphreys, 1996). In addition, theorems include non-existence statements, i.e., that there can be no examples of objects with a given definition. For example: there are no solutions to the equation $a^n + b^n = c^n$ for positive integers a, b and c with $n > 2$ (Fermat's Last Theorem (Singh, 1997)).

Finally, a proof demonstrating the truth of the theorem statement completes the quartet. The proof is usually a sequence of logical inferences proceeding from the premises of the theorem statement to the conclusion (although other strategies exist, as discussed in Section 2.2). Any newly defined concepts which were not explicitly mentioned in the theorem statement will probably appear in the proof, as the concepts may have been extracted from the proof in order to make it easier to understand.

In addition to concepts, theorems and proofs, authors sometimes present open conjectures which, like theorems, are statements about concepts, but which lack a proof, such as the open conjecture that there are no odd perfect numbers. These are provided in the hope that someone will one day provide a proof and turn the conjecture into a theorem. Algorithms are another type of ground mathematics. These provide a faster way to calculate examples of a concept than a naive method using the definition alone, e.g., the sieve of Eratosthenes for finding prime numbers. Finally, examples of concepts often appear in journal papers, particularly when examples are rare, or when finding the next in a sequence has historical interest, as with the largest prime number.

2.2 Meta-mathematical Knowledge

Information about how mathematical explorations have been undertaken, and how to undertake them in general, is rarely found in published form. There are, of course, exceptions to this, and mathematicians such as Poincaré have written about how to do mathematics (as discussed in (Ghiselin, 1996)), with similar expositions from (Lakatos, 1976) and (Pólya, 1988).

There has also been research that makes explicit some ways to solve mathematical problems. Such problems are often posed as exercises in mathematics texts and usually involve either the solution of a fairly simple, but illustrative, theorem, or the construction of a particular example with some desired properties. For instance, Paul Zeitz suggests the 'plug and chug' method: if possible, make calculations which are relevant to the problem (e.g., put numbers into a formula), and examine the output in the hope of finding a pattern which leads to a Eureka step and the eventual solution (Zeitz, 1999). This approach is discussed in more detail in (Colton, 2000).

Some work has also been undertaken to make explicit certain strategies for proving theorems, and there are some known approaches such as proof by induction, proof by contradiction, and proof by reductio ad absurdum. In particular, the area of proof planning has made more formal the notion of a proof strategy and inductive theorem provers utilise these ideas (Bundy, 1988).

2.3 Discovery in Pure Mathematics

Before discussing computational approaches to discovery in pure mathematics, we must first discuss what constitutes a discovery. Naively, any algorithm, concept, conjecture, theorem or proof new to mathematics is a discovery. However, we need to qualify this: a new result in mathematics must be important in some way. For instance, it is fairly easy to define a new concept in number theory. However, if there are no interesting provable properties of the concept and no obvious applications to previous results, the concept may be an invention, but it is unlikely to be accepted as a genuine discovery.

Similarly, any conjecture, theorem, proof or algorithm must be interesting in the context within which it is discovered. An important result may have application to the domain in which it was discovered; for example the definition of a concept may simplify a proof. Similarly, a conjecture, which if true, may demonstrate the truth of many other results. An important result may also be something without obvious application, but which expresses some unusual[1] connection between seemingly unrelated areas. The question of interestingness in mathematics is discussed more extensively in (Colton et al., 2000b) and (Colton, 2002b).

In addition to finding a new result, discoveries can be made about previous results. For instance, while the nature of pure mathematics makes it less likely to find errors than in other sciences, another type of discovery is the identification of an error of reasoning in a proof. For example, Heawood discovered a flaw in Kempe's 1879 proof of the four colour theorem,[2] which had been accepted for 11 years. A more recent example was the discovery that Andrew Wiles' original proof of Fermat's Last Theorem was flawed (but not, as it turned out, fatally flawed, as Wiles managed to fix the problem (Singh, 1997)).

[1] The mathematician John Conway is much quoted as saying that a good conjecture must be 'outrageous' (Fajtlowicz, 1999).

[2] This theorem states that every map needs only four colours to ensure that no two touching regions are coloured the same.

Similarly, papers are often published with a shorter proof of a known result than any previously found. For instance, Appel and Haken's original proof of the four colour theorem (Appel & Haken, 1977) was criticised because it required a computer to verify around 1500 configurations, and as such was not repeatable by a human. Fortunately, Robertson et al. have discovered a much simplified proof (Robertson et al., 1996).

In predicting that a computer would discover a new theorem (which may involve new concepts, algorithms, etc.), Newell and Simon restricted their discussion to discoveries at the object level. It is true that most mathematical discoveries occur at the object level, but discoveries at the meta-level are certainly possible. In particular, a theorem may be important not because of the relationship it expresses among the concepts in the theorem statement, but rather because of an ingenious step in the proof that is applicable to many other problems. Similarly, a new generic way to form concepts – such as a geometric or algebraic construction – can also be viewed as a discovery. For instance, in the long term, Galois' thesis on the insolubility of polynomials was more important because he introduced the notion of groups – an algebraic construct which arises in a multitude of other domains – than the actual result, although that in itself was a major breakthrough, as (Stewart, 1989) explains.

To summarise, mathematical discoveries include three main activities:

- discovering an important (interesting/applicable) concept, open conjecture, theorem, proof or algorithm;
- revising a previous result, by for instance, the identification of a flaw or the simplification of a proof;
- deriving a new method, in particular a proof strategy or a construction technique for concepts.

2.4 Approaches to Computational Discovery

In the next five sections, we discuss some computational approaches that have led to discoveries in pure mathematics. Firstly, we deal with computer algebra packages, the most common programs employed in pure mathematics. Following this, we look at the various uses of automated theorem provers which have led to discoveries. The next category covers a broad range of general and ad-hoc techniques designed to find examples of concepts. In the fourth category, we examine how the use of mathematical databases can facilitate discovery. The final category covers a multitude of systems designed to invent concepts, make conjectures, and in general form a theory about a domain, rather than to pursue specific results. We look at the first two categories in terms of the types of problems solved. However, for the final three categories, because the techniques are more ad-hoc, we sub-categorise them in terms of the techniques themselves, rather than the problems the techniques are used to solve.

3 Computer Algebra Systems

Computer algebra systems are designed to perform complex mathematical calculations, including algebraic manipulations, polynomial solution, differentiation, and so on. Such systems include Maple (Abell & Braselton, 1994), Mathematica (Wolfram, 1999) and GAP (Gap, 2000). The systems are usually accompanied by large libraries of functions – many of which are written by the mathematicians who make use of them – which cover a wide range of domains.

Four common ways in which computer algebra systems are used to facilitate discovery in pure mathematics are:

- Verifying a result with examples before a proof is attempted;
- Providing examples from which a pattern can hopefully be induced in order to state a conjecture;
- Filling in the specifics of a theorem statement;
- Proving theorems.

The first three applications are so common that we can illustrate them with examples from our own experience. We illustrate the fourth application of computer algebra to discovery with an example from the work of Doron Zeilberger.

3.1 Giving Empirical Weight

Computer algebra packages can be very useful for adding empirical weight to a conjecture before attempting to prove the conjecture. That is, whenever a plausible conjecture arises and it is possible to generate counterexamples, some effort is usually expended trying to find such a counterexample before an attempt to prove the conjecture is made. For example, using techniques described in Section 7, we made the conjecture that perfect numbers are never refactorable (with a refactorable number being such that the number of divisors is itself a divisor). Using the fast routines for integers in the GAP computer algebra package, we verified this result for all perfect numbers from 1 to 10^{54}. This gave us the confidence to attempt a proof, and we eventually proved this theorem (Colton, 1999). Of course, the alternate outcome is possible: a counterexample can be found, but we leave discussion of this until Section 5.

3.2 Presenting Data for Eureka Steps

Another way in which computer algebra systems can facilitate discovery is to produce examples from complex calculations and present the data in a way that lets the user notice patterns and possibly make a conjecture. To the best of our knowledge, there are no data-mining tools within computer algebra packages that could automate the pattern spotting part of the process. However, they are often equipped with visualisation packages, which can certainly help to highlight patterns in output. Furthermore, computer algebra systems include programmable languages, so the output can be tailored to enhance the chances of identifying a pattern.

A fairly trivial, but illustrative, example occurred when we used the Maple computer algebra system to help solve a problem posed by Paul Zeitz (Zeitz, 1999):

> Show that integers of the form $n(n+1)(n+2)(n+3)$ are never square numbers.

Following Zeitz's 'plug and chug' advice, we simply used Maple to calculate the value of the function $f(n) = n(n+1)(n+2)(n+3)$, for the numbers 1 to 4, giving:

$$f(1) = 24, f(2) = 120, f(3) = 360, f(4) = 840$$

As predicted by Zeitz, we noticed that the output was always one less than a square number, so proving this would solve the problem. We also used Maple in the proof, by guessing that $n(n+1)(n+2)(n+3)$ could be written as a quadratic squared minus one. Then we tried different quadratics until Maple agreed upon an equality between $(n^2+3n+1)^2-1$ and $n(n+1)(n+2)(n+3)$. We could, of course, have done all this entirely without Maple, but it illustrates how computer algebra systems can be used to find, and, in some cases, verify patterns in data. We discuss a different approach to Zeitz's problem in (Colton, 2000).

3.3 Specifying Details in Theorems

A third way in which computer algebra systems can aid discovery is by filling in details in theorem statements. For example, we became interested in *divisor graphs* of integers, which are constructed by taking an integer n and making the divisors of n the nodes of a graph. Then, we joined any two distinct nodes with an edge if one divided the other. Figure 1 show three examples of divisor graphs.

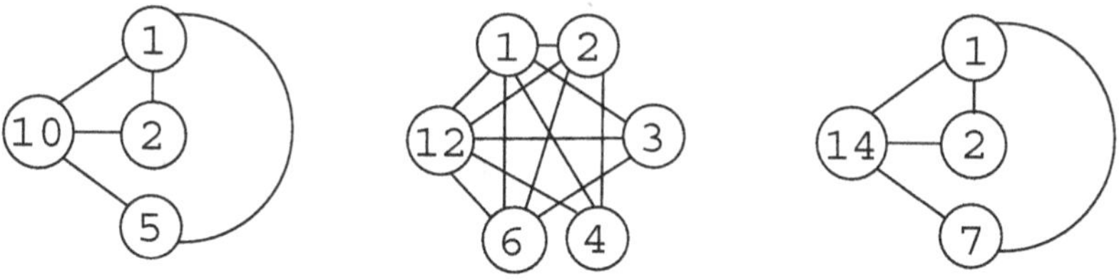

Fig. 1. Divisor graphs for the numbers 10, 12 and 14

As both 1 and n divide all the other divisors, it follows that, for every integer, the result is a connected graph. We became interested in the question: which integers produce planar divisor graphs? A planar graph can be drawn on a piece of paper in such a way that no edge crosses another, such as the first and third graphs in Figure 1. Kuratowski's theorem (Kuratowski, 1930) is used to determine whether or not a graph is planar.

To answer our question, we first realised that the divisor graph of an integer is dependent only on its prime signature: if an integer n can be written as $n = p_1^{k_1} p_2^{k_2} \dots p_m^{k_m}$, for primes $p_1 < p_2 < \dots < p_m$, then the prime signature of n is the list $[k_1, k_2, \dots, k_m]$. Hence, as depicted in Figure 1, the numbers $10 = 2 \times 5$ and $14 = 2 \times 7$ have the same divisor graph because they have the same prime signature: $[1, 1]$. Furthermore, we determined that if an integer n is divisible by a non-planar integer, then n itself will be non-planar, because its divisor graph will have a non-planar subgraph.

These two facts enabled us to approach our problem by looking at the divisor graphs of numbers with different prime signatures and seeing whether the divisor graphs are planar. As most prime signatures will produce non-planar divisor graphs, we reasoned that we could determine a boundary on those prime signatures producing planar divisor graphs. We wrote a small Maple program to construct the divisor graph for an integer, then used the built-in `isplanar` function to determine whether the graph was planar. Rather than thinking too hard about the exact strategy, we simply printed out 'planar' or 'non-planar' for the numbers 1 to 100, and looked at the occurrences of non-planar divisor graphs.

We found that the number 30 produced a non-planar divisor graph. The number 30 can be written as pqr for three distinct primes p, q and r. Hence, using the above reasoning about prime signatures and graphs with non-planar subgraphs being non-planar themselves, we concluded that all integers divisible by three or more primes must be non-planar. Hence we focussed on integers with 1 or 2 prime divisors and we easily identified a boundary for the planar divisor graphs: integers of the form p^4 produced non-planar divisor graphs, as did integers of the form p^2q^2.

Hence, we answered our question, and could state the theorem as follows:

> Only the number 1 and integers of the form: p, p^2, p^3, pq and p^2q for distinct primes p and q produce planar divisor graphs.

We see that, whereas the overall idea for the theorem was ours, using Maple let us easily fill in the specifics of the theorem statement. Note also that the proof of the theorem evolved alongside the actual statement. That is, there was no time when we stated a conjecture and tried to prove it. This method of discovery is common in mathematics, but publications are usually stated with the theorem and proof very much separate. Further details of this and related theorems are found in (Colton, 2002b), appendix C.

3.4 Proving Theorems

As highlighted by Doron Zeilberger's invited talk at the IJCAR-2001 conference, Levi Ben Gerson's treatise of 1321 had around 50 theorems with rigorous proofs proving what are now routine algebraic identities, such as: $(a + b)^2 = a^2 + 2ab + b^2$ (which took a page and a half to prove). Computer algebra systems can now "prove" much more elaborate identities than these with simple rewriting techniques. Zeilberger argues that, if mathematicians are happy to embrace such theorems as routine enough for a computer to carry out the proof, then we should embrace all such computer generated proofs. That is, rather than being skeptical of theorems that require a computer to prove them (such as the four colour theorem, which we discuss in Section 5.5), he encourages such proofs, arguing that they are an order of magnitude more difficult than theorems that a mere human can prove (but still trivial, as they can be proved).

As an example, we can take Zeilberger's proof of Conway's Lost Cosmological Theorem (Ekhad & Zeilberger, 1997). This starts with the sequence of integers $1, 11, 21, 1211, 111221, \ldots$, which is obtained by describing in words the previous term, e.g., one, one one, two ones, one two (and) one one, etc. Remarkably, Conway proved that the length of the $(n+1)th$ term divided by the length of the nth term tends to a constant $\lambda = 1.303577269\ldots$ known as *Conway's constant.* The generalised cosmological theorem states that starting with any non-trivial integer other than 22, and constructing the sequence in the same manner, will give the same ratio between lengths. Conway lamented that his proof of this theorem, and another by Guy, were lost (Conway, 1987).

Zeilberger chose not to prove this theorem with pen and paper, but rather to write a Maple package called HORTON (after John Horton Conway). The proof relies on the fact that most numbers can be split into halves that develop independently of each other as the sequence proceeds. There are a finite number of integers (atoms) that cannot be split in this way. By ranging over all possibilities for atoms, the HORTON program showed that all numbers eventually decompose into atoms after a certain number of steps, which proved the conjecture (Ekhad & Zeilberger, 1997). This proof is similar in nature to the proof of the four colour theorem described in Section 5.5, but was undertaken using a computer algebra systems, rather than by writing software specifically for the problem.

4 Automated Theorem Proving

One of the original goals of computer science and artificial intelligence was to write a program that could prove mathematics theorems automatically, and many systems have been implemented which can prove fairly complicated theorems. The ability to prove theorems is a powerful skill and automated theorem provers have performed a variety of discovery tasks in pure mathematics. These include proving established conjectures, improving proofs, finding new axiomatisations, and discovering new theorems. We concentrate here on discoveries made using deductive methods. We leave discussion of (counter)example construction methods – which have also solved theorems – until Section 5.

4.1 Proving Established Conjectures

The most extensive application of automated theorem proving to pure mathematics has been undertaken by the research team at Argonne laboratories, using various theorem provers, including EQP and, more recently, Otter (McCune, 1990).

Certainly the most famous theorem to be proved by an automated theorem prover is the Robbins Algebra conjecture, which McCune et al. solved using the EQP theorem prover (McCune, 1997). Herbert Robbins proposed that commutative, associative algebras are Boolean if they satisfy the extra condition that $n(n(x) + y) + n(n(x) + n(y)) = x$. These algebras became known as Robbins algebras and the question of whether they are Boolean defeated the attempts of mathematicians and logicians for more than 60 years.

In addition to providing the solution in 1996, automated reasoning techniques were also used during the development of this problem. In particular, on the advice of Wos, Winker used a combination of automated techniques and a standard mathematical approach to find two simpler conditions on the algebras, which, if true, would show that they are Boolean. The EQP program eventually proved that the second condition does in fact hold for Robbins algebras. The search for the proof took around eight days on an RS/6000 processor and used around 30 megabytes of memory.

EQP's successor, the Otter program (McCune, 1990), has also had much success discovering proofs to theorems in pure mathematics. In addition to finding new axiomatisations of algebras, as discussed in Section 4.3, Otter has been used to prove research theorems in algebraic geometry and cubic curves, lattice theory, Boolean algebras and quasigroup theory. A particularly fruitful partnership between McCune (Otter's author and principal user) and the mathematician Padmanabhan has developed. Padmanabhan has supplied many theorems relevant to his research that Otter has proved for the first time. Much of this work was written up in (McCune & Padmanabhan, 1996). A web page describing the discoveries due to the Argonne provers can be found at: `http://www-unix.mcs.anl.gov/AR/new_results/`.

4.2 Improving Proofs

One of the first applications of automated theorem proving was the use of Newell, Shaw and Simon's Logic Theory Machine (Newell et al., 1957) to prove theorems from Whitehead and Russell's Principia Mathematica. The program proved 38 of the 52 theorems they presented to it, and actually found a more elegant proof to theorem 2.85 than provided by Whitehead and Russell. (MacKenzie, 1995) points out that, on hearing of this, Russell wrote to Simon in November 1956:

> 'I am delighted to know that Principia Mathematica can now be done by machinery ... I am quite willing to believe that everything in deductive logic can be done by machinery.'

Newell, Shaw and Simon submitted an article about theorem 2.85, co-authored by the Logic Theory Machine, to the *Journal of Symbolic Logic*. However, it was refused publication as it was co-authored by a program.

More recently, Larry Wos has been using Otter to find smaller proofs of theorems than the current ones. To this end, he uses Otter to find more succinct methods than those originally proposed. This often results in detecting double negations and removing unnecessary lemmas, some of which were thought to be indispensable. (Wos, 1996) presents a methodology using a strategy known as resonance to search for elegant proofs with Otter. He gives examples from mathematics and logic, and also argues that this work also implications for other fields such as circuit design.

(Fleuriot & Paulson, 1998) have studied the geometric proofs in Newton's Principia and investigated ways to prove them automatically with the Isabelle interactive theorem prover (Paulson, 1994). To do this, they formalised the Principia in both Euclidean geometry and non-standard analysis. While working through one of the key results (proposition 11 of book 1, the Kepler problem) they discovered an anomaly in the reasoning. Newton was appealing to a cross-multiplication result which wasn't true for infinitesimals or infinite numbers. Isabelle could therefore not prove the result, but Fleuriot managed to derive an alternative proof of the theorem that the system found acceptable.

4.3 Finding Axiom Schemes

Another interesting project undertaken by the Argonne team aims to find different axioms systems for well known algebras, such as groups and loops (McCune, 1992) (McCune, 1993). In many cases, it has been possible to reduce the axioms to a single axiom. For instance, group theory can be expressed in terms of the multiplication and inverse operators in a single axiom:

$$\forall\, x, y, z, u \in G,\ \ (x(y(((zz')(uy)')x))') = u,$$

where a' indicates the inverse of a. These results were achieved by using Otter to prove the equivalence of the standard group axioms with exhaustively generated formulae. Similar results have been found for different operators in group theory and for Abelian groups, odd exponent groups and inverse loops. (Kunen, 1992) has subsequently proved that there are no smaller single axioms schemes than those produced by Otter.

4.4 Discovering Theorems

With a powerful theorem prover, it is possible to speculate that certain statements are theorems, and discard those which the prover does not prove. As the prover is so efficient, an exhaustive search of theorems can be undertaken, without having to invent concepts or worry about notions of interestingness as is the case with the programs described in Section 7.

(Chou, 1985) presents improvements on Wu's method for proving theorems in plane geometry (Wu, 1984). In Chapter 4, he describes three approaches to using his prover to find new theorems: (i) ingenious guessing using geometric intuition (ii) numerical searching and (iii) a systematic approach based on the Ritt-Wu decomposition algorithm. Using the first approach, he found a construction based on Pappus' Theorem which led to a colinearity result believed to be new. Using the second method – suggesting additional points and lines within given geometric configurations – he started with a theorem of Gauss (that the midpoints of the three diagonals of a complete quadrilateral are collinear) and constructed a theorem about taking subsets of five lines from six, which he believed to be of importance. With the third method, he discovered a generalisation of Simson's theorem. (Bagai et al., 1993) provide a more general approach to automated exploration in plane geometry, although it appears that no new theorems resulted from this work.

5 Example Construction

The construction of examples of concepts has advanced pure mathematics in at least these three ways:

- by discovering counterexamples to open conjectures;
- by solving non-existence conjectures, either by finding a counterexample or by exhausting the search space to prove the conjecture;
- by finding larger (more complex) examples of certain objects, such as the largest primes.

We break down our overview of example construction according to the general techniques used to solve problems, rather than the type of problem solved.

5.1 Constraint Satisfaction Solving

Specifying a problem in terms of constraint satisfaction has emerged as a powerful, general purpose, technique (Tsang, 1993). To do this, the problem must be stated as a triple of: variables, domains for the variables, and constraints on the assignment of values from the domain to each variable. The solution of the problem comes with the assignment of a value (or a range of values) to each of the variables in such a way that none of the constraints are broken. There are various strategies for the assignment of variables, propagation of constraints, and backtracking in the search.

This approach has been applied to problems from pure mathematics, in particular quasigroup existence problems. For instance, the FINDER program (Slaney, 1992) has solved many quasigroup existence problems. (Slaney et al., 1995) used FINDER along with two different automated reasoning programs called DDPP (a Davis Putnam implementation, as discussed in Section 5.2) and MGTP to solve numerous quasigroup existence problems. For example, they found an idempotent type 3 quasigroup (such that $\forall\ a, b (a * b) * (b * a) = a$) of size 12, settling that existence problem. They had similar results for quasigroups of type 4 and solved many other existence questions, both by finding counterexamples and by exhausting the search to show that no quasigroups of given types and sizes exist.

Open quasigroup problems have also been solved with computational methods by a number of other researchers.[3] Constraint satisfaction techniques have been

[3] With details at this web page:
`http://www.cs.york.ac.uk/~tw/csplib/combinatorial.html`

applied to other problems from combinatorial mathematics, such as Golomb rulers (Dewdney, 1985), and Ramsey numbers (Graham & Spencer, 1990). However, optimised, specialised algorithms usually out-perform the constraint satisfaction approach. For instance, while the constraint approach works well for Golomb rulers, all the actual discoveries have been made by specialised algorithms.

5.2 The Davis Putnam Method

This method is used for generating solutions to satisfiability problems (Davis & Putnam, 1960), (Yugami, 1995). It works by searching for an assignment of variables that satisfies all clauses in a formula expressed in conjunctive normal form. The procedure uses unit propagation to improve performance and works by choosing a variable in a clause containing a single literal, and assigning a value that satisfies the clause.

The MACE program (McCune, 2001) uses the Davis-Putnam method to generate models as counterexamples to false conjectures and has also been employed to solve some existence problems. As discussed in (McCune, 1994), MACE found a pair of orthogonal Mendelsohn triple systems of order 9. Also, MACE found a quasigroup of type 3 of order 16 with 8 holes of size 2. Furthermore, MACE has solved the existence problem of size 17 quasigroups of type 6 (such that $f(f(x,y),y) = f(x,f(x,y)))$ by finding an example. Another Davis Putnam program which has been used to solve more than 100 open questions about quasigroups in design theory is SATO (Zhang et al., 1996), (Zhang & Hsiang, 1994). Also, as mentioned above, the DDPP program is an implementation of the Davis-Putnam method.

5.3 The PSLQ Algorithm

The PSLQ algorithm, as described in (Bailey, 1998), is able to efficiently suggest new mathematical identities of the form $a_1x_1 + a_2x_2 + \ldots + a_nx_n = 0$ by finding non-trivial coefficients a_i if supplied with real numbers x_1 to x_n.

One application of the algorithm is to find whether a given real number, α, is algebraic. To do this, the values $\alpha, \alpha^2, \ldots, \alpha^n$ are calculated to high precision and the PSLQ algorithm then searches for non trivial values a_i such that

$$a_1\alpha + a_2\alpha^2 \ldots + a_n\alpha^n = 0$$

This functionality finds application in discovering Euler sums, which has led to a remarkable new formula for π:

$$\pi = \sum_{i=0}^{\infty} \frac{1}{16^i}\left(\frac{4}{8i+1} - \frac{2}{8i+4} - \frac{1}{8i+5} - \frac{1}{8i+6}\right).$$

Note that the formula was actually discovered by hand and the numbers found by computation. This formula is interesting as it can be used to calculate the nth hexadecimal digit of π *without* calculating the first $n-1$ digits, as discussed in (Bailey et al., 1997). Until this discovery, it was assumed that finding the nth digit of π was not significantly less expensive than finding the first $n-1$ digits. The new algorithm can calculate the millionth hexadecimal digit of π in less than two minutes on a personal computer.

5.4 Distributed Discovery Projects

With the increase in internet usage in the last decade, many distributed attempts to find ever larger examples of certain concepts have been undertaken. In particular, the latest effort to find the largest prime number is the Great Internet Mersenne Prime Search (GIMPS), which is powered by parallel technology running free software available at `www.mersenne.org`. Since 1996, the GIMPS project has successfully found the four most recent record primes. The record stands with a prime number with over two million digits and there are predictions that GIMPS will find a billion-digit prime number by 2010.

The search for prime numbers has an interesting history from a computational point of view. In particular, in 1961 in a single session on an IBM7090, Hurwitz found two Mersenne primes which beat the previous record. However, due to the way in which the output was presented, Hurwitz read the largest first. There is now a debate as to whether the smaller of the two was ever the largest known prime: it had been discovered by the computer before the larger one, but the human was only aware of it after he had found a larger one. Opinions[4] are split as to whether a discovery only witnessed by a computer should have the honour of being listed historically as the largest prime number known at a given time.

While finding the next largest prime is not seen as the most important activity in pure mathematics, it is worth remembering that one of the first programs worked on by Alan Turing (actually written by Newman and improved by Turing (Ribenboim, 1995)) was to find large prime numbers. Moreover, finding primes may one day lead to the solution of an important number theory conjecture by Catalan: that 8 and 9 are the only consecutive powers of integers.

There are many similar distributed attempts, some of which have come to a successful conclusion. Examples include a search to find ten consecutive primes in arithmetic progression (`www.ltkz.demon.co.uk/ar2/10primes.htm`) and a distributed computation to find the quadrillionth bit of π (which is a 0), as described at `http://www.cecm.sfu.ca/projects/pihex/pihex.html`.

5.5 Ad-Hoc Construction Methods

Individual programs tailored by mathematicians to solve particular problems are ubiquitous. One famous instance, which can be considered under the umbrella of example generation, is the four colour theorem. As mentioned in Section 2.1 above, this theorem has a colourful history (Saaty & Kainen, 1986) and was eventually solved in 1976 with the use of a computer to check around 1500

[4] See `www.utm.edu/cgi-bin/caldwell/bubba/research/primes/cgi/discoverer/` for some opinions on the notion of discovery with respect to prime numbers.

configurations in an avoidable set (see (Saaty & Kainen, 1986) for details). To solve the conjecture, Appel and Haken used around 1200 hours of computing time to check the configurations, looking for (and not finding) a configuration which would break the four colour theorem. This was the first major theorem to be proved by a computer that could not be verified directly by a human. As discussed above, simplifications have since been made to the proof that has made it less controversial and the truth of the theorem is now generally accepted (Robertson et al., 1996).

Another set of ad-hoc computational methods were used to solve the existence problem of finite projective planes of order 10. Clement Lam eventually proved that there are none, but only with the help of many mathematicians who wrote numerous programs during the 1980s. The complexities of this problem and its eventual solution are beyond the scope of this survey, but full details are given in (Lam et al., 1989).

Finally, we mentioned in Section 3 that mathematicians use computer algebra programs to find counterexamples to conjectures they originally think to be true. As an example of this, in appendix C of (Colton, 2002b), we discuss the following conjecture:

> A refactorable number is such that the number of divisors is itself a divisor (Colton, 1999). Given a refactorable number, n, then define the following function: $f(n) = |\{(a,b) \in \mathbf{N} \times \mathbf{N} : ab = n \text{ and } a \neq b\}|$. Then $f(n)$ divides n if and only if n is a non-square.

We attempted to find a counterexample to this claim using the GAP computer algebra system, but found none between 1 and 1,000,000. After abortive attempts to actually prove the conjecture, we began to look for counterexamples again. We eventually found three counterexamples: 36360900, 79388100 and 155600676. Hence, in this case, we disproved a conjecture by finding a counterexample. Unless the theorem is of importance, these kinds of results are rarely published, but they still represent computer discoveries in mathematics.

6 Mathematical Databases

The representation of mathematical knowledge and its storage in large databases is a priority for many researchers. For example, the MBASE project (Kohlhase & Franke, 2000) aims to create a database of concepts, conjectures, theorems and proofs, to be of use to the automated reasoning community, among others. Simply accessing databases of mathematical information can lead to discoveries. Such events occur when the item(s) returned by a database search differ radically from those expected. Because the data is mathematical, there is a chance that the object returned from the search is related in some mathematical way to the object you were actually looking for.

One particularly important database is the Online Encyclopedia of Integer Sequences,[5] which contains over 75,000 integer sequences, such as prime numbers

[5] `http://www.research.att.com/~njas/sequences`

and square numbers. They have been collected over 35 years by Neil Sloane, with contributions from hundreds of mathematicians. The Encyclopedia is very popular, receiving over 16,000 queries every day. The first terms of each sequence are stored, and the user queries the database by providing the first terms of a sequence they wish to know more about. In addition to the terms of the sequence, a definition is given and keywords are assigned, such as 'nice' (intrinsically interesting) and 'core' (fundamental to number theory or some other domain). Sloane has recorded some times when using the Encyclopedia has led to a conjecture being made. For instance, in (Sloane, 1998), he relates how a sequence that arose in connection with a quantization problem was linked via the Encyclopedia with a sequence that arose in the study of three-dimensional quasicrystals.

Another important database is the Inverse Symbolic Calculator,[6] which, given a decimal number, attempts to find a match to one in its database of over 50 million taken largely from mathematics and the physical sciences. Other databases include the GAP library of around 60 million groups, the Mizar library of formalised mathematics (Trybulec, 1989), the Mathworld online Encyclopedia,[7] and the MathSciNet citation and review server,[8] which contains reviews for more than 10,000 mathematics articles and references for over 100,000. (Colton, 2001b) provides a more detailed survey of mathematical databases

7 Automated Theory Formation

We collect here some ad-hoc techniques generally designed to *suggest* conjectures and theorems rather than prove them. This usually involves an amount of invention (concept formation), induction (conjecture making), and deduction (theorem proving), which taken together amount to forming a theory.

7.1 The Graffiti Program

The Graffiti program (Fajtlowicz, 1988) makes conjectures of a numerical nature in graph theory. Given a set of well known, interesting graph theory invariants, such as the diameter, independence number, rank, and chromatic number, Graffiti uses a database of graphs to empirically check whether one sum of invariants is less than another sum of invariants. The empirical check is time consuming, so Graffiti employs two techniques, called the **beagle** and **dalmation** heuristics, to discard certain trivial or weak conjectures before the empirical test (as described in (Larson, 1999)). If a conjecture passes the empirical test and Fajtlowicz cannot prove it easily, it is recorded in (Fajtlowicz, 1999), and he forwards it to interested graph theorists.

As an example, conjecture 18 in (Fajtlowicz, 1999) states that, for any graph G:

$$cn(G) + r(G) \leq md(G) + fmd(G),$$

[6] `http://www.cecm.sfu.ca/projects/ISC/`

[7] `http://mathworld.wolfram.com`

[8] `http://www.ams.org/mathscinet`

where $cn(G)$ is the chromatic number of G, $r(G)$ is the radius of G, $md(G)$ is the maximum degree of G and $fmd(G)$ is the frequency of the maximum degree of G. This conjecture was passed to some graph theorists, one of whom found a counterexample. The conjectures are useful because calculating invariants is often computationally expensive and bounds on invariants may bring computation time down. Moreover, these types of conjecture are of substantial interest to graph theorists, because they are simply stated, yet often provide a significant challenge to resolve – the mark of an important theorem such as Fermat's Last.

In terms of adding to mathematics, Graffiti has been extremely successful. The conjectures it has produced have attracted the attention of scores of mathematicians, including many luminaries from the world of graph theory. There are over 60 graph theory papers published which investigate Graffiti's conjectures.[9]

7.2 The AutoGraphiX Program

(Caporossi & Hansen, 1999) have recently implemented an algorithm to find linear relations between variables in polynomial time. This has been embedded in AutoGraphiX (AGX), an interactive program used to find extremal graphs for graph invariants (Caporossi & Hansen, 1997). AGX has been employed to refute three conjectures of Graffiti and has also been applied to automatic conjecture making in graph theory. Given a set of graph theory invariants calculated for a database of graphs in AGX, the algorithm is used to find a basis of affine relations on those invariants. For example, AGX was provided with 15 invariants calculated for a special class of graphs called colour-constrained trees. The invariants included:

$$\begin{aligned} \alpha &= \text{the stability number} \\ D &= \text{the diameter} \\ m &= \text{the number of edges} \\ n_1 &= \text{the number of pending vertices} \\ r &= \text{the radius} \end{aligned}$$

The algorithm discovered a new linear relation between the invariants:

$$2\alpha - m - n_1 + 2r - D = 0,$$

which Caporossi and Hansen have proved for all colour-constrained trees (Caporossi & Hansen, 1999).

7.3 The HR Program

HR is a theory formation program designed to undertake discovery tasks in domains of pure mathematics such as group, graph and number theory (Colton,

---

[9] See `http://cms.dt.uh.edu/faculty/delavinae/research/wowref.htm` for a list of the papers involving Graffiti's conjectures.

2002b), (Colton et al., 1999). The system is given some objects of interest from a domain, such as graphs or integers or groups, and a small set of initial concepts, each supplied with a definition and examples. From this, HR constructs a theory by inventing new concepts using general production rules to build them from one (or two) old concepts. The new concepts are produced with correct examples and a predicate definition that describes some relation between the objects in the examples.

HR uses the examples of the invented concepts to make empirical conjectures about them. For instance, if it finds that the examples of a new concept are exactly the same those of an old one, it conjectures that the definitions of the two concepts are logically equivalent. In finite algebraic systems such as group theory, HR uses the Otter theorem prover (McCune, 1990) to prove the conjectures it makes. If Otter fails to prove a conjecture, HR invokes the MACE model generator (McCune, 1994) to attempt to find a counterexample. Any counterexamples found are incorporated into the theory, thus reducing the number of further false conjectures generated.

We have used HR in domains such as anti-associative algebras (with only one axiom – no triple of elements is associative). This made us aware of theorems which were new to us; for example, there must be two different elements on the diagonal of the multiplication tables and that anti-associative algebras cannot be quasigroups or have an identity (in fact, no element can have a local identity). More results from HR's application to discovery tasks are presented in Chapter 12 of (Colton, 2002b). More recently, as discussed in (Colton & Miguel, 2001), we have applied HR to the invention of additional constraints to improve efficiency when solving constraint satisfaction problems. For instance, HR conjectured and proved that Qg3-quasigroups are anti-Abelian; i.e., for each a, b such that $a \neq b$, $a * b \neq b * a$. HR also discovered a symmetry on the diagonal of Qg3-quasigroups, namely that $\forall\, a, b\, (a * a = b \rightarrow b * b = a)$.

7.4 The NumbersWithNames Program

As discussed in (Colton, 1999) and (Colton et al., 2000a), one of the original applications of HR to mathematical discovery was the invention of integer sequences worthy of inclusion in the Encyclopedia of Integer Sequences. To be included, they must be shown to be interesting, so we also used HR to supply conjectures about the sequences it invented, some of which we proved. To augment the supply of conjectures, we enabled HR to search the Encyclopedia to find sequences which were empirically related to the ones it had invented. Such relationships include one conjecture being a subsequence (or supersequence) of another, and one sequence having no terms in common with another.

We have extracted this functionality into a program called NumbersWithNames, which HR accesses. NumbersWithNames contains a subset of 1000 sequences from the Encyclopedia, such as prime numbers, and square numbers, which are of sufficient importance to have been given names. The internet interface[10] lets the user choose one of the sequences or input a new one. The program

[10] This is available at: `http://www.machine-creativity.com/programs/nwn`

then makes conjectures about the sequence by relating it to those in its database. It also attempts to make conjectures by forming related sequences and searching the Encyclopedia with respect to them. The new sequences are formed by combining the given sequence with 'core' sequences in the database. NumbersWithNames orders the conjectures it makes in terms of a plausibility measure which calculates the probability of the conjecture occurring if the sequence had been chosen at random along the number line.

Some examples of conjectures made using this approach are given in (Colton, 1999), (Colton et al., 2000a) and appendix C of (Colton, 2002b). An example is the theorem that if the sum of the divisors of an integer n is a prime number, then the number of divisors of n will also be a prime number. Another recent appealing example is that perfect numbers are pernicious. That is, if we write a perfect number in binary, there will be a prime number of 1s (the definition of pernicious numbers). More than this, the 1s will be first, followed by zeros (another relation found by the program). For example, the first three perfect numbers are 6, 28 and 496, and when written in binary, these are 110, 11100 and 111110000. This unobvious conjecture – which we proved – is typical of those found by NumbersWithNames.

8 Summary

We have described what constitutes knowledge in pure mathematics and surveyed some computational techniques which have led to new knowledge in this field. We do not claim to have covered all mathematical discoveries aided by computer, as there are hundreds or perhaps thousands of individual programs crafted to assist in the discovery process, many of which we never hear about. However, from the evidence presented, we can draw some general conclusions:

- Computational discovery occurs at the object-level. While proof planning and automated theorem proving in general help us understand proof strategies, no new ones suitable for mathematicians have resulted from this. Similarly, although some work in meta-level concept formation (Colton, 2001a) has been undertaken, and theory formation in general gives us a better understanding of the exploration process, no new concept formation or conjecture making techniques have come to light.
- Most discoveries identify new results rather than improving old results, although improving proofs is developing into a useful application of theorem proving.
- Most types of object-level mathematics have been discovered by computer, including concepts, examples of concepts, open conjectures, theorems and proofs. We have not given evidence of any algorithms being discovered by computer, but neither do we rule this out.
- While theory formation programs are designed to carry out most activities in mathematics, computer algebra systems appear to be the most equipped to do this in a way that might lead to genuine discovery. We have given evidence of computer algebra systems being used to generate examples, to

prove and disprove conjectures, to fill in specifics in theorem statements, and to make new conjectures.
- Most major discoveries have been made via specialised programs, or by much manipulation of the problem statement into a form by which automated techniques were applicable. For example, while a general theorem prover solved Robbin's algebra, this was only after much human (and automated) theory formation about the problem.
- Most of the important discoveries have played on the ability of computers to perform massive calculations with high accuracy. However, the discoveries made by theory formation programs required relatively less computing power.

Given this summarization, we can return to Newell and Simon's prediction and suggest ways in which we can increase and improve the use of computers in the mathematical discovery process.

9 Discussion

To recap, Newell and Simon predicted that, by 1968, a computer would have discovered and proved an important theorem in mathematics. By 2002, however, it is difficult to claim that a computer has both discovered and proved a theorem which the mathematical community would describe as important. Certainly, the smallest axiomatisations of groups discussed, the new geometry theorems discussed in Section 4 and the results in quasigroup theory as discussed in Section 7 were both discovered and proved by computer. Furthermore, as it is difficult to assess discoveries at the time they are made (there may be hidden applications or implications), these results may be considered important one day.

Moreover, automated theorem provers have proved important theorems, in particular the Robbins Algebra theorem, but they did not discover them. Conversely, the NumbersWithNames program and Graffiti have discovered, but not proved theorems in number theory and graph theory, respectively, which have interested mathematicians. The theorems from Graffiti are particularly important, as they provide bounds for numerical invariants that increase efficiency in calculations.

In looking for possible reasons for the lack of success, we refer back to the title of Newell and Simon's paper in which they make their predictions: "Heuristic Problem Solving: The Next Advance in Operations Research". The key word here is, of course, *heuristic*. When asked about theorem provers, Simon (personal communication, 2002) lamented that the heuristic approach in theorem proving had dwindled, replaced largely by complete methods. While guaranteed to find proofs eventually, complete methods have difficulty in proving hard results because of combinatorial explosions in their searches.

Simon certainly had a point and his early work shows the undeniable utility of heuristics in artificial intelligence. Certainly, if automated theorem provers are to prove important results in pure mathematics, heuristic searches will play an important role. However, this criticism may be a little unfair on automated theorem provers, some of which do use heuristic techniques. More importantly, theorem provers have found greater application in hardware and software verification and safety critical systems – where it is imperative that the proof be formal and complete – than in pure mathematics.

This suggests another possible reason why there has been less success than we expected: a lack of interest from the mathematical community. It has become clear through personal communication with mathematicians and through organisation of workshops such as (Colton & Sorge, 2000), that many mathematicians believe that, for the moment, they have all the computer tools they want. Only a few pure mathematicians use automated theorem provers on a regular basis, including Padmanabhan, who has used Otter on many occasions. However, we should note that Padmanabhan has never actually invoked Otter: he passes theorems to Otter's author, Bill McCune, who tunes settings, runs Otter, and emails proofs back. Without mathematicians using and improving automated tools, it seems unlikely that an important theorem will be discovered and proved in the near future.

Perhaps the major reason why Newell and Simon's prediction has not come true is the implicit assumption that the computer acts autonomously – somehow it finds an important conjecture and manages to prove it. We have repeatedly mentioned that many discoveries are made possible only through the use of software packages (in particular computer algebra systems), but that the computer only facilitates the discovery: it cannot be said to have made the discovery, as usually the result is known (at least sketchily) to the user before the computer is employed.

For autonomous discovery to occur, more research must be put into automated theory formation. Much more research has been undertaken into automated theorem proving than into concept formation and conjecture making. Hence, computers are more able to prove results suggested by their user than to intelligently suggest interesting new ones. While many theorem proving systems have been built by entire teams, the more general question of automated theory formation has only been addressed in an ad-hoc manner by individual researchers. As mentioned in Section 1.1, examples of AI research in this area include the AM, Eurisko, GT, ARE, IL, DC, SCOT, MCS and Cyrano programs. While some of these built on previous work (e.g., Cyrano was based on AM), they were all isolated projects, and with the exception of SCOT and MCS, they are no longer current.

10 Conclusions

We have supplied evidence that a number of computational techniques have led to discoveries in pure mathematics, and argued that computer algebra systems are the most widely used programs among research mathematicians, many of whom believe they have all the computational tools they require at the moment. Taken together, these insights suggest that we embed more discovery techniques into computer algebra systems, such as theorem proving in the Theorema system (Buchberger et al., 1997). Alternatively, we could enable computer algebra systems to utilise other computational processes via some network of systems, as in the MathWeb project (Franke et al., 1999) and the Calculemus project.[11] Certainly, to encourage more autonomous discoveries, we need more integrated systems, as in the HR project and the SCOTT theorem prover (Hodgson & Slaney, 2001), which combines the example generation power of FINDER with the deductive power of Otter. Furthermore, if we embed general techniques such as resolution theorem proving, constraint solving, and the Davis Putnam procedure into computer algebra systems, this may enhance the mathematician's productivity, as they would have to spend less time hand crafting programs to complete a task.

The mathematicians Andrew Wiles and Doron Zeilberger represent two extremes of computer use in pure mathematics. Wiles famously proved Fermat's last theorem in 1995 (Singh, 1997), and was reported to have used a computer only to write up his results. Zeilberger, on the other hand, recommends that we teach students of mathematics to program rather than to prove and – drawing on Chaitin's results on the limits of mathematics (Chaitin, 1998) – argues that the mathematics we can actually prove is trivial. He states that we can only view non-trivial mathematics through computer experimentation, and we should rely on quasi-proof rather than formal deduction to determine whether or not to accept a theorem. Zeilberger regularly publishes papers co-authored with his computer, which he calls Shalosh B. Ekhad.

It is difficult to gauge the extent to which an average mathematician uses a computer. There are still mathematicians who do not use computers at all, and many who use them for email and/or typesetting only. However, computer algebra systems such as Maple and Mathematica are being employed increasingly by mathematicians for research and teaching purposes. Furthermore, due to the number of undergraduate and postgraduate computer algebra courses, it seems likely that every new university-trained mathematician will be literate in this topic. Moreover, computer algebra systems have, without doubt, enriched pure mathematics, and there are scores of theorems that would not have been stated or proved without computer algebra systems and other computational techniques.

With projects such as the Center for Experimental Mathematics at Simon Fraser University and journals like the *Journal of Experimental Mathematics*, computer enhanced discovery is now recognised as a worthy approach to pure mathematics. Newell and Simon's 1958 predictions were optimistic, but they were certainly not unachievable. Their prediction, for instance, that a computer would beat the world chess champion did eventually come true. We hope that, with a new generation of mathematicians and the increased power of the mathematical software available to them, Newell and Simon will soon be vindicated in their optimism about automated mathematical discovery.

---

[11] See `http://www.eurice.de/calculemus` for details of the Calculemus project.

Acknowledgments: I would like to thank Derek Sleeman and the Departments of Computing Science and Mathematics at the University of Aberdeen for inviting me to talk about computational discovery in pure mathematics, which provided the backbone for the work presented here. Ideas presented here have also arisen from conversations at the 2000 CADE workshop on the role of automated deduction in mathematics, the 2001 IJCAR workshop on future directions in automated deduction and the 2001 IJCAI Workshop on Knowledge Discovery from Distributed, Dynamic, Heterogeneous, Autonomous Sources. I would like to thank Alan Bundy and Toby Walsh for their continued input to this work and I am very grateful to Ursula Martin, Paul Cairns, Ian Gent, Geoff Sutcliffe, Larry Wos, and Doron Zeilberger for suggesting some mathematical discoveries made by computer. Finally, I would like to thank Herbert Simon for meeting with me and discussing many interesting topics, some of which have inspired this paper. This work was supported by EPSRC grant GR/M98012.

References

Abell, M., Braselton, J.: Maple V handbook. Academic Press, San Diego, CA (1994)

Appel, K., Haken, W.: Every planar map is four colorable. Illinois Journal of Mathematics 21, 429–567 (1977)

Bagai, R., Shanbhogue, V., Żytkow, J., Chou, S.: Automatic theorem generation in plane geometry. In: Proceedings of the Seventh International Symposium on Methodologies for Intelligent Systems, Trondheim, Norway, pp. 415–424 (1993)

Bailey, D.: Finding new mathematical identities via numerical computations. ACM SIGNUM 33(1), 17–22 (1998)

Bailey, D., Borwein, M., Borwein, P., Plouffe, S.: The quest for pi. Mathematical Intelligencer 19(1), 50–57 (1997)

Buchberger, B., Jebelean, T., Kriftner, F., Marin, M., Tomuta, E., Vasaru, D.: A survey of the theorema project. In: International Symposium on Symbolic and Algebraic Computation, Maui, HI, pp. 384–391 (1997)

Bundy, A.: The use of explicit plans to guide inductive proofs. In: Proceedings of the Ninth Conference on Automated Deduction. Argonne, IL (1988)

Caporossi, G., Hansen, P.: Variable neighbourhood search for extremal graphs. 1. The AutoGraphiX system (Technical Report Les Cahiers du GERAD G-97-41). École des Hautes Études Commerciales, Montréal (1997)

Caporossi, G., Hansen, P.: Finding relations in polynomial time. In: Proceedings of the Sixteenth International Joint Conference on Artificial Intelligence, Stockholm, Sweden, pp. 780–785 (1999)

Chaitin, G.: The limits of mathematics. Springer, Berlin (1998)

Chou, S.: Proving and discovering geometry theorems using Wu's method (Technical Report 49). Computing Science, University of Austin at Texas (1985)

Colton, S.: Refactorable numbers - a machine invention. Journal of Integer Sequences 2 (1999), `http://www.research.att.com/~njas/sequences/JIS`

Colton, S.: Automated plugging and chugging. Computation and Automated Reasoning, A.K. Peters, Natick, MA, pp. 247–248 (2000)

Colton, S.: Experiments in meta-theory formation. In: Proceedings of the AISB-01 Symposium on Creativity in Arts and Science. York, UK (2001a)

Colton, S.: Mathematics: A new domain for data mining. In: Proceedings of the IJCAI-01 Workshop on Knowledge Discovery from Distributed, Dynamic, Heterogeneous, Autonomous Sources. Seattle, WA (2001b)

Colton, S.: Automated theory formation applied to mutagenesis data. In: Proceedings of the First British-Cuban Workshop on Bioinformatics (2002a)

Colton, S.: Automated theory formation in pure mathematics. Springer, Berlin (2002b)

Colton, S., Bundy, A., Walsh, T.: HR: Automatic concept formation in pure mathematics. In: Proceedings of the Sixteenth International Joint Conference on Artificial Intelligence, pp. 786–791. Stockholm, Sweden (1999)

Colton, S., Bundy, A., Walsh, T.: Automatic invention of integer sequences. In: Proceedings of the Seventeenth National Conference on Artificial Intelligence, Austin, TX, pp. 558–563 (2000a)

Colton, S., Bundy, A., Walsh, T.: On the notion of interestingness in automated mathematical discovery. International Journal of Human Computer Studies 53(3), 351–375 (2000b)

Colton, S., Miguel, I.: Constraint generation via automated theory formation. In: Proceedings of the Seventh International Conference on the Principles and Practice of Constraint Programming, Paphos, Cyprus, pp. 572–576 (2001)

Colton, S., Sorge, V. (eds.): Proceedings of the Cade-00 workshop on the role of automated deduction in mathematics. Pittsburgh, PA (2000)

Conway, J.: The weird and wonderful chemistry of audioactive decay. In: Open Problems in Communication and Computation, pp. 173–188, Springer, Berlin (1987)

Davis, M., Putnam, H.: A computing procedure for quantification theory. Associated Computing Machinery 7, 201–215 (1960)

Dewdney, A.: Computer recreations. Scientific American, December issue, pp. 16–26 (1985)

Ekhad, S., Zeilberger, Z.: Proof of conway's lost cosmological theorem. Electronic Announcements of the American Mathematical Society 3, 78–82 (1997)

Epstein, S.: On the discovery of mathematical theorems. In: Proceedings of the Tenth International Joint Conference on Artificial Intellignce, pp. 194–197. Milan, Italy (1987)

Fajtlowicz, S.: On conjectures of Graffiti. Discrete Mathematics 72, 113–118 (1988)

Fajtlowicz, S.: The writing on the wall (Unpublished preprint) (1999) Available from `http://math.uh.edu/~clarson/`

Fajtlowicz, S.: Computer generated conjectures in mathematical chemistry. Dimacs Working Group Meeting on Computer-Generated Conjectures from Graph Theoretic and Chemical Databases I (2001)

Fleuriot, J., Paulson, L.: A combination of nonstandard analysis and geometry theorem proving, with application to newton's principia. In: Proceedings of the Fifteenth International Conference on Automated Deduction. Lindau, Germany (1998)

Franke, A., Hess, S., Jung, C., Kohlhase, M., Sorge, V.: Agent-oriented integration of distributed mathematical services. Journal of Universal Computer Science 5, 156–187 (1999)

Gap. GAP reference manual. The GAP Group, School of Mathematical and Computational Sciences, University of St. Andrews (2000)

Ghiselin, B. (ed.): The creative process. University of California Press, Berkeley (1996)
Gould, R.: Graph theory. San Francisco, CA: Benjamin Cummings (1988)
Graham, R., Spencer, J.: Ramsey theory. Scientific American, July issue, pp. 112–117 (1990)
Haase, K.: Discovery systems. In: Proceedings of the Seventh European Conference on Artificial Intelligence, Brighton, UK, pp. 546–555 (1986)
Hardy, G., Wright, E.: The theory of numbers. Oxford University Press, Oxford (1938)
Hodgson, K., Slaney, J.: System description: SCOTT-5. In: Proceedings of the First International Joint Conference on Automated Reasoning, Siena, Italy, pp. 443–447 (2001)
Humphreys, J.: A course in group theory. Oxford University Press, Oxford (1996)
Kohlhase, M., Franke, A.: MBase: Representing knowledge and context for the integration of mathematical software systems. Journal of Symbolic Computation 11, 1–37 (2000)
Kunen, K.: Single axioms for groups. Journal of Automated Reasoning 9(3), 291–308 (1992)
Kuratowski, G.: Sur la problème des courbes gauches en topologie. Fund. Math 15–16 (1930)
Lakatos, I.: Proofs and refutations: The logic of mathematical discovery. Cambridge University Press, Cambridge (1976)
Lam, C., Thiel, L., Swiercz, S.: The nonexistence of finite projective planes of order 10. Canadian Journal of Mathematics 41, 1117–1123 (1989)
Langley, P.: The computer-aided discovery of scientific knowledge. In: Proceedings of the First International Conference on Discovery Science, Fukuoka, Japan, pp. 25–39 (1998)
Larson, C.: Intelligent machinery and discovery in mathematics (Unpublished preprint) (1999), available from `http://math.uh.edu/~clarson/`
Lenat, D.: AM: Discovery in mathematics as heuristic search. Knowledge-Based Systems in Artificial Intelligence. McGraw-Hill Advanced Computer Science Series (1982)
Lenat, D.: Eurisko: A program which learns new heuristics and domain concepts. Artificial Intelligence 21, 61–98 (1983)
MacKenzie, D.: The automation of proof: A historical and sociological exploration. IEEE Annals of the History of Computing 17(3), 7–29 (1995)
McCune, W.: The OTTER user's guide (Technical Report ANL/90/9). Argonne National Laboratories (1990)
McCune, W.: Automated discovery of new axiomatizations of the left group and right group calculi. Journal of Automated Reasoning 9(1), 1–24 (1992)
McCune, W.: Single axioms for groups and abelian groups with various operations. Journal of Automated Reasoning 10(1), 1–13 (1993)
McCune, W.: A Davis-Putnam program and its application to finite first-order model search (Technical Report ANL/MCS-TM-194). Argonne National Laboratories (1994)
McCune, W.: Solution of the Robbins problem. Journal of Automated Reasoning 19(3), 263–276 (1997)
McCune, W.: Mace 2.0 reference manual and guide (Technical Report ANL/MCS-TM-249). Argonne National Laboratories (2001)
McCune, W., Padmanabhan, R.: Automated deduction in equational logic and cubic curves. Springer, Heidelberg (1996)
Morales, E.: DC: a system for the discovery of mathematical conjectures. Master's thesis, University of Edinburgh (1985)

Newell, A., Shaw, J., Simon, H.: Empirical explorations of the logic theory machine: A case study in heuristic. In: Proceedings of the Western Joint Computer Conference, Los Angeles, CA, pp. 218–239 (1957)

Paulson, L.: Isabelle: A generic theorem prover. Springer, Heidelberg (1994)

Pistori, H., Wainer, J.: Automatic theory formation in graph theory. In: Proceedings of the Argentine Symposium on Artificial Intelligence, Buenos Aires, Argentina, pp. 131–140 (1999)

Pólya, G.: How to solve it. Princeton University Press, Princeton (1988)

Ribenboim, P.: The new book of prime number records, 3rd edn. Springer, Heidelberg (1995)

Robertson, N., Sanders, D., Seymour, P., Thomas, R.: A new proof of the four-color theorem. Elecronic Resources of the American Mathematical Society 2, 17–25 (1996)

Saaty, T., Kainen, P.: The four-color problem: Assaults and conquest. Dover Publications, Mineola (1986)

Shen, W.: Functional transformations in AI discovery systems (Technical Report CMU-CS-87-117). Computer Science Department, CMU (1987)

Simon, H., Newell, A.: Heuristic problem solving: The next advance in operations research. Operations Research 6(1), 1–10 (1958)

Sims, M.: IL: An Artificial Intelligence approach to theory formation in mathematics. Doctoral dissertation, Rutgers University (1990)

Singh, S.: Fermat's last theorem. Fourth Estate (1997)

Slaney, J.: FINDER (finite domain enumerator): Notes and guide (Technical Report TR-ARP-1/92). Australian National University Automated Reasoning Project (1992)

Slaney, J., Fujita, M., Stickel, M.: Automated reasoning and exhaustive search: Quasigroup existence problems. Computers and Mathematics with Applications 29, 115–132 (1995)

Sloane, N.J.A.: My favorite integer sequences. In: Proceedings of the International Conference on Sequences and Applications, Singapore, pp. 103–130 (1998)

Stewart, I.: Galois theory. Chapman & Hall, Sydney (1989)

Trybulec, A.: Tarski grothendieck set theory. Journal of Formalised Mathematics, Axiomatics (1989)

Tsang, E.: Foundations of constraint satisfaction. Academic Press, San Diego (1993)

Valdés-Pérez, R.: Generic tasks of scientific discovery. In: Working Notes of the AAAI Spring Symposium on Systematic Methods of Scientific Discovery (1995)

Valdés-Pérez, R.: Principles of human computer collaboration for knowledge discovery in science. Artificial Intelligence 107(2), 335–346 (1999)

Wolfram, S.: The mathematica book, 4th edn. Wolfram Media/Cambridge University Press (1999)

Wos, L.: The automation of reasoning: An experimenter's notebook with OTTER tutorial. Academic Press, San Diego (1996)

Wu, W.: Basic principles of mechanical theorem proving in geometries. Journal of System Sciences and Mathematical Sciences 4(3), 207–235 (1984)

Yugami, N.: Theoretical analysis of Davis-Putnam procedure and propositional satisfiability. In: Proceedings of the Fourteenth International Joint Conference on Artificial Intelligence, Montreal, Canada, pp. 282–288 (1995)

Zeitz, P.: The art and craft of problem solving. John Wiley & Sons, New York (1999)

Zhang, H., Bonacina, M., Hsiang, H.: PSATO: a distributed propositional prover and its application to quasigroup problems. Journal of Symbolic Computation 21, 543–560 (1996)

Zhang, H., Hsiang, J.: Solving open quasigroup problems by propositional reasoning. In: Proceedings of the International Computer Symposium. Hsinchu, Taiwan (1994)

Zhang, J.: MCS: Model-based conjecture searching. In: Proceedings of the Sixteenth Conference on Automated Deduction, Trento, Italy, pp. 393–397 (1999)

Part II
Computational Scientific Discovery in Biomedicine

Automatic Computational Discovery of Chemical Reaction Networks Using Genetic Programming

John R. Koza[1], William Mydlowec[2], Guido Lanza[2], Jessen Yu[2], and Martin A. Keane[3]

[1] Biomedical Informatics, Department of Medicine, Department of Electrical Engineering
Stanford University, Stanford, California, USA
koza@stanford.edu
[2] Genetic Programming Inc.
Los Altos, California, USA
bill@pharmix.com, guido@pharmix.com, jyu@cs.stanford.edu
[3] Econometrics Inc.
Chicago, Illinois, USA
makeane@ix.netcom.com

Abstract. The concentrations of substances participating in networks of chemical reactions are often modeled by non-linear continuous-time differential equations. Recent work has demonstrated that genetic programming is capable of automatically creating complex networks (such as analog electrical circuits and controllers) whose behavior is modeled by linear and non-linear continuous-time differential equations and whose behavior matches prespecified output values. This chapter demonstrates that it is possible to automatically induce (reverse engineer) a network of chemical reactions from observed time-domain data. Genetic programming starts with observed time-domain concentrations of substances and automatically creates both the topology of the network of chemical reactions and the rates of each reaction of a network such that the behavior of the automatically created network matches the observed time-domain data. Specifically, genetic programming automatically created a network of four chemical reactions that consume glycerol and fatty acid as input, use ATP as a cofactor, and produce diacyl-glycerol as the final product. The network was created from 270 data points. The topology and sizing of the entire network was automatically created using the time-domain concentration values of diacyl-glycerol (the final product). The automatically created network contains three key topological features, including an internal feedback loop, a bifurcation point where one substance is distributed to two different reactions, and an accumulation point where one substance is accumulated from two sources.

1 Introduction

A living cell can be viewed as a dynamical system in which a large number of different substances react continuously and non-linearly with one another. In order to understand the behavior of a continuous non-linear dynamical system with numerous

S. Džeroski and L. Todorovski (Eds.): Computational Discovery, LNAI 4660, pp. 205–227, 2007.

interacting parts, it is usually insufficient to study behavior of each part in isolation. Instead, the behavior must usually be analyzed as a whole (Tomita et al., 1999).

The concentrations of substrates, products, and catalysts (e.g., enzymes) participating in chemical reactions are often modeled by non-linear continuous-time differential equations, such as the Michaelis-Menten equations (Voit, 2000).

Considerable amounts of time-domain data are now becoming available concerning the concentration of biologically important chemicals in living organisms. Such data include both gene expression data (often obtained from microarrays) and data on the concentration of substances participating in networks of chemical reactions (Ptashne, 1992; McAdams & Shapiro, 1995; Loomis & Sternberg, 1995; Arkin et al., 1997; Yuh et al., 1998; Laing et al., 1998; Mendes & Kell, 1998; D'haeseleer et al., 1999).

The question arises as to whether it is possible to start with observed time-domain concentrations of substances and automatically create both the topology of the network of chemical reactions and the rates of each reaction that produced the observed data—that is, to automatically reverse engineer the network from the data.

Genetic programming (Koza et al., 1999a) is a method for automatically creating a computer program whose behavior satisfies certain high-level requirements. Recent work has demonstrated that genetic programming can automatically create complex networks that exhibit prespecified behavior in areas where the network's behavior is governed by differential equations (both linear and non-linear).

For example, genetic programming is capable of automatically creating both the topology and sizing (component values) for analog electrical circuits (e.g., filters, amplifiers, computational circuits) composed of transistors, capacitors, resistors, and other components merely by specifying the circuit's output ¾ that is, the output data values that would be observed if one already had the circuit. This reverse engineering of circuits from data is performed by genetic programming even though there is no general mathematical method for creating the topology and sizing of analog electrical circuits from the circuit's desired (or observed) behavior (Koza et al.,1999a). Seven of the automatically created circuits infringe on previously issued patents. Others duplicate the functionality of previously patented inventions in a novel way.

As another example, genetic programming is capable of automatically creating both the topology and sizing (tuning) for controllers composed of time-domain blocks such as integrators, differentiators, multipliers, adders, delays, leads, and lags merely by specifying the controller's effect on the to-be-controlled plant (Koza et al., 1999d; Koza et al., 2000). This reverse engineering of controllers from data is performed by genetic programming even though there is no general mathematical method for creating the topology and sizing for controllers from the controller's behavior. Two of the automatically created controllers infringe on previously issued patents.

As yet other examples, genetic programming, has been used to create a better-than-classical quantum algorithm for the Deutsch-Jozsa "early promise" problem (Spector et al., 1998) and the Grover's database search problem (Spector et al., 1999a), a quantum algorithm for the depth-2 AND/OR query problem that is better than any previously published result (Spector et al., 1999b; Barnum et al., 2000), and a quantum algorithm for the depth-1 OR query problem that is better than any previously published result (Barnum et al., 2000).

As yet another example, it is possible to automatically create antennas composed of a network of wires merely by specifying the antenna's high-level specifications (Comisky et al., 2000).

Our approach to the problem of automatically creating both the topology and sizing of a network of chemical reactions involves

1. establishing a representation for a network of chemical networks involving symbolic expressions (S-expressions) and program trees that can be progressively bred by means of genetic programming, and
2. defining a fitness measure that measures how well the behavior of an individual network matches the observed data.

We then use the above representation and fitness measure in a run of genetic programming. During the run, the evaluation of the fitness of each individual in the population involves

1. converting each individual program tree in the population into an analog electrical circuit representing the network of chemical reactions,
2. converting each analog electrical circuit into a netlist so that the circuit can be simulated,
3. obtaining the behavior of the network of chemical reactions by simulating the corresponding electrical circuit, and
4. using the network's behavior and characteristics to calculate its fitness.

The implementation of our approach entails working with five different representations for a network of chemical reactions, namely

- **Reaction Network:** Biochemists often use this representation (shown in Figure 1) to represent a network of chemical reactions. In this representation, the blocks represent chemical reactions and the directed lines represent flows of substances between reactions.
- **System of Non-Linear Differential Equations:** A network of chemical reactions can also be represented as a system of non-linear differential equations (as shown in Section 6).
- **Program Tree:** A network of chemical reactions can also be represented as a program tree whose internal points are functions and external points are terminals (as shown in Section 4). This representation enables genetic programming to breed a population of programs in a search for a network of chemical reactions whose time-domain behavior concerning concentrations of final product substance(s) closely matches the observed data.
- **Symbolic Expression:** A network of chemical reactions can also be represented as a symbolic expression (S-expression) in the style of the LISP programming language. This representation is used internally by the run of genetic programming.
- **Analog Electrical Circuit:** A network of chemical reactions can also be represented as an analog electrical circuit (as shown in Figure 3 in Section 6.4). Representation of a network of chemical reactions as a circuit facilitates simulation of the network's time-domain behavior.

Section 2 states an illustrative problem. Section 3 briefly describes genetic programming. Section 4 presents a method of representing networks of chemical reactions with program trees. Section 5 presents the preparatory steps for applying genetic programming to the illustrative problem. Section 6 presents the results. Section 7 states the conclusions. Section 8 describes possible future work.

2 Statement of the Illustrative Problem

Our goal in this chapter is to automatically induce (reverse engineer) *both* the topology and sizing of a network of chemical reactions.

The *topology* of a network of chemical reactions comprises (1) the number of substrates consumed by each reaction, (2) the number of products produced by each reaction, (3) the pathways supplying the substrates (either from external sources or other reactions) to the reactions, and (4) the pathways dispersing the reaction's products (either to other reactions or external outputs).

The *sizing* of a network of chemical reactions consists of the numerical values representing the rates of each reaction.

We chose, as an illustrative problem, a network that incorporates three key topological features. These features include an internal feedback loop, a bifurcation point (where one substance is distributed to two different reactions), and an accumulation point (where one substance is accumulated from two sources). The particular chosen network is part of a phospholipid cycle, as presented in the E-CELL cell simulation model (Tomita et al., 1999). The network's external inputs are glycerol and fatty acid. The network's final product is diacyl-glycerol. The network's four reactions are catalyzed by Glycerol kinase (EC2.7.1.30), Glycerol-1-phosphatase (EC3.1.3.21), Acylglycerol lipase (EC3.1.1.23), and Triacylglycerol lipase (EC3.1.1.3). Figure 1 shows this network of chemical reactions with the correct rates of each reaction in parenthesis.

3 Background on Genetic Programming

Genetic programming is a method for automatically creating a computer program whose behavior satisfies user-specified high-level requirements (Koza, 1992; Koza, 1994a; Koza, 1994b; Koza et al., 1999a; Koza et al., 1999b; Koza & Rice, 1992). Genetic programming is an extension of the genetic algorithm (Holland, 1975) in which a population of programs is bred.

Genetic programming starts with a large population of randomly created computer programs (program trees) and uses the Darwinian principle of natural selection, crossover (sexual recombination), mutation, gene duplication, gene deletion, and certain mechanisms of developmental biology to breed a population of programs over a series of generations. Although there are many mathematical algorithms that solve problems by producing a set of numerical values, a run of genetic programming can create both a graphical structure (topology) and numerical values. That is, genetic programming also produces the structure in which the numerical values reside.

Genetic programming breeds computer programs to solve problems by executing the following three steps:

1) Generate an initial population of compositions (typically random) of the functions and terminals of the problem.
2) Iteratively perform the following substeps (referred to herein as a generation) on the population of programs until the termination criterion has been satisfied:
   A) Execute each program in the population and assign it a fitness value using the fitness measure.
   B) Create a new population of programs by applying the following operations. The operations are applied to program(s) selected from the population with a probability based on fitness (with reselection allowed).
      i) Reproduction: Copy the selected program to the new population.
      ii) Crossover: Create a new offspring program for the new population by recombining randomly chosen parts of two selected programs.
      iii) Mutation: Create one new offspring program for the new population by randomly mutating a randomly chosen part of the selected program.
      iv) Architecture-altering operations: Select an architecture-altering operation from the available repertoire of such operations and create one new offspring program for the new population by applying the selected architecture-altering operation to the selected program.
3) Designate the individual program that is identified by result designation (e.g., the best-so-far individual) as the result of the run of genetic programming. This result may be a solution (or an approximate solution) to the problem.

The genetic operations are applied to the individual(s) selected from the population on the basis of fitness.

The reproduction operation copies a selected program to the new population.

The mutation operation acts (asexually) on one selected individual. It deletes a randomly chosen subtree and replaces it with a subtree that is randomly grown in the same way that the individuals of the initial random population were originally created.

The crossover operation acts (sexually) on a pair of selected individuals. Its creates an offspring program for the new population by exchanging a randomly chosen subtree from one parent with a randomly chosen subtree from the other parent.

The architecture-altering operations are patterned after gene duplication and gene deletion in nature. The architecture-altering operation of subroutine creation acts on one selected individual by adding a subroutine. The architecture-altering operation of subroutine deletion acts on one selected individual by deleting a subroutine. These operations are described in detail in Koza et al. (1999a).

The individual programs that are evolved by genetic programming are typically multi-branch programs consisting of one or more result-producing branches and zero, one, or more automatically defined functions (subroutines). Architecture-altering operations enable genetic programming to automatically determine the number of automatically defined functions, the number of arguments that each possesses, and the nature of the hierarchical references, if any, among automatically defined functions.

Additional information on genetic programming can be found in books such as (Banzhaf et al., 1998); books in the series on genetic programming (such as (Langdon, 1998)); in the *Advances in Genetic Programming* series (Spector et al., 1999c); in the proceedings of the Genetic Programming Conferences (Koza et al., 1998); in the proceedings of the annual Genetic and Evolutionary Computation Conferences (Spector et al., 2001); in the proceedings of the annual Euro-GP conferences (Foster et al., 2002); and in the *Genetic Programming and Evolvable Machines* journal.

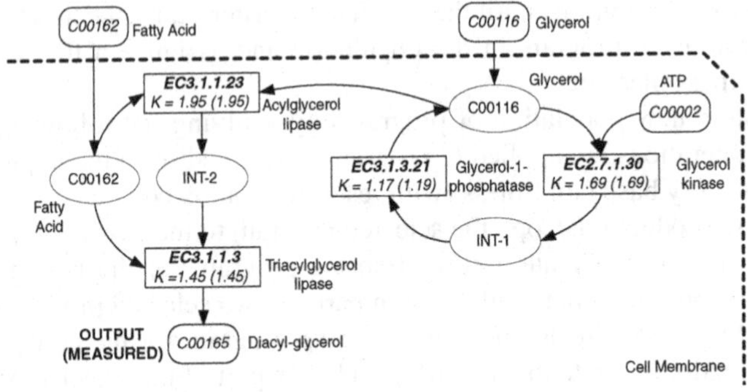

Fig. 1. Network of chemical reactions involved in the phospholipid cycle

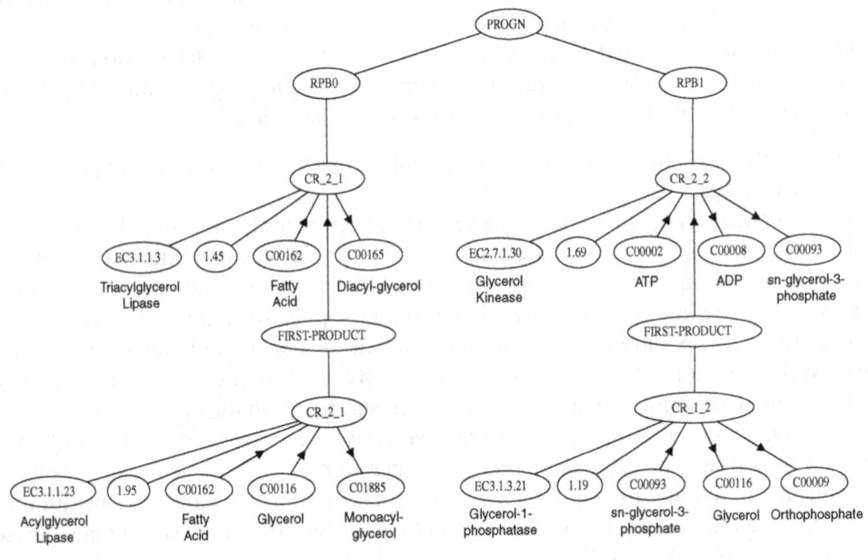

Fig. 2. Program tree corresponding to network of Figure 1

4 Representation of Chemical Reaction Networks

We first describe a method for representing a network of chemical reactions as a program tree suitable for use with genetic programming. Each program tree represents an interconnected network of chemical reactions involving various substances. A chemical reaction may consume one or two substances and produce one or two substances. The consumed substances may be external input substances or intermediate substances produced by reactions. The chemical reactions, enzymes, and substances of a network may be represented by a program tree that contains

- internal nodes representing chemical reaction functions,
- internal nodes representing selector functions that select the reaction's first versus the reaction's second (if any) product,
- external points (leaves) representing substances that are consumed and produced by a reaction,
- external points representing enzymes that catalyze a reaction, and.
- external points representing numerical constants (reaction rates).

Each program tree in the population is a composition of functions from the function set and terminals from the terminal set.

4.1 Repertoire of Functions

There are four chemical reaction functions and two selector functions.

The first argument of each chemical reaction (CR) function identifies the enzyme that catalyzes the reaction. The second argument specifies the reaction's rate. In addition, there are two, three, or four arguments specifying the substrate(s) and product(s) of the reaction. Table 1 shows the number of substrate(s) and product(s) and overall arity for each of the four chemical reaction functions.

Table 1. Four chemical reaction functions

Function	Substrates	Products	Arity
CR_1_1	1	1	4
CR_1_2	1	2	5
CR_2_1	2	1	5
CR_2_2	2	2	6
CR_2_2	2	2	6

Each chemical reaction function returns a list composed of the reaction's one or two products. The one-argument FIRST function returns the first of the one or two products produced by the chemical reaction function designated by its argument. The one-argument SECOND function returns the second of the two products (or the first product, if the reaction produces only one product).

The rates of chemical reactions are modeled by rate laws. The concentrations of substrates, products, intermediate substances, and catalysts participating in reactions are modeled by various rate laws, including first-order rate laws, second-order rate laws, power laws, and the Michaelis-Menten equations (Voit, 2000).

For example, the Michaelis-Menten rate law for a one-substrate chemical reaction is

$$\frac{d[P]}{dt} = \frac{k_2[E]_0[S]_t}{[S]_t + K_m}.$$

In this rate law, [P] is the concentration of the reaction's product P; $[S]_t$ is the concentration of substrate S at time *t*; $[E]_0$ is the concentration of enzyme E at time 0 (and at all other times); and K_m is the Michaelis constant defined as

$$K_m = \frac{k_{-1} + k_2}{k_1}.$$

However, when the constants K_m and k_2 are considerably greater than the concentrations of substances, it is often satisfactory to use a pseudo-first-order rate law such as

$$\frac{d[P]}{dt} = \frac{k_2[E]_0[S]_t}{K_m} = k_{new}[E]_0[S]_t\,,$$

where K_{new} is a constant defined as

$$k_{new} = \frac{k_2}{K_m}.$$

Note that the above pseudo-first-order rate law involves only multiplication.

The Michaelis-Menten rate law for a two-substrate chemical reaction is

$$Rate_t = \frac{[E]_0}{\frac{1}{K_0} + \frac{1}{K_A[A]_t} + \frac{1}{K_B[B]_t} + \frac{1}{K_{AB}[A]_t[B]_t}}.$$

However, when $k_{-1} \sim 0$ and $k_{-1} << k_1 << k_2$, it is often satisfactory to use a pseudo-second-order rate law such as

$$Rate_t = k_1[A][B][E]$$

The runs in this chapter use a first-order and second-order rate law.

For additional details on this and other aspects of this chapter, see our more lengthy Stanford University technical report (Koza et al., 2000).

4.2 Repertoire of Terminals

Some terminals represent substances (input substances, intermediate substances created by reactions, or output substances). Other terminals represent the enzymes that catalyze the chemical reactions. Still other terminals represent numerical constants for the rate of the reactions.

4.3 Constrained Syntactic Structure

The trees are constructed in accordance with a constrained syntactic structure. The root of every result-producing branch must be a chemical reaction function. The enzyme that catalyzes a reaction always appears as the first argument of its chemical reaction function. A numerical value representing a reaction's rate always appears as the second argument of its chemical reaction function. The one or two input arguments to a chemical reaction function can be either a substance terminal or selector function (`FIRST` or `SECOND`). The result of having a selector function as an input argument is to create a cascade of reactions. The one or two output arguments to a chemical reaction function must be a substance terminal. The argument to a one-argument selector function (FIRST or SECOND) is always a chemical reaction function.

4.4 Example

Figure 2 shows a program tree that corresponds to the network of chemical reactions of Figure 1. The program tree is presented in the style of the LISP programming language. The program tree (Figure 2) has two result-producing branches, RPB0 and RPB1. These two branches are connected by means of the LISP connective PROGN function.

As can be seen, there are four chemical reaction functions in Figure 2. The first argument of each chemical reaction function is constrained to be an enzyme and the second argument is constrained to be a numerical rate. The remaining arguments are substances, such as externally supplied input substances, intermediate substances produced by reactions within the network, and the final output substance produced by the network. The remaining arguments of each chemical reaction function are marked, purely as a visual aid to the reader, by an arrow.

There is a two-substrate, one-product chemical reaction function CR_2_1 in the lower left part of Figure 2. For this reaction, the enzyme is Acylglycerol lipase (EC3.1.1.23) (the first argument of this chemical reaction function); its rate is 1.95 (the second argument); its two substrates are fatty acid (C00162) (the third argument) and Glycerol (C00116) (the fourth argument); and its product is Monoacyl-glycerol (C01885) (the fifth argument).

There is a FIRST-PRODUCT function between the two chemical reaction functions in the left half of Figure 2. The FIRST-PRODUCT function selects the first of the two products of the lower CR_2_1 function. The line in the program tree from the lower chemical reaction function to the FIRST-PRODUCT function and the line between the FIRST-PRODUCT function and the higher CR_2_1 reaction means that when this tree is converted into a network of chemical reactions, the first (and, in this case, only) substance produced by the lower CR_2_1 reaction is a substrate to the higher reaction. In particular, the product of the lower reaction function (i.e., an intermediate substance called Monoacyl-glycerol) is the second of the two substrates to the higher chemical reaction function (i.e., the fourth argument of the higher function). Thus, although there is no return value for any branch or for the program tree as a whole, the return value(s) of all but the top chemical reaction function of a particular branch (as well the return values of a FIRST-PRODUCT function and a SECOND-PRODUCT function) define the flow of substances in the network of chemical reactions represented by the program tree.

Notice that the fatty acid (C00162) substance terminal appears as a substrate argument to both of these chemical reaction functions (in the left half of Figure 2 and also in the left half of Figure 1). The repetition of a substance terminal as a substrate argument in a program tree means that when the tree is converted into a network of chemical reactions, the available concentration of this particular substrate is distributed to two reactions in the network. That is, the repetition of a substance terminal as a substrate argument in a program tree corresponds to a bifurcation point where one substance is distributed to two different reactions in the network of chemical reactions represented by the program tree. There is another bifurcation point in this network of chemical reactions where Glycerol (C00116) appears as a substrate argument to both the two-substrate, one-product chemical reaction function CR_2_1 (in the lower left of Figure 2 and in the upper left part of Figure 1) and the two-substrate, two-product chemical reaction function CR_2_2 (in the upper right part of Figure 2 and in the upper right part of Figure 1).

Glycerol (C00116) has two sources in this network of chemical reactions. First, it is externally supplied (shown at the top right of Figure 1). Second, this substance is the product of the one-substrate, two-product chemical reaction function CR_1_2 (in the middle of Figure 1 and in the lower right of Figure 2). When a substance in a network has two or more sources (by virtue either of being externally supplied, by virtue of being a product of a reaction of a network, or any combination thereof), the substance is accumulated. When the program tree is converted into a network, all the sources of this substance are pooled. That is, there is an accumulation point for the substance.

Also, Glycerol (C00116) appears as part of an internal feedback loop consisting of two reactions, namely

- the one-substrate, two-product chemical reaction function CR_1_2 catalyzed by EC3.1.3.21 (in the middle of Figure 1 and in the lower right of Figure 2) and
- the two-substrate, two-product chemical reaction function CR_2_2 catalyzed by EC2.7.1.30 (in the upper right part of Figure 2 and in the right part of Figure 1).

The presence of an internal feedback loop is established in this network because of the following two features of this program tree:

- There exists a substance, namely sn-Glycerol-3-Phosphate (C00093) such that this substance
  - is a product (sixth argument) that is produced by the two-substrate, two-product chemical reaction function CR_2_2 (catalyzed by EC2.7.1.30) in the upper right part of Figure 2, and
  - is also a substrate that is consumed by the one-substrate, two-product chemical reaction function CR_1_2 (catalyzed by EC3.1.3.21) in the lower right part of Figure 2 that lies beneath the CR_2_2 function.
- There exists a second substance, namely glycerol (C00116), that
  - is a product that is produced by the chemical reaction function CR_1_2 (catalyzed by EC3.1.3.21) and
  - is a substrate that is consumed by the chemical reaction function CR_2_2 (catalyzed by EC2.7.1.30).

In summary, the network of Figure 2 contains the following three noteworthy topological features:

- an internal feedback loop in which Glycerol (C00116) is both consumed and produced in the loop,
- two bifurcation points (one where Glycerol is distributed to two different reactions and one where and fatty acid is distributed to two different reactions), and
- an accumulation point where one substance, namely Glycerol, is accumulated from two sources.

5 Preparatory Steps

Before applying genetic programming to a problem involving the automatic synthesis of a network of chemical reactions, six major preparatory steps are required, namely (1) determine the architecture of the program trees, (2) identify the functions, (3) identify the terminals, (4) define the fitness measure, (5) choose control parameters for the run, and (6) choose the termination criterion and method of result designation.

5.1 Program Architecture

Each program tree in the initial random population (generation 0) has one result-producing branch. In subsequent generations, the architecture-altering operations may insert and delete result-producing branches to and from particular individual program trees in the population. Each program tree may have up to four result-producing branches.

5.2 Function Set

The function set, F, consists of six functions.

F = {CR1_1, CR1_2, CR2_1, CR2_2, FIRST, SECOND}.

5.3 Terminal Set

The terminal set, T, is

T = {$\Re$, C00116, C00162, C00002, C00165, INT_1, INT_2, INT_3, EC2_7_1_30, EC3_1_3_21, EC3_1_1_23, EC3_1_1_3}.

$\Re$ denotes a perturbable numerical value from 0.0 and 2.0. In the initial random generation (generation 0) of a run, each perturbable numerical value is set, individually and separately, to a random value in the chosen range. Each perturbable numerical value is subsequently perturbed during the run.

In the illustrative problem herein, C00116 is the concentration of glycerol. C00162 is the concentration of fatty acid. These two substances are inputs to the illustrative overall network of interest herein. C00002 is the concentration of the cofactor ATP. C00165 is the concentration of diacyl-glycerol. This substance is the final product of the illustrative network herein.

INT_1, INT_2, and INT_3 are the concentrations of intermediate substances 1, 2, and 3 (respectively). This set makes more intermediate substances available than are needed for the problem at hand.

EC2_7_1_30, EC3_1_3_21, EC3_1_1_23, and EC3_1_1_3 are enzymes.

5.4 Fitness Measure

In order to evaluate the fitness of an individual program tree in the population, the program tree is converted into a directed graph representing a network of chemical reactions. The result-producing branches are executed from left to right. The functions in a particular result-producing branch are executed in a depth-first manner. One reactor (representing the concentration of the substances participating in the reaction) is inserted into the network for each chemical reaction function that is encountered in a branch. The reactor is labeled with the reaction's enzyme and rate. A directed line entering the reactor is added for each of the reaction's one or two substrate(s). A directed line leaving the reactor is added for each of the reaction's one or two product(s). The first product of a reaction is selected whenever a `FIRST` function is encountered in a branch. The second product is selected whenever a `SECOND` function is encountered in a branch.

After the network is constructed, the behavior and characteristics of the network are ascertained by simulation. There are several alternative ways to perform the simulation, including, for example, conversion of the network into an electrical circuit and simulation of the resulting electrical circuit by the SPICE3 simulator (Quarles et al., 1994). We use the SPICE3 simulator and provide SPICE with subcircuit definitions necessary to implement chemical reaction equations as electrical circuits.

Figure 3 shows the electrical circuit corresponding to the network of Figure 1. The triangles in the Figure represent integrators.

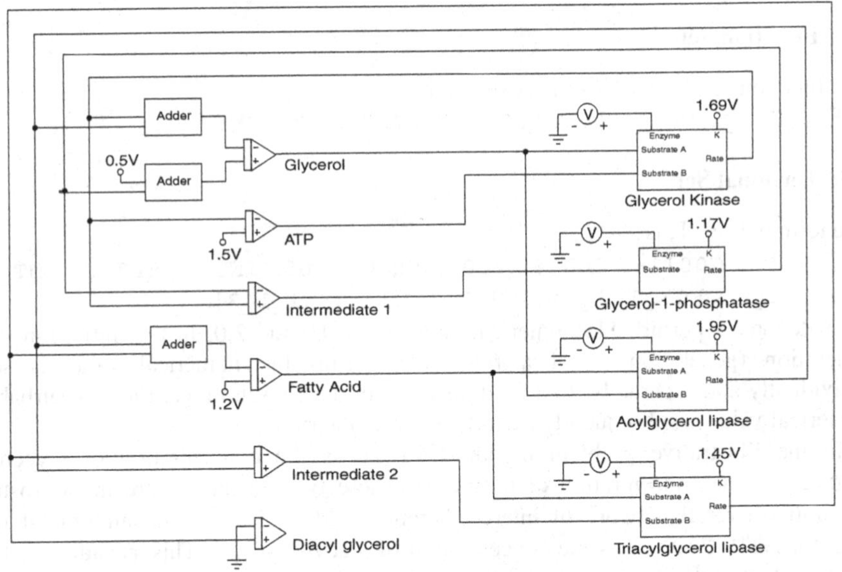

Fig. 3. Electrical circuit corresponding to the chemical reaction network of Figure 1

Each individual chemical reaction network is exposed to nine time-domain signals (table 2) representing the time-varying concentrations of four enzymes (EC2.7.1.30, EC3.1.3.21, EC3.1.1.23, and EC3.1.1.3) over 30 half-second time steps. Each of these time series patterns has been structured so as to vary the concentrations between 0 and 2.0 in a pattern to which a living cell might conceivably be exposed.

Each individual chemical reaction network is exposed to nine test cases. Thus, there are 270 fitness cases (9 test cases, each consisting of 30 time steps).

Each of the nine test cases (Table 2) is constructed by choosing four different time series from the above set of six time series as the concentration for the four enzymes (EC2.7.1.30, EC3.1.3.21, EC3.1.1.23, and EC3.1.1.3).

The slope-up time series for an enzyme starts at a concentration of 0.5 at time step 0 and increases linearly to a concentration of 1.75 at time step 30.

The slope-down time series for an enzyme starts at a concentration of 1.55 and decreases linearly to a concentration of 1.0 at time step 30.

The step-down time series starts at a concentration of 1.95 at time step 0, continues at that level until time step 6, steps down to a concentration of 1.5 at time step 7, and then remains at the lower level.

The step-up time series starts at a concentration of 0.4 at time step 0, continues at that level until time step 12, steps up to a concentration of 1.65 at time step 13, and then remains at the lower level.

The sawtooth time series starts at a concentration of 2.0 at time step 0, linearly decreases to a concentration of 1.0 at time step 5, and linearly increases to a concentration of 2.0 at time step 10. This pattern is then repeated between times steps 10 and 20 and between time steps 20 and 30.

The knock-out time series starts at a concentration of 1.2 at time step 0 and then drops to a concentration of 0 for all remaining time steps.

Table 2. Variations in the levels of the four enzymes

Signal	**EC2.7.1.30**	**EC3.1.3.21**	**EC3.1.1.23**	**EC3.1.1.3**
1	Slope-Up	Sawtooth	Step-Down	Step-Up
2	Slope-Down	Step-Up	Sawtooth	Step-Down
3	Step-Down	Slope-Up	Slope-Down	Step-Up
4	Step-Up	Slope-Down	Step-Up	Step-Down
5	Sawtooth	Step-Down	Slope-Up	Step-Up
6	Sawtooth	Step-Down	Knock-Out	Slope-Up
7	Sawtooth	Knock-Out	Slope-Up	Step-Down
8	Knock-Out	Step-Down	Slope-Up	Sawtooth
9	Step-Down	Slope-Up	Sawtooth	Knock-Out

There are a total of 270 data points. The data was obtained from the E-CELL cell simulation model (Tomita et al., 1999; Voit, 2000).

The concentrations of all intermediate substances and the network's final product are 0 at time step 0.

For the runs in this paper, Glycerol (`C00116`), Fatty acid (`C00162`), and ATP (`C00002`) are externally supplied at a constant rate (table 3). That is, these values are not subject to evolutionary change during the run.

Table 3. Rates for three externally supplied substances

Substance	**Rate**
Glycerol (`C00116`)	0.5
Fatty acid (`C00162`)	1.2
ATP (`C00002`)	1.5

The fitness of an individual network of chemical reactions is measured in terms of how closely the concentration of the network's end product matches the observed concentration of diacyl-glycerol (C00165). Fitness is the sum, over the 270 data points (fitness cases), of the absolute value of the difference between the concentration of the end product of the individual reaction network and the observed concentration of diacyl-glycerol (C00165). The smaller the fitness, the better.

An individual that cannot be simulated by SPICE is assigned a high penalty value of fitness (10^8).

The number of hits is defined as the number of fitness cases (0 to 270) for which the concentration of the measured substances is within 5% of the observed data value.

5.5 Control Parameters for the Run

The population size, *M*, is 100,000. A generous maximum size of 500 points (for functions and terminals) was established for each result-producing branch. The percentages of the genetic operations for each generation are 58.5% one-offspring crossover on internal points of the program tree other than perturbable numerical values, 6.5% one-offspring crossover on points of the program tree other than perturbable numerical values, 1% mutation on points of the program tree other than perturbable numerical values, 20% mutation on perturbable numerical values, 10% reproduction, 3% subroutine creation, and 2% subroutine deletion. The above values (and our other choices of minor parameters) are the same as we have applied to a broad range of problems from a number of different fields (Koza et al., 1999a).

5.6 Termination

The run was manually monitored and manually terminated when the fitness of many successive best-of-generation individuals appeared to have reached a plateau.

5.7 Implementation on Parallel Computing System

We used a home-built Beowulf-style (Sterling et al., 1999; Koza et al., 1999a) parallel cluster computer system consisting of 1,000 350 MHz Pentium II processors (each accompanied by 64 megabytes of RAM). The system has a 350 MHz Pentium II computer as host. The processing nodes are connected with a 100 megabit-per-second Ethernet. The processing nodes and the host use the Linux operating system. A distributed version of the genetic programming algorithm with unsynchronized generations and semi-isolated subpopulations was used. The subpopulation size was 500 at each of 1,000 processing nodes. As each processor (asynchronously) completes a generation, four small batches of emigrants from that processing node (subpopulation) are dispatched to each of its four toroidally adjacent processors. The 1,000 processors are hierarchically organized. There are 5 × 5 = 25 high-level groups (each containing 40 processors). If the adjacent node belongs to a different group, the migration rate is 2% and emigrants are selected based on fitness. If the adjacent node belongs to the same group, emigrants are selected randomly and the migration rate is 5% (10% if the adjacent node is in the same physical box).

6 Results

The population for the initial generation (generation 0) of a run of genetic programming is created at random. This initial random population is a blind random search of the space of possible chemical reaction networks. The results of generation 0 also provide a useful baseline for comparing the results of subsequent generations.

Although the results of blind random search are invariably poor for non-trivial problems, some individuals are better than others. The fitness of the best individual (Figure 4) from generation 0 of our one (and only) run of this problem is 86.4. This individual scores 126 hits (out of 270). In the best individual from generation 0, substance `C00162` (fatty acid) is used as an input substance to this network of chemical reactions; however, glycerol (`C00116`) and ATP (`C00002`) are not. Two of the four available reactions (EC 3.1.1.23 and EC 3.1.1.3) are used. However; a third reaction (EC 3.1.3.21) consumes a non-existent intermediate substance (INT_2) and the fourth reaction (EC 2.7.1.30) is not used at all. This network contains one important topological feature, namely the bifurcation of `C00162` to two different reactions. However, this network does not contain any of the other important topological features of the correct network.

Generation 1 (and each subsequent generation of a run of genetic programming) is created from the population of the preceding generation by performing the reproduction, crossover, mutation, and architecture-altering operations on individuals selected probabilistically from the population on the basis of fitness.

Over successive generations, both the average fitness of all individuals in the population as a whole and the fitness of the best individual in the population tend to improve. This run (which is typical of runs of genetic programming for many different problems from a variety of problem domains) shows how the evolutionary process exploits the small differential advantages possessed by the more fit individuals in the population at each generation.

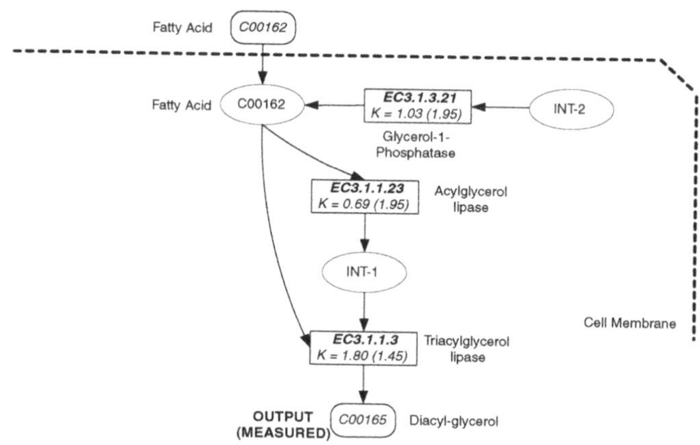

Fig. 4. Best of generation 0

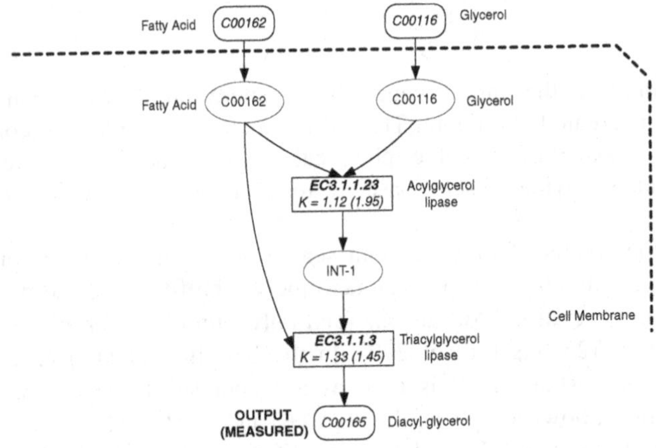

Fig. 5. Best of generation 10

In generation 10, the fitness of the best individual (Figure 5) is 64.0. This individual scores 151 hits. This network is superior to the best individual of generation 0 in that it uses both C00162 (fatty acid) and glycerol (C00116) as external inputs. Although this individual is an improvement over the best individual of generation 0, this network is defective in that it does not use ATP (C00002) and in that it contains only two of the four available reactions.

In generation 25, the fitness of the best individual (Figure 6) is 14.3. This individual scores 224 hits. This network is an improvement over the previous best-of-generation individuals in that it contains all four of the available reactions and in that it contains two topological features not previously seen. First, this network contains an internal feedback loop in which one substance (glycerol C00116) is consumed by one reaction (catalyzed by enzyme EC 2.7.1.30), produced by another reaction (catalyzed by enzyme EC 3.1.3.21), and then supplied as a substrate to the first reaction. Second, this network contains a place where there is an addition of quantities of one substance. Specifically, glycerol (C00116) comes from the reaction catalyzed

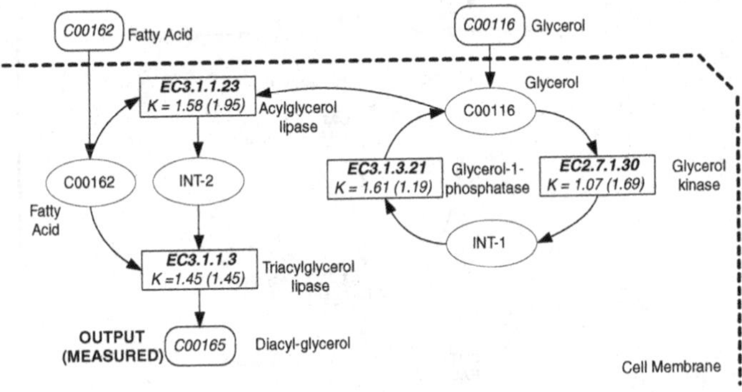

Fig. 6. Best of generation 25

by enzyme EC 3.1.3.21 and is also externally supplied. This network also contains two substances (`C00116` and `C00162`) where a substance is bifurcated to two different reactions.

In generation 120, the fitness of the best individual is 2.33. This network is an improvement over the previous best-of-generation individuals in that the cofactor ATP (`C00002`) appears as an input. In fact, this network has the same topology as the correct network (figure 1). However, the numerical values (sizing) are not yet correct (as indicated by the fact that it scores only 255 out of 270 hits).

The best-of-run individual appears in generation 225 (Figure 1). The rates that are outside the parenthesis in Figure 1 are the rates of the best individual from generation 225 of the run of genetic programming. Its fitness is almost zero (0.054). This individual scores 270 hits (out of 270). In addition to having the same topology as the correct network, the rate constants of three of the four reactions match the correct rates (to three significant digits) while the fourth rate differs by only about 2% from the correct rate (i.e., the rate of EC 3.1.3.21 is 1.17 compared with 1.19 for the correct network).

In the best-of-run network from generation 225, the rate of production of the network's final product, diacyl-glycerol (`C00165`), is given by

$$\frac{d[C00165]}{dt} = 1.45[C00162][INT\_2][\mathrm{EC}\,3.1.1.3].$$

Note that genetic programming has correctly determined that the reaction that produces the network's final product diacyl-glycerol (`C00165`) has two substrates and one product; it has correctly identified enzyme EC3.1.1.3 as the catalyst for this final reaction; it has correctly determined the rate of this final reaction as 1.45; and it has correctly identified the externally supplied substance, fatty acid (C00162), as one of the two substrates for this final reaction. None of this information was supplied *a priori* to genetic programming.

Of course, genetic programming has no way of knowing that biochemists call the intermediate substance (`INT_2`) by the name Monoacyl-glycerol (`C01885`) (as indicated in Figure 1). It has, however, correctly determined that an intermediate substance is needed as one of the two substrates of the network's final reaction and that this intermediate substance should, in turn, be produced by a particular other reaction (described next).

In the best-of-run network from generation 225, the rate of production and consumption of the intermediate substance `INT_2` is given by

$$\frac{d[INT\_2]}{dt} = 1.95[C00162][C00116][\mathrm{EC}\,3.1.1.23] - 1.45[C00162][INT\_2][\mathrm{EC}\,3.1.1.3].$$

Again, genetic programming has correctly determined that the reaction that produces the intermediate substance (`INT_2`) has two substrates and one product; it has correctly identified enzyme EC3.1.1.23 as the catalyst for this reaction; it has correctly determined the rate of this reaction as 1.95; and it has correctly identified two externally supplied substances, fatty acid (C00162) and glycerol (`C00116`), as the two substrates for this reaction.

In the best-of-run network from generation 225, the rate of production and consumption of the intermediate substance INT_1 in the internal feedback loop is given by

$$\frac{d[INT\_1]}{dt} = 1.69[C00116][C00002][\text{EC } 2.7.1.30] - 1.17[INT\_1][\text{EC } 3.1.3.21].$$

Note that the numerical rate constant of 1.17 in the above equation is slightly different from the correct rate (as shown in Figure 1).

Here again, genetic programming has correctly determined that the reaction that produces the intermediate substance (INT_1) has two substrates and one product; it has correctly identified enzyme EC2.7.1.30 as the catalyst for this reaction; it has almost correctly determined the rate of this reaction to be 1.17 (whereas the correct rate is 1.19, as shown in Figure 1); and it has correctly identified two externally supplied substances, glycerol (C00116) and the cofactor ATP (C00002), as the two substrates for this reaction.

Genetic programming has no way of knowing that biochemists call the intermediate substance (INT_1) by the name sn-Glycerol-3-Phosphate (C00093) (as indicated in Figure 1). Genetic programming has, however, correctly determined that an intermediate substance is needed as the single substrate of the reaction catalyzed by Glycerol-1-phosphatase (EC3.1.3.21) and that this intermediate substance should, in turn, be produced by the reaction catalyzed by Glycerol kinase (EC2.7.1.30).

In the best-of-run network from generation 225, the rate of supply and consumption of ATP (C00002) is

$$\frac{d[ATP]}{dt} = 1.5 - 1.69[C00116][C00002][\text{EC } 2.7.1.30]$$

The rate of supply and consumption of fatty acid (C00162) is

$$\frac{d[C00162]}{dt} = 1.2 - 1.95[C00162][C00116][\text{EC } 3.1.1.23] - 1.45[C00162][INT\_2][\text{EC } 3.1.1.3].$$

The rate of supply, consumption, and production of glycerol (C00116) is

$$\frac{d[C00116]}{dt} = 0.5 + 1.17[INT\_1][\text{EC } 3.1.3.21] - 1.69[C00116][C00002][\text{EC } 2.7.1.30] - 1.95[C00162][C00116][\text{EC } 3.1.1.23]$$

Notice the internal feedback loop in which C00116 is both consumed and produced.

In summary, driven by the time-domain concentration values of the final product C00165 (diacyl-glycerol), genetic programming created both the topology and sizing for an entire network of chemical reactions whose time-domain behavior closely matches that of the naturally occurring pathway, including

- the total number of reactions in the network,
- the number of substrate(s) consumed by each reaction,
- the number of product(s) produced by each reaction,
- an indication of which enzyme (if any) acts as a catalyst for each reaction,
- the pathways supplying the substrate(s) (either from external sources or other reactions in the network) to each reaction,

- the pathways dispersing each reaction's product(s) (either to other reactions or external outputs),
- the number of intermediate substances in the network,
- emergent topological features such as
  - internal feedback loops,
  - bifurcation points,
  - accumulation points, and
- numerical rates (sizing) for all reactions.

7 Future Work

Numerous directions for future work are suggested by the work described herein.

7.1 Improved Program Tree Representation

The run consumed 170 hours of computer time. Although the representation used herein yielded the desired results, the authors believe that alternative representations for the program tree (i.e., the function set, terminal set, and constrained syntactic structure) would significantly improve the efficiency of the search.

7.2 Minimum Amount of Data Needed

The work in this chapter has not addressed the important question of the minimal number of data points necessary to automatically create a correct network or the question whether the requisite amount of data is available in practical situations.

7.3 Opportunities to Use Knowledge

There are numerous opportunities to incorporate and exploit preexisting knowledge about chemistry and biology in the application of the methods described in this chapter.

The chemical reactions functions used in this chapter (i.e., `CR_1_1`, `CR_1_2`, `CR_2_1`, `CR_2_2`) are intentionally open-ended in the sense that they permit great flexibility and variety in the networks that can be created by the evolutionary process. However, there is a price, in terms of efficiency of the run, that is paid for this flexibility and generality. Alternative chemical reaction functions that advantageously incorporate preexisting knowledge might be defined and included in the function set.

For example, a particular substrate, a particular product, or both might be made part of the definition of a new chemical reaction function. For example, a variant of the `CR_2_2` chemical reaction function might be defined in which ATP is hard-wired as one of the substrates and ADP is hard-wired as one of products. This new chemical reaction function would have only one free substrate argument and one free product argument. This new chemical reaction function might be included in the function set in addition to (and conceivably in lieu of) the more general and open-ended `CR_2_2` chemical reaction function. This new chemical reaction function would exploit the well-known fact that there are a number of biologically important and biologically common reactions that employ ATP as one of its two substrates and produce ADP as one of its products.

Similarly, a particular enzyme might be made part of the definition of a new chemical reaction function. That is, a chemical reaction function with k substrates and j products might be defined in which a particular enzyme is hard-wired. This new chemical reaction function would not possess an argument for specifying the enzyme. This new chemical reaction function would exploit knowledge of the arity of reactions catalyzed by a particular enzyme.

Also, a known rate might be made part of the definition of a new chemical reaction function. This approach might be particularly useful in combination with other alternatives mentioned above.

7.4 Designing Alternative Metabolisms

Mittenthal et al. (1998) presented a method for generating alternative biochemical networks. They illustrated their method by generating diverse alternatives to the non-oxidative stage of the pentose phosphate pathway. They observed that the naturally occurring pathway is especially favorable in several respects to the alternatives that they generated. Specifically, the naturally occurring pathway has a comparatively small number of steps, does not use any reducing or oxidizing compounds, requires only one ATP in one direction of flux, and does not depend on recurrent inputs.

Mendes & Kell (1998) have also suggested that novel metabolic pathways might be artificially constructed.

It would appear that genetic programming could also be used to generate diverse alternatives to naturally occurring networks. Conceivably, realizable alternative metabolisms might emerge from such evolutionary runs.

In one approach, the fitness measure in a run of genetic programming might be oriented toward duplicating the final output(s) of the naturally occurring network (as was done in this chapter). However, instead of harvesting only the individual from the population with the very best value of fitness, individuals that achieve a slightly poorer value of fitness could be examined to see if they simultaneously possess other desirable characteristics.

In a second approach, the fitness measure in a run of genetic programming might be specifically oriented to factors such as the network's efficiency or use or non-use of certain specified reactants or enzymes.

In a third approach, the fitness measure in a run of genetic programming might be specifically oriented toward achieving novelty. Genetic programming has previously been used as an invention machine by employing a two-part fitness measure that incorporates both the degree to which an individual in the population satisfies the certain performance requirements and the degree to which the individual does not possess the key characteristics of previously known solutions (Koza et al., 1999c; Koza et al., 1999a).

8 Conclusion

We have demonstrated the principle that genetic programming can automatically create a network of chemical reactions involving four chemical reactions that takes in glycerol and fatty acid as input, uses ATP as a cofactor, and produces diacyl-glycerol as its final product. The automatically created network contains three key topological features, including an internal feedback loop, a bifurcation point where one substance is distributed to two different reactions, and an accumulation point where one substance is accumulated from two sources. This example demonstrates the principle that it is possible to reverse engineer a network of chemical reactions.

Acknowledgement: Douglas B. Kell of the University of Wales made helpful comments on a draft of this material.

References

Arkin, A., Shen, P., Ross, J.: A test case of correlation metric construction of a reaction pathway from measurements. Science. 277, 1275–1279 (1997)

Banzhaf, W., Nordin, P., Keller, R.E., Francone, F.D.: Genetic Programming – An Introduction. Morgan Kaufmann and Heidelberg dpunkt, San Francisco, CA (1998)

Barnum, H., Bernstein, H.J., Spector, L.: Quantum circuits for OR and AND of ORs. Journal of Physics A: Mathematical and General 33, 8047–8057 (2000)

Comisky, W., Yu, J., Koza, J.: Automatic synthesis of a wire antenna using genetic programming. In: Late Breaking Papers at the 2000 Genetic and Evolutionary Computation Conference, Las Vegas, NV, pp. 179–186 (2000)

D'haeseleer, P., Wen, X., Fuhrman, S., Somogyi, R.: Linear modeling of mRNA expression levels during CNS development and injury. In: Proceedings of the Pacific Symposium on Biocomputing, pp. 41–52. World Scientific, Island of Hawaii, HI (1999)

Foster, J.A., Lutton, E., Miller, J., Ryan, C., Tettamanzi, A.G.B. (eds.): Proceedings of the Fifth European Conference on Genetic Programming, Kinsale, Ireland. Springer, Heidelberg (2002)

Holland, J.H.: Adaptation in Natural and Artificial Systems: An Introductory Analysis with Applications to Biology, Control, and Artificial Intelligence, 2nd edn. The MIT Press, Cambridge, MA (1992)

Koza, J.R.: Genetic Programming: On the Programming of Computers by Means of Natural Selection. MIT Press, Cambridge, MA (1992)

Koza, J.R.: Genetic Programming II: Automatic Discovery of Reusable Programs. MIT Press, Cambridge, MA (1994a)

Koza, J.R.: Genetic Programming II Videotape: The Next Generation. MIT Press, Cambridge, MA (1994b)

Koza, J.R., Banzhaf, W., Chellapilla, K., Deb, K., Dorigo, M., Fogel, D.B., Garzon, M.H., Goldberg, D.E., Iba, H., Riolo, R. (eds.): Proceedings of the Third Annual Conference on Genetic Programming. Morgan Kaufmann, Madison, WI (1998)

Koza, J.R., Bennett, F.H., Andre, D., Keane, M.A.: Genetic Programming III: Darwinian Invention and Problem Solving. Morgan Kaufmann, San Francisco, CA (1999a)

Koza, J.R., Bennett, F.H, Andre, D., Keane, M.A., Brave, S.: Genetic Programming III Videotape: Human-Competitive Machine Intelligence. Morgan Kaufmann, San Francisco, CA (1999b)

Koza, J.R., Bennett, F.H., Stiffelman, O.: Genetic programming as a Darwinian invention machine. In: Proceedings of the Second European Workshop on Genetic Programming, Göteborg, Sweden, pp. 93–108. Springer, Heidelberg (1999c)

Koza, J.R., Keane, M.A., Yu, J., Bennett, F.H., Mydlowec, W.: Automatic creation of human-competitive programs and controllers by means of genetic programming. Genetic Programming and Evolvable Machines 1, 121–164 (2000)

Koza, J.R., Keane, M.A., Yu, J., Bennett, F.H., Mydlowec, W., Stiffelman, O.: Automatic synthesis of both the topology and parameters for a robust controller for a non-minimal phase plant and a three-lag plant by means of genetic programming. In: Proceedings of the Thirtyeighth Conference on Decision and Control, Phoenix, AZ, pp. 5292–5300 (1999d)

Koza, J.R., Mydlowec, W., Lanza, G., Yu, J., Keane, M.A.: Reverse Engineering and Automatic Synthesis of Metabolic Pathways from Observed Data Using Genetic Programming. Stanford Medical Informatics Technical Report SMI-2000-0851 (2000)

Koza, J.R., Rice, J.P.: Genetic Programming: The Movie. MIT Press, Cambridge, MA (1992)

Laing, S., Fuhrman, S., Somogyi, R.: REVEAL: A general reverse engineering algorithm for inference of genetic network architecture. In: Proceedings of the Pacific Symposium on Biocomputing, pp. 18–29. World Scientific, Maui, HI (1998)

Langdon, W.B.: Genetic Programming and Data Structures: Genetic Programming + Data Structures = Automatic Programming! Kluwer, Amsterdam (1998)

Loomis, W.F., Sternberg, P.W.: Genetic networks. Science 269, 649 (1995)

McAdams, H.H., Shapiro, L.: Circuit simulation of genetic networks. Science 269, 650–656 (1995)

Mendes, P., Kell, D.B.: Non-linear optimization of biochemical pathways: Applications to metabolic engineering and parameter estimation. Bioinformatics 14, 869–883 (1998)

Mittenthal, J.E., Ao, Y., Bertrand, C., Scheeline, A.: Designing metabolism: Alternative connectivities for the pentose phosphate pathway. Bulletin of Mathematical Biology 60, 815–856 (1998)

Ptashne, M.: A Genetic Switch: Phage λ and Higher Organisms, 2nd edn. Cell Press and Blackwell Scientific Publications, Cambridge, MA (1992)

Quarles, T., Newton, A.R., Pederson, D.O., Sangiovanni-Vincentelli, A.: SPICE 3 Version 3F5 User's Manual. Department of Electrical Engineering and Computer Science, University of California. Berkeley, CA (1994)

Spector, L., Barnum, H., Bernstein, H.J.: Genetic programming for quantum computers. In: Proceedings of the Third Annual Conference on Genetic Programming, pp. 365–373. Morgan Kaufmann, Madison, WI (1998)

Spector, L., Barnum, H., Bernstein, H.J.: Quantum computing applications of genetic programming. In: Advances in Genetic Programming 3, pp. 135–160. MIT Press, Cambridge, MA (1999a)

Spector, L., Barnum, H., Bernstein, H.J., Swamy, N.: Finding a better-than-classical quantum AND/OR algorithm using genetic programming. In: Proceedings of the 1999 Congress on Evolutionary Computation, pp. 2239–2246. IEEE Press, Washington, DC (1999b)

Spector, L., Langdon, W.B., O'Reilly, U., Angeline, P. (eds.): Advances in Genetic Programming 3. MIT Press, Cambridge, MA (1999c)

Spector, L., Goodman, E., Wu, A., Langdon, W.B., Voigt, H.-M., Gen, M., Sen, S., Dorigo, M., Pezeshk, S., Garzon, M., Burke, E. (eds.): Proceedings of the Genetic and Evolutionary Computation Conference. Morgan Kaufmann Publishers, San Francisco, CA (2001)

Sterling, T.L., Salmon, J., Becker, D.J., Savarese, D.F.: How to Build a Beowulf: A Guide to Implementation and Application of PC Clusters. MIT Press, Cambridge, MA (1999)

Tomita, M., Hashimoto, K., Takahashi, K., Shimizu, T.S., Matsuzaki, Y., Miyoshi, F., Saito, K., Tanida, S., Yugi, K., Venter, J.C., Hutchison, C.A.: E-CELL: Software environment for whole cell simulation. Bioinformatics 15, 72–84 (1999)

Voit, E.O.: Computational Analysis of Biochemical Systems. Cambridge University Press, Cambridge (2000)

Yuh, C.-H., Bolouri, H., Davidson, E.H.: Genomic cis-regulatory logic: Experimental and computational analysis of a sea urchin gene. Science. 279, 1896–1902 (1998)

Discovery of Genetic Networks Through Abduction and Qualitative Simulation

Blaž Zupan[1,3], Ivan Bratko[1,2], Janez Demšar[1], Peter Juvan[1], Adam Kuspa[3,4], John A. Halter[5], and Gad Shaulsky[3]

[1] Artificial Inteligence Laboratory
Faculty of Computer and Information Science, University of Ljubljana, Slovenia
[2] Department of Intelligent Systems
Jožef Stefan Institute, Ljubljana, Slovenia
[3] Department of Biochemistry and Molecular Biology
Baylor College of Medicine, Houston, Texas, USA
[4] Department of Molecular and Human Genetics
Baylor College of Medicine, Houston, Texas, USA
[5] PM&R and Division of Neuroscience
Baylor College of Medicine, Houston, Texas, USA

Abstract. GenePath is an automated system for reasoning about genetic networks, wherein a set of genes have various influences on one another and on a biological outcome. It acts on a set of experiments in which genes are knocked-out or overexpressed, and the outcome of interest is evaluated. Implemented in Prolog, GenePath uses abductive inference to elucidate network constraints based on background knowledge and experimental results. Two uses of the system are demonstrated: synthesis of a consistent network from abduced constraints, and qualitative reasoning-based approach that generates a set of networks consistent with the data. In practice, as illustrated by an example on aggregation of a soil amoeba *Dictyostelium discoideum*, a combination of constraint satisfaction and qualitative reasoning produces a small set of plausible networks.

1 Introduction

Biologists often develop genetic networks to reason on relations between genes and biological processes. To do this, their main tools are mutations. Initially, mutations help define and catalogue genes that participate in a biological mechanism. Relationships between the genes are then determined using combinations of mutations in two or more genes.

The methods used to relate genes and biological processes are most often fairly straightforward and require only a short time relative to the time required to obtain the experimental data, as preparing and experimenting with a single mutant can, typically, take about two weeks in the wet lab. However, accounting for all the data becomes complicated when the amount of data increases since the number of possible genetic networks grows combinatorially with the number

S. Džeroski and L. Todorovski (Eds.): Computational Discovery, LNAI 4660, pp. 228–247, 2007.

of genes. Even ordering a set of about ten genes – in contemporary genetics this may be considered as a moderately-sized problem requiring quite laborious experimental work – may be a tedious task.

For this purpose, we developed GenePath, an intelligent assistant in cataloguing genes and experimental data and deriving models of gene interactions. Although in principle capable of handling hundreds of genes at a time, it is primarily intended to work with current problems in functional genomics that usually consider a much smaller number of genes. To fulfill its mission of intelligent assistance, its uniqueness lies in ability to communicate the results with the geneticist in a way that explains how and why were these constructed. GenePath can provide the evidence for everything it derives, allows what-if analysis in which a geneticist can ask the system what would happen if certain experiments were added or removed, and finally, can provide explanations for relations it found from the data.

This chapter first gives some background on biology and introduces qualitative genetic networks. Next, we present a data set that will be used in examples throughout the chapter. The data we use comes from a contemporary real-life problem which studies the behavior of a soil amoeba *Dictyostelium discoideum*. The next two sections show a GenePath's architecture and present it's methods for derivation of genetic networks. Following is a section with a short description of GenePath's web-based user interface. Discussion describes some other test cases, and discusses the scalability and limitations of GenePath.

2 Some Background on Biology and on Qualitative Genetic Networks

Genes are the basic units of genetic material. They are coded in double-stranded DNAs and reside within chromosomes of each living cell. Generally, each gene encodes a specific protein. In the process of protein synthesis a gene is first transcribed to a single-stranded RNA molecule (mRNA), which is then translated to a protein. Expression of genes, *e.g.*, the amount of mRNA molecules transcribed from DNA, is regulated, so that under different environments and conditions genes may or may not be expressed, and the level of expression may vary. Through evolution, nature has established great variety of complicated regulatory mechanisms involving alterations in transcription, translation and other biochemical processes. For the purposes of his paper, we will assume that a single gene translation product can affect the process of transcription of another gene.

A geneticist's main tool to investigate biological phenomena is inducing mutations. There are two types of mutations most often used. In gene inactivation, geneticists remove a gene from the DNA (this is also called a "knock-out") or use some other means to prevent a gene from being expressed. An opposite mutation that is often harder to induce in the laboratory is activation of a gene (also referred to as *overexpression*); this results n a gene being significantly more expressed than under normal conditions. Although one can mutate several genes at the time, geneticists mostly use only single and double mutants since inducing a greater number of mutations can be technically difficult and unreliable.

Experiments that are used by GenePath observe a specific mutant and record an outcome of the experiments in terms of phenotypes. A phenotype is an observed state of biological process being studied in experiment. For example, we can observe whether the mutant cells grow or not, so in this case the biological process is "cell growth" and the possible phenotypes may be "cells grow" and "cells don't grow". We will also refer to a phenotype as an *outcome* of an experiment.

To present relations between genes and the biological processes, biologists often use genetic networks. A genetic network is a graph whose nodes correspond to genes and biological processes, the latter always occurring as terminal nodes. Arcs correspond to influences of genes on other genes and on biological processes. Genetic networks thus describe some biological phenomenon from the viewpoint of interactions between genes. They provide a high-level view and disregard most details on how one gene regulates the expression of another.

In this paper we consider the most commonly used and reported type of genetic networks called *qualitative genetic networks*. In qualitative genetic networks, the interaction between elements of the network is simplified to sole inhibition (negative influence) or excitation (positive influence), while the magnitude of the influence is not shown.

Consider for example a genetic network shown in Figure 1. It includes four genes (*regA*, *pufA*, *pkaR* and *pkaC*) and a single biological process (*agg*, aggregation of soil amoeba *D. discoideum* described later). Arcs marked with ⊣ denote inhibition, and those with → denote excitation. Hypothetically, keeping all other expressions constant, expression of *pufA* should rise under the magnification of expression of *regA*, or lower under the reduction of expression of *regA*. Conversely, and according to the network, expression of *pkaC* should decrease under increase of expression of *pkaR* (or *pufA*). Notice that the qualitative network does not denote to what extent or in which direction the expression of one gene should change to induce the change of expression in other genes. It depicts only the type of relation. Perhaps even more importantly, genetic networks encode the genetic pathways, e.g., chains (paths) of gene regulation. For instance, from Figure 1 it is clear that gene *pkaC* is regulated by gene *regA*, but the regulation is not direct as it goes through another gene (*pufA*). On the other hand, genes *pufA* and *pkaR* directly influence the expression of *pkaC*.

The notion of "directness" is based on the closed-world assumption: genetic networks consider only a (small) subset of the cell's genes and ignore all other genes. For instance, the network from Figure 1 depicts only relations between genes *regA*, *pufA*, *pkaR*, *pkaC* and aggregation. Therefore, the arrow between *regA* and *pufA* only states that none of the other considered genes mediates between *regA* and *pufA*. There may be other cell's genes – ones that are not included in the study and the network – on the path from gene *regA* to gene *pufA*.

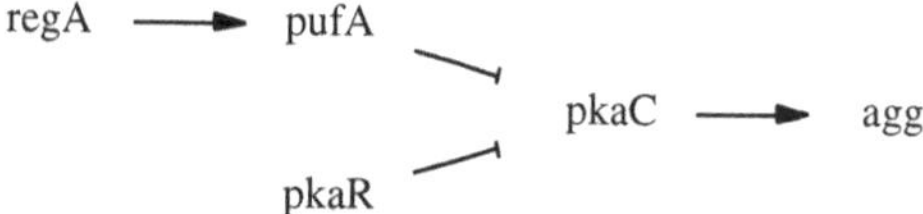

Fig. 1. An example of genetic network

Geneticists infer genetic networks by comparing the phenotypes of different strains of mutated organisms, or by comparing mutant organisms to organisms without any artificially induced mutations (so-called *wild-type* organism). Experiments that we consider in this paper are tuples of mutations and outcomes. Mutations are specified as a set of one or more genes with information on the type of mutation (activation or inactivation of a gene). The outcome of an experiment is assessed through the description of a phenotype (e.g., cells grow, or cells do not grow, etc).

3 A Data Set on Aggregation of *D. discoideum*

A data set that will be used to illustrate some of GenePath's functions is given in Table 1. The data comes from the studies of the soil amoeba *Dictyostelium discoideum*. In food-rich environment the amoeba behaves like an ordinary single-cell amoeba. Upon starvation, individual amoebae start to emit a chemical signal and aggregate. When the concentration of cells is high enough, the cells differentiate into a basal disk, stalk and sorus. The resulting multicellular organism moves toward the surface; the spores wait there to be transmitted to a new, food-rich environment where they begin a new cycle as individual cells. The transition from ordinary single cells to a multicellular structure was in the past and at present of great interest for biologists interested in the evolution and development of multicellular organisms (Zimmer, 1998; Souza et al., 1998; Souza et al., 1999).

Our example data set focuses on cell aggregation. It includes seven genes: *yakA*, *pufA*, *gdtB*, *pkaR*, *pkaC*, *acaA*, and *regA*. Seventeen experiments were carried out: in each experiment (except for the wild type) selected genes were either knocked out or overexpressed (denoted with "-" or "+", respectively). Observed aggregation was either normal ("+"), excessive ("++"), reduced ("±") or absent ("-"). For instance, experiment 2 from Table 1 depicts that knocking out *yakA* results in no aggregation, while *Dictyostelia* with removed *pufA* aggregate faster. Notice that the aggregation level is an ordinal variable, with the order being from no aggregation to excessive aggregation: -, ±, +, ++.

Background knowledge may be given in the form of known parts of a network. For our example, let us assume that we already know that *acaA* inhibits *pkaR* and *pkaR* inhibits *pkaC*, which in turn excites aggregation. This can be formally denoted with a single path of

$$acaA \dashv pkaR \dashv pkaC \rightarrow \text{aggregation}$$

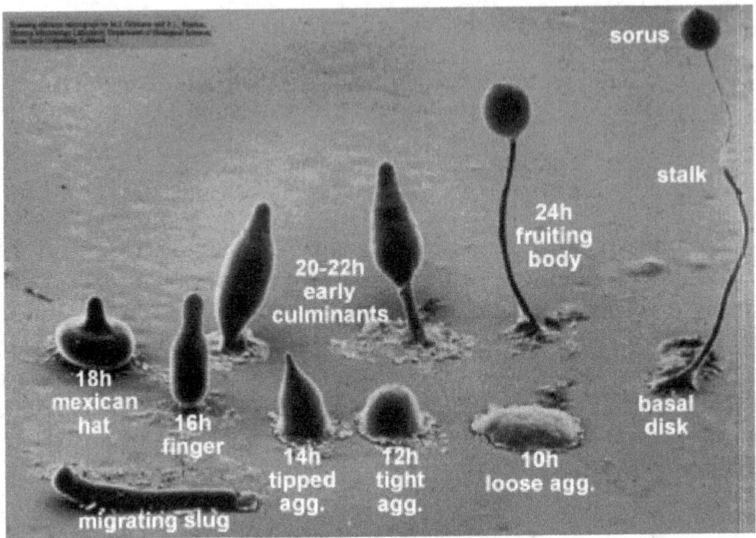

Fig. 2. Aggregation and differentiaion of *Dictyostelium* (Courtesy of R. Blanton, M. Grimson, Texas Tech. Univ.)

Table 1. Genetic data set: aggregation of *Dictyostelium*

#	Genotype	Aggregation
1	wild-type	+
2	yakA-	-
3	pufA-	++
4	gdtB-	+
5	pkaR-	++
6	pkaC-	-
7	acaA-	-
8	regA-	++
9	acaA+	++
10	pkaC+	++
11	pkaC-, regA-	-
12	yakA-, pufA-	++
13	yakA-, pkaR-	+
14	yakA-, pkaC-	-
15	pkaC-, yakA+	-
16	yakA-, pkaC+	++
17	yakA-, gdtB-	±

where $\rightarrow$ denotes excitation and $\dashv$ inhibition (for qualitative definition of these relations, see Section 5.2). A complete network is expected to be consistent with relations from the background knowledge.

4 GenePath's Architecture

GenePath is designed as geneticist's intelligent assistant in performing experiments, reasoning about the outcomes and designing new experiments. It receives information in form of experimental data, such as given in Table 1 and background knowledge in form of the already known relations between genes. Its goal is to derive a genetic network that explains the behavior of the mutants.

GenePath allows the user to examine the experimental evidence and the logic that were used to determine each relationship between genes. Furthermore, GenePath can generate several networks that are consistent with the data and propose further experiments that may help determine the most likely network. As such, GenePath is not intended to give ultimate answers and to be used as a black box; rather, it is focused on supporting of process of cataloging and exploring genetic experiments, and derivation of genetic networks.

GenePath performs abductive inference. Given *Background Knowledge* and *Observations*, the goal of abduction (Flach, 1994) is to find an explanation such that

$$Background\ Knowledge \cup Explanation \models Observations$$

That is, *Observations* logically follow from *Background Knowledge* and *Explanation.* In GenePath, *Observations* are a set of genetic experiments, *Background Knowledge* is defined by geneticists as *a priori* known relations in the network, and *Explanation* is derived in form of a genetic network.

GenePath implements a framework for reasoning about genetic experiments and hypothesizing genetic networks. It consists of the following entities:

- *genetic data*, *i.e.*, experiments with mutations and corresponding outcomes,
- *background knowledge* in the form of known gene-to-gene and gene-to-outcome relations,
- *expert-defined patterns*, which are used by GenePath to abduce gene relations from genetic experiments,
- *abductive inference engine* that, by matching encoded patterns with genetic data, obtains constraints over genetic network,
- *network synthesis methods*, that form networks (hypotheses) that are consistent with genetic data, abduced constraints and background knowledge.

The architecture of GenePath is shown in Figure 3. GenePath starts with abduction of relations between genes and outcomes (step (a) in Figure 3). Additional relations directly obtained from background knowledge are added to the collection of abduced relations (b). They can be used directly to obtain a single genetic network (c). A set of possible genetic networks can be derived through qualitative modelling directly from the data (d). In a combined approach (e), the abduced relations are used as constraints to filter networks generated from data according to a qualitative semantics of genetic networks.

GenePath is implemented in the Prolog computer language (Bratko, 2001). Since it is a declarative language based on logic, Prolog is an excellent environment for this purpose: Prolog makes it easy to consistently encode the data,

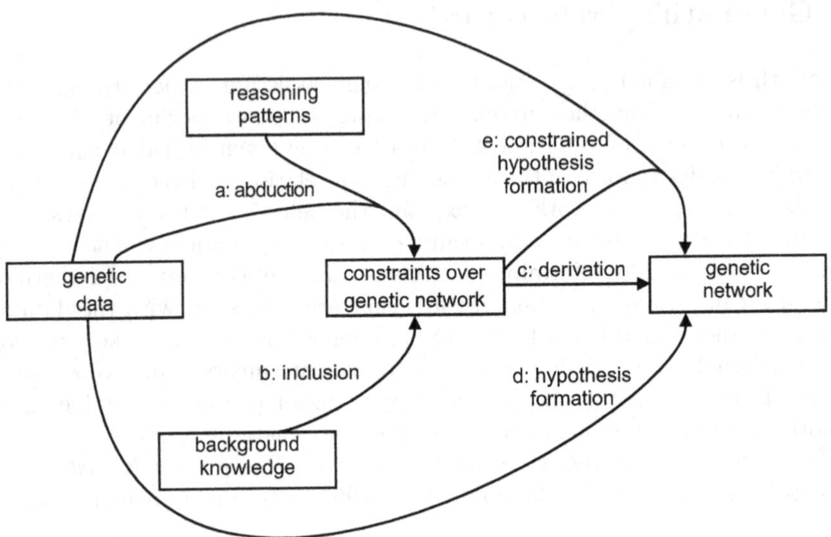

Fig. 3. GenePath's architecture

background knowledge and expert-defined patterns, and allows for a relatively straightforward integration of different GenePath's components.

5 GenePath's Approaches to Genetic Network Discovery

The core component of GenePath is construction of genetic networks. GenePath can use a set of expert defined patterns, which, when applied to the data, identify relations between genes that can serve as constraints on the network. GenePath constructs the genetic network either from such constraints directly, or combines constraints with a qualitative reasoning algorithm to identify a number of possible networks.

5.1 Derivation of Networks Using Expert-Defined Patterns

In this approach, genetic networks are constructed in a two-step process that mimics the construction of networks as done manually by geneticists. In the first step, gene-to-gene and gene-to-outcome relations are discovered and in the second they are used to construct a network.

Relating the genes. Geneticists often construct genetic networks by first establishing the relations between pairs of genes, and between genes and phenotypes. GenePath uses following types of relations:

1. *influences*: gene `A` either excites or inhibits the phenotype or a downstream gene (*i.e.*, gene closer to the biological process); the relation can be either direct or indirect;

2. *parallel*: both gene A and gene B influence the phenotype, but are on separate (parallel) paths;
3. *epistatic*: gene A precedes gene B in a path for phenotype (both genes are therefore on the same path and gene B is in the network positioned closer to the biological process than gene A);
4. *not_influences*: gene A does not influence the phenotype.

To find whether a gene is in a certain relation to another gene or a phenotype, geneticists look for evidence in a form of (typically) two or three experiments with appropriate mutations of the gene(s) and appropriate outcomes. We have crafted several such patterns and coded them as Prolog clauses. Some relations can be detected by different patterns while for others we defined only a single pattern. For instance, there are two patterns for detecting epistasis, but only one pattern for detecting a parallel relation. To illustrate how the patterns are composed and matched with the data, we will discuss two of them in detail. The others are defined in similar fashion and are presented in detail in (Zupan et al., 2003).

One of the most important patterns for discovering the epistatic relation is defined as follows. Let the data include three experiments. Let the first two show that separate mutations of genes A and B result in outcomes OA and OB, respectively. In the third experiment, both genes are mutated. If the phenotype of the double mutant equals OB, the mutation of B overrides the mutation of A. By this we conclude that gene B is epistatic to gene A. In other words, A precedes B on the pathway to the phenotype.

For our data from Table 1, we can, for instance, apply this pattern to experiments 8, 6 and 11, where A=*regA* and B=*pkaC*. Strains with genes *regA* and *pkaC* knocked-out (experiments 8 and 6) aggregate excessively and do not aggregate, respectively. A double mutant does not aggregate, so knocking-out *pkaC* cancels (overrides) the effect of knocking out *regA*. This suggests that *regA* precedes *pkaC* on the pathway to aggregation. The effects of mutations are opposite, so we conclude that *regA* (directly or indirectly) inhibits *pkaC*.

There are two other triplets of experiments that match this pattern. From experiments 2, 10 and 16 we conclude that *yakA* excites *pkaC*. From experiments 2, 3 and 12 we conclude that *yakA* inhibits *pufA*.

For another illustration of patterns, consider the pattern for gene parallelism. Like the above pattern for epistasis, this also requires three experiments: if the first two show that single mutations of genes A and B result in different outcomes OA and OB, respectively, and a double mutant with both genes mutated exhibit a third phenotype OAB different from phenotypes of single mutants, then we conclude that the genes are on parallel pathways. For instance, *yakA* and *pkaR* are considered to act in parallel pathways because the outcomes of the single gene mutations in experiments 2 and 5 are different from each other and from the phenotype of the double gene mutation in experiment 13. Similarly, experiments 2, 4 and 17 indicate for parallelism of *yakA* and *gdtB*.

Genepath uses Prolog's search engine to find the experiments that match the patterns and form a set of gene-to-gene and gene-outcome relations. Table 2 lists the relations found for the experiments from Table 1. The relations discovered by GenePath can be back-tracked: the geneticist can ask the system for the evidence for each relation. Since patterns resemble the geneticist's reasoning, the provided evidence is easy to understand. For instance, for the relation between genes *regA* and *pkaC* (*regA* ⊣ *pkaC*), GenePath gives the following explanation:

pkaC acts after regA because the outcomes of the single gene mutations in experiments 6 and 8, respectively, are different from each other and the outcome of the double gene mutation in experiment 11 is the same as for the single gene mutation in pkaC (experiment 6).

Table 2. A list of all relations derived from experiments in Table 1. The list does not include relations given as background knowledge.

influence	epistasis	parallelism	non-influence
yakA → agg	*yakA* ⊣ *pufA*	*yakA*, *pkaR*	(none found)
pufA ⊣ agg	*regA* ⊣ *pkaC*	*yakA*, *gdtB*	
gdtB ⊣ agg	*yakA* → *pkaC*		
pkaR ⊣ agg	*acaA* → *pkaC*		
pkaC → agg			
acaA → agg			
regA ⊣ agg			

Synthesis of Genetic Network. To construct a network from the abduced relations, GenePath first tests their mutual consistency. If conflicts are found – such as, for instance, a gene reported to not influence the phenotype and at the same time involved in epistasis for that biological process – they are shown to the domain expert and resolved either by either removal of one of the relations in conflict or through appropriate revision of the data (if an error has been found there). For the paper's example, no conflicts in the abduced relations were detected. Relations abduced from the data about aggregation of *Dictyostelium* were examined by the domain expert who confirmed that they are also consistent with his knowledge. An exception was the gene *gdtB*, which was found by GenePath to inhibit aggregation. Relying on the expert's comment, this constraint and *gdtB* were removed from further analysis.[1]

The construction of the genetic network is based on identification of *directly* related genes. Genes `A` and `B` are assumed to be in direct relation if they are not on parallel paths and if no other gene is found that precedes `A` and is at the same time preceded by `B`. For a start, GenePath considers all the genes that influence the outcome and removes the genes that influence any other gene in the group.

[1] To the biologically oriented reader: the effect of mutating *gdtB* alone is not manifested in the aggregation step of development, but is revealed when aggregation is severely reduced by the absence of *yakA*. This is a common finding in genetics. We expect that biochemical studies will be required to decipher the function of *gdtB* and its relationship to *yakA*.

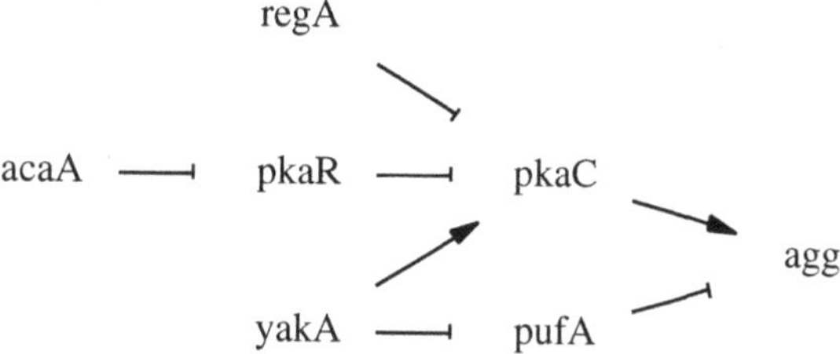

Fig. 4. Genetic network derived through constraint satisfaction

The remaining genes are inserted into the network so as to directly influence the biological process.

From our data set, three genes in direct relation were found: *yakA* → *pkaC*, *regA* ⊣ *pkaC*, and *yakA* ⊣ *pufA*. The genetic network resulting from the background knowledge and GenePath's findings is as shown in Figure 4.

The constructed genetic network was examined by the domain expert (GS). He immediately realized that while the network was in general consistent with his domain knowledge, the direct influence of *pufA* to aggregation was not expected. Reviewing relations that GenePath found revealed that the reason for that was an undetermined relation between *pufA* and *pkaC*. It was also observed that relation of *regA* with most of other genes was not determined. GenePath failed to find the correct placement of *regA* and *pufA* in the network due to insufficient data that would allow GenePath to infer required relations.

5.2 Construction of Networks Through Qualitative Reasoning

Alternatively to the approach described above, GenePath can in effect search through all potential networks and expose those that are consistent with the data. To distinguish this part of GenePath from its abductive inference part, we will refer to it as GeneHyp ("genetic hypothesizer").

GeneHyp assumes that each node of genetic network (gene or biological process) is in some state. The states are *qualitative*: `pos`, `zero`, and `neg`. These states are qualitative abstractions of the real valued levels of the gene expressions and biological processes. The corresponding meanings are:

- `zero` = normal or neutral, same as in wild type
- `pos` = overexpressed, active
- `neg` = inhibited, underexpressed, inactive

The approach implemented in GeneHyp can be outlined as follows. The method starts by determining how genes influence the outcome. Next, it constructs a candidate network that accounts for these influences and checks if the network is consistent with the data. A list of candidate networks is examined and those consistent with the data are shown to the user, ordered by geneticists-defined preferential bias.

Influences between genes and the outcome are determined directly from single mutation experiments. If in a single gene experiment, a gene A and biological process BP change in the same direction, then the gene is said to *positively influence* the biological process. So there must be a path in the network between A and BP with the overall positive influence of A on BP. For example, for the path

GeneA → GeneB ⊣ GeneC ⊣ BP

the chain of influences is: pos, neg, neg. Namely, GeneA positively influences GeneB, GeneB negatively influences GeneC etc. The path's overall effect is pos (the product of three qualitative values pos × neg × neg = pos). Such a positive influence is represented in GeneHyp by the Prolog term:

```
path( GeneA, pos, BP)
```

A negative influence of some Gene on BP is defined analogously and denoted by:

```
path( Gene, neg, BP)
```

A model generated by GeneHyp has to explain all the influences found in the single gene experiments, and it also has to be consistent with all the other (multiple gene mutation) experiments. Typically this gives rise to a large number of models that are logically consistent with the experimental data. Finally, GeneHyp only outputs those models that are consistent with the experimental data and the abduced constraints (those inferred by geneticists' abductive patterns, Section 4.1). It should be noticed that these patterns, defined by the geneticists, in fact define the geneticists' bias, that is the experts' preferences over the set of all possible models that are logically consistent with the data.

As GenePath, GeneHyp is implemented in Prolog. According to the outline above, the top level predicate of GeneHyp is:

```
explain( Model)  :-  % Model explains experiments
  influences( Influences),          % Influences detected in data
  construct( Influences, Model),    % Model accounts for Influences
  verify( Model),                   % Model consistent with the experiments
  genepath_consistent( Model).      % Consistent with abduced constraints
```

Influences are constructed directly from single mutation experiments, as described above. To construct a model M that accounts for all the influences Is, the predicate

```
construct( Is, M)
```

does the following: for each influence path(Node1,Sign,Node2) it is ensured that there is a path in M from Node1 to Node2 with cumulative influence Sign. The model M is constructed gradually, starting from the initial network that contains no arcs. Then for each influence path(N1,Sign,N2) do:

if there is a path in the current network from `N1` to `N2` with
the cumulative influence `Sign` then do nothing,
else either add to the current network `arc(N1,N2,Sign)`,
or add `arc(N1,N,Sign1)` for some node `N` in the current network, and
call the `construct` predicate recursively so that it also takes care
of another influence `path(N,Sign2,N2)` where `Sign1`×Sign2=Sign

Obviously, explaining the observed influences by constructing corresponding network models is a combinatorial process that involves search. This search is controlled by the following mechanisms that were elicited from the geneticists and reflect their preferential bias: (1) Among the models consistent with the data, look for models that minimize the total number of arcs. (2) Among the models with equal total number of arcs, first look for models with the smallest number of nodes' inputs.

These search preferences reflect the following criteria: first, small models in terms of the number of arcs are preferred; second, among the models with equal number of arcs, networks with longer paths are preferred. According to the latter criterion, networks with small number of long paths are preferred to models with large number of short paths. In other words, the degree of parallelism in the network is to be minimized.

The constructed models are checked for consistency with all the experimental data by carrying out a qualitative simulation (Weld & de Kleer, 1990). For the purpose of this simulation we defined a qualitative semantics of genetic networks as follows. `A` → `B` means: if A increases and all other things in the network stay equal, B also increases. For multiple direct influences (e.g., when the state of gene `C` directly depends on genes `A` and `B`), the meaning of the combined influence of A and B on C was chosen to be the qualitative summation. As usual in qualitative physics, this is defined as in Table 3.

Table 3. Rules for qualitative summation

A	B	C=A+B
pos	pos	pos
pos	zero	pos
pos	neg	Anything

A	B	C=A+B
zero	pos	pos
zero	zero	zero
zero	neg	neg

A	B	C=A+B
neg	pos	Anything
neg	zero	neg
neg	neg	neg

The qualitative simulation-based approach was applied to our sample data set. Several thousands of networks comprising six to eight arcs and consistent with the data set were derived – an obviously too large a number to be analyzed by the expert. We therefore needed to resort to an approach that further constrains the number of constructed networks. This is outlined below.

5.3 A Combined Approach to Network Construction

The qualitative reasoning approach by GeneHyp usually proposes a set of hypothesized genetic networks that is too large to be practical for a geneticist.

To further limit the number of proposed networks, we propose a combined approach that uses GeneHyp but uses additional relations abductively inferred by GenePath using GenePath's abductive inference patterns as constraints. This substantially narrows the search space, and for our example data set of Table 1 results in only four networks of low complexity (six edges in the network) that are consistent with abduced constraints and with experimental data (Figure 5).

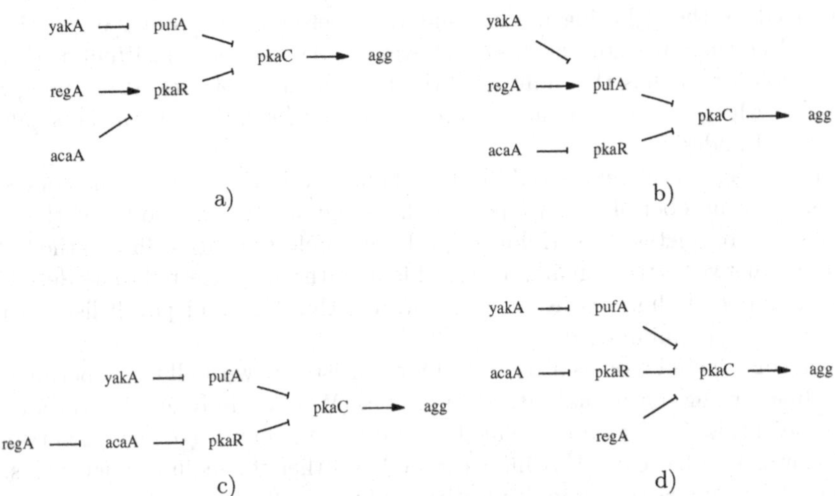

Fig. 5. Genetic networks proposed by combined approach

The close observation shows that the only difference between the networks is in the role of *regA*: it can (a) excite *pkaR*, (b) excite *pufA*, (c) inhibit *acaA* or (d) inhibit *pkaC*. Models c) and d) were discarded because they are inconsistent with biochemical knowledge (additional background knowledge which was not given to the system). Although we preferred the network from Figure 5.a which is completely consistent with our knowledge and experience in the domain, we did not have experimental evidence in our sample data set (Table 1) to discard the network from Figure 5.b. Additional experiments with appropriate *Dictyostelium* mutants would be required to decide in favor of one or the other.

6 Implementation and Web-Based Interface

While the Prolog programming language proved an excellent environment for prototyping and testing both GenePath and its qualitative reasoning part GeneHyp, attempts to convince (even) participating geneticists to access the programs in their native environment were futile (as expected). Since our goal is to develop a broadly accessible and easy-to-use intelligent assistant for explorative analysis of genetic data, we consequently designed a web-based interface, where Prolog's core resides on the server, and is accessed through a server's Visual Basic interface to provide a pure HTML-based client-server solution. Currently, a web-based interface is available only for the abductive inference part, and can be freely accessed at `http://genepath.org`.

A web-based interface comprises several screens that allow a geneticist to enter genes and outcomes, define the background knowledge and provide a list of genetic experiments. The data can be saved to a local file on a client, making it available for future uploads and changes. GenePath uses a single window for analysis of results (Figure 6), where a hypothesized network is displayed together with a choice of abduced relations. Importantly, the user can request, by a mouse click, evidence for specific relation, obtaining a window that describes the reasoning by which this constraint was abduced in plain English (Figure 7).

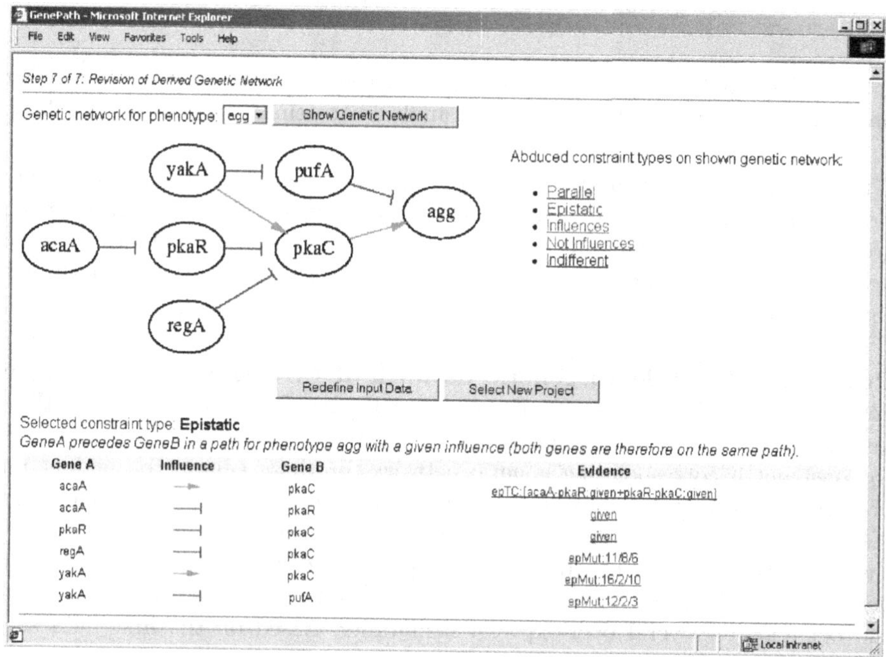

Fig. 6. A web-based interface showing the results of analysis

7 Discussion

7.1 On Other Experiments with GenePath

Besides a *Dictyostelium* aggregation genetic network, in its two year-long existence GenePath has been tried on a number of other problems. Here, we briefly describe some of them, while directing interested reader to GenePath's web site

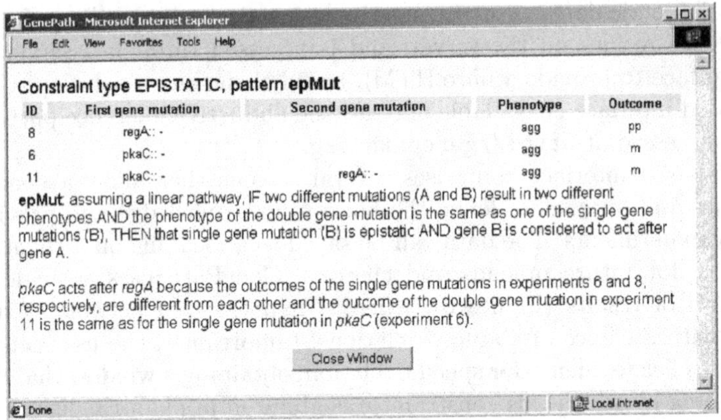

Fig. 7. Explanation of the abduced relation regA⊣ pkaC

(`http://genepath.org`), where she can find the data sets of these and other problems freely available and ready for analysis or demonstration of GenePath.

Growth and development of *Dictyostelium*. Using 28 experiments, GenePath found a correct genetic network for the *Dictyostelium*'s transition from growth to development upon starvation. The genetic network includes 5 genes: *acaA*, *pkaC*, *yakA*, *pufA* and *pkaR*, which are arranged in two pathways (Souza et al., 1998; Souza et al., 1999).

Programmed cell death genetic network of *C. elegans*. During normal development of *C. elegans* 131 of its 1090 cells die. Programmed cell death is known to be regulated by genes *egl1*, *ced3*, *ced4* and *ced9*; GenePath task was to construct a corresponding genetic network for these genes. To conduct an objective experiment and avoid any fitting of the method, one of the co-authors (GS, geneticist) gave another author (BZ, computer scientist) only the data, not the expected solution. The data for this problem came in data sets; the first data set included only knock-out mutations, while the second also described the behavior of several over-expression mutants. Based on the first data set, GenePath was able to find the correct pathway, except that it could not order genes *ced3* and *ced4*. This ambiguity was then correctly resolved with the second data set.

Participating geneticist confirmed the correctness of the solution. Even more, the observation that the data set is sufficient to derive the pathway except for genes *ced3* and *ced4* was also stated in the paper from which the data originated (Metzstein et al., 1998). This clearly confirms the similarity between GenePath's and geneticists' reasoning.

Blind experiment. In its most severe test, 79 experiments involving 16 genes were presented to GenePath. Data was encoded to eliminate the possibility of help from an informed geneticist: genes were not named but rather given an ID number, the origin of the data was unknown. At the start, GenePath pointed out for inconsistencies in the data and suggested a single gene that was a source of them. After that gene was excluded from the analysis, GenePath developed a single solution in the form of a (linear) path that, when presented to the owner of the data, was confirmed to be correct and consistent with the one published (Riddle et al., 1981). Interestingly, the same reported for inconsistencies by GenePath was also found problematic in the original publication from which the data was taken, and its exact function is still a subject of investigation.

7.2 On Utility of Abduced Relations

The case with aggregation of Dictyostelium and other cases tested by GenePath indicate that the constraints obtained through abduction are very useful. Since they are based on expert-defined patterns, they speak the expert's language and may thus be easy to interpret. While the relation-based approach to network construction does not (formally) guarantee a solution consistent with the data — it only guarantees consistency with relations abduced from data, it was often found that where there were enough experiments, this approach found a sole and correct solution. However, especially when used in the domains where genetic networks are yet unknown, the approach that combines abduction of relations and qualitative modelling should be the one of choice, since it usually results in several alternative networks. The geneticist can than either sort out the networks according to his domain knowledge, or he can use GenePath to assist him at reasoning on networks and planning experiments that would narrow down the choice of networks to a single model.

7.3 Scalability

With respect to scalability, the abductive part of GenePath should scale well: most GenePath's patterns are defined over a couple of genes and although the corresponding search space grows exponentially with the number of genes considered, geneticists would usually investigate regulatory networks that include at most a few dozen of genes. For both problems examined in this paper, GenePath constructed the resulting genetic networks within few seconds of CPU time (Pentium III, 1000 MHz). The response time was equally fast even for our biggest real-life data set with 16 genes and 79 experiments mentioned above.

To test GenePath on data sets of the size beyond currently available ones in classical genetics, we have computer-generated several large genetic networks, constructed experiments from them and gave them to GenePath for reconstruction. The networks we have generated were constrained in the number of parallel paths. Experiments were generated so that we made sure that all necessary epistasis relations could be determined from the data. The results of these experiments are given in Table 4 and show that GenePath effectively handles data that contains hundreds of genes and experiments. Even for the largest case investigated (1000 genes and about 3000 experiments) the response time was reasonable. Experiments further show that the time to find relations from the data is negligible in comparison with the time needed to construct genetic networks from abduced constraints. Notice that these experiments are rather unrealistic and largely exceed in size the biggest problems examined by contemporary classical genetics.

Table 4. GenePath's response time on a set of computer-generated data sets. (#Genes = number of genes in experiments, #Exp = number of experiments (about 1/3 of these are double mutants), #Par = number of parallel paths in the network, $T_{relations}[s]$ = CPU time for abduction of relations, $T_{network}[s]$ = CPU time for construction of genetic network)

#Genes	#Exp	#Par	$T_{relations}[s]$	$T_{network}[s]$
1000	2998	50	43	2262
500	1259	25	8	308
500	1205	0	8	353
250	650	25	3	50
250	633	10	3	37
250	615	0	2	46
100	264	10	1	4
100	261	5	<1	4
100	256	0	<1	2

7.4 On Limitations of the Current Approach

The present version of GenePath does not handle cycles, and hence does not propose networks that would include any cyclic component. In principle, cycles could be dealt with a GenePath's part that proposes network constraints – for instance, GenePath may be able to derive that gene A excites gene B, gene B excites gene C, and that gene C inhibits gene A, thus potentially leading to a network where genes A, B, and C would be on a cyclic path. Our present problem with handling cycles are rather in their interpretation: effects of cycles should be observed in time, and our data set exclude any notion of the time effects. Furthermore, without knowing the time scales in which a gene influences another genes, there may be simply too many possible hypothetical networks that include cycles and are consistent with data. We also believe that not being able to deal with cycles does not severely limit the utility of GenePath, as majority of studies in classical genetics that use mutant data do reason mostly on acyclic networks. Still, we do seek a solution to this problem, but anticipate that we will also need to incorporate some additional data that will give notion on temporal relationships.

All the examples we have shown in this paper included a biological process that could be characterized in qualitative terms (like *Dictyostelium* “aggregates”, “does not aggregate”, “has a weak aggregation”, etc.). To reason on the nature of relationships (excitement or inhibition) GenePath requires that qualitative outcomes are ordered. This approach could equally deal with numerical outcomes (like aggregation expressed in percents, where a 100% would denote a wild-type aggregation): the only reason this was not tried yet in GenePath was that we had no data of this type available.

8 Related Work

GenePath uses a genetic logic similar to that described by Avery and Wasserman (Avery & Wasserman, 1992) for determining gene order by epistasis in regulatory networks. It uses abduction (see, for instance, (Kakas et al., 1998)) as an inference mechanism, and principles of expert systems and artificial intelligence for combining expert and domain knowledge and data.

The best-known AI system intended for application in classical genetics is Mark Stefik's MOLGEN (Stefik, 1981). MOLGEN is an expert system for planning gene-cloning experiments in molecular genetics. Stefik (Stefik, 1981) gives a detailed example of how MOLGEN reconstructed a solution to a rat-insulin problem solved previously by Ullrich et al. (Ullrich et al., 1977). However, due to the limitations of its knowledge base and possibly other difficulties, MOLGEN was never applied to find a solution to a previously unsolved genetic experiment problem.

In contrast to MOLGEN, GenePath is not intended to plan experiments, but to interpret experimental data and suggest new experiments that would, if carried out, provide most useful new information about the regulatory mechanism under study. Also, GenePath uses different AI techniques than MOLGEN. GenePath performs abductive inference, whereas the main AI mechanism in MOLGEN is hierarchical means-ends design of plans.

In terms of application, the problem of genetic network construction from data has recently gained substantial interest, but in modelling most of the related work focuses on analysis of gene expression data (Dutilh, 1999). For instance, Friedman at al. (Peer et al., 2000) use Bayesian networks to discover and Shrager et al. (Shrager et al., 2002) use heuristic search to revise genetic networks, and Akutsu et al. (Akutsu et al., 2000) infer genetic networks in the form of Boolean or qualitative networks. Like GenePath, most contemporary systems infer networks which are directional and include both excitation and inhibition links.

Data considered by GenePath comes from classical genetics, *i.e.*, each experiment observes the effect of the mutation to some biological process. The motivation for having a program that supports explorative research in this area is twofold. First, most of research in functional genomics is still based on classical phenotypes. And second, there is a potential bridge between classical genomics and expression data analysis where a whole gene expression array would be considered as a mutant's phenotype (Hughes et al., 2000). It is beyond the scope of this paper to speculate on how gene relations could be inferred from expression arrays phenotype, yet we have to note that even through such mechanisms the genetic networks would remain small due to the requirement for the construction and experimentation with single and double mutants. Namely, epistasis, a cornerstone relation for construction of genetic networks, can by contemporary genetics be proved only through the use of mutants, and to gather a higher volume of relevant data would require bypassing the bottleneck of constructing and analyzing a large number of mutants in parallel (see, for instance, (Ross-Macdonald et al., 1999; Winzeler et al., 1999)). Such experiments are likely to be performed in other organisms in the near future (Kuspa et al., 2001), so the need for automated methods for organizing genes in genetic pathways is evident.

9 Conclusion

GenePath has been a subject of experimental verification for over a year, and is now approaching the point where it can assist geneticists, scholars and students in discovering and reasoning on genetic networks. There are a few improvements under development: an enhancement of a web-based interface that would show a range of hypothesized networks instead of a single one, and an enhancement that allows GenePath to propose experiments. These developments should further enable GenePath to mature into a seamless intelligent assistant for explorative genetic data analysis and reasoning on genetic networks.

References

Akutsu, T., Miyano, S., Kuhara, S.S.: Inferring qualitative relations in genetic networks and methabolic pathways. Bioinformatics 16, 727–734 (2000)

Avery, L., Wasserman, S.: Ordering gene function: the interpretation of epistasis in regulatory hierarchies. Trends in Genetics 8, 312–316 (1992)

Bratko, I.: Prolog programming for artificial intelligence, 3rd edn. Addison-Wesley, Reading, MA (2001)

Dutilh, B.: Analysis of data from microarray experiments, the state of the art in gene network reconstruction. Literature thesis, Theoretical biology and Bioinformatics, Utrecht University (1999)

Flach, P.: Simply logical: Intelligent reasoning by example. John Wiley & Sons, New York (1994)

Hughes, T.R., Marton, M.J., Jones, A.R., Roberts, C.J., Stoughton, R., Armour, C.D., Bennett, H.A., Coffey, E., Dai, H., He, Y.D., Kidd, M.J., King, A.M., Meyer, M.R., Slade, D., Lum, P.Y., Stepaniants, S.B., Shoemaker, D.D., Gachotte, D., Chakraburtty, K., Simon, J., Bard, M., Friend, S.H.: Functional discovery via a compendium of expression profiles. Cell 102, 109–126 (2000)

Kakas, A.C., Kowalski, R.A., Toni, F.: Handbook of logic in Artificial Intelligence and Logic Programming. In: Chapter The role of abduction in logic programming, vol. 5, pp. 235–324. Oxford University Press, Oxford, UK (1998)

Kuspa, A., Sucgang, R., Shaulsky, G.: The promise of a protist: the dictyostelium genome project. Functional and Integrative Genomics 1, 279–293 (2001)

Metzstein, M.M., Stanfield, G.M., Horvitz, H.R.: Genetics of programmed cell death in C. elegans: past, present and future. Trends in Genetics 14, 410–416 (1998)

Peer, D., Nachman, I., Linial, M., Friedmann, N.: Using Bayesian networks to analyze whole genome expression data. Journal of Computational Biology 7, 601–620 (2000)

Riddle, D.L., Swanson, M.M., Albert, P.S.: Interacting genes in nematode dauer larva formation. Nature 14, 668–671 (1981)

Ross-Macdonald, P., Coelho, P.S., Roemer, T., Agarwal, S., Kumar, A., Jansen, R., Cheung, K.H., Sheehan, A., Symoniatis, D., Umansky, L., Heidtman, M., Nelson, F.K., Iwasaki, H., Hager, K., Gerstein, M., Miller, P., Roeder, G.S., Snyder, M.: Large-scale analysis of the yeast genome by transposon tagging and gene disruption. Nature 402, 413–418 (1999)

Shrager, J., Langley, P., Pohorille, A.: Guiding revision of regulatory models with expression data. In: Proceedings of the Pacific Symposium on Biocomputing, Lihue, HI, vol. 7, pp. 486–497 (2002)

Souza, G.M., da Silva, A.M., Kuspa, A.: Starvation promotes dictyostelium development by relieving pufa inhibition of pka translation through the yaka kinase pathway. Development 126, 3263–3274 (1999)

Souza, G.M., Lu, S.J., Kuspa, A.: YakA, a protein kinase required for the transition from growth to development in Dictyostelium. Development 125, 2291–2302 (1998)

Stefik, M.: Planning with constraints (MOLGEN: Part 1). Artificial Intelligence 16, 111–140 (1981)

Ullrich, A., Shine, J., Chirgwin, J., Pictet, R., Tischer, E., Rutter, W.J., Goodman, H.M.: Rat insulin genes: construction of plasmids containing the coding sequences. Science 196, 1313–1319 (1977)

Weld, D.S., de Kleer, J.: Readings in Qualitative Reasoning about Physical Systems. Morgan Kaufmann, San Mateo, CA (1990)

Winzeler, E.A., Shoemaker, D.D., Astromoff, A., Liang, H., Anderson, K., Andre, B., Bangham, R., Benito, R., Boeke, J.D., Bussey, H., Chu, A.M., Connelly, C., Davis, K., Dietrich, F., Dow, S.W., El Bakkoury, M., Foury, F., Friend, S.H., Gentalen, E., Giaever, G., Hegemann, J.H., Jones, T., Laub, M., Liao, H., Liebundguth, N., Lockhart, D.J., Lucau-Danila, A., Lussier, M., M'Rabet, N., Menard, P., Mittmann, M., Pai, C., Rebischung, C., Revuelta, J.L., Riles, L., Roberts, C.J., Ross-MacDonald, P., Scherens, B., Snyder, M., Sookhai-Mahadeo, S., Storms, R.K., Veronneau, S., Voet, M., Volckaert, G., Ward, T.R., Wysocki, R., Yen, G.S., Yu, K., Zimmermann, K., Philippsen, P., Johnston, M., Davis, R.W.: Functional characterization of the S. cerevisiae genome by gene deletion and parallel analysis. Science 285, 901–906 (1999)

Zimmer, C.: The slime alternative — what the behavior of amoebas may tell us. Discover, 86–93 (May 1998)

Zupan, B., Demšar, J., Bratko, I., Juvan, P., Halter, J.A., Kuspa, A., Shaulsky, G.: GenePath: a system for automated construction of genetic networks from mutant data. Bioinformatics 19, 383–389 (2003)

Learning Qualitative Models of Physical and Biological Systems

Simon M. Garrett[1], George M. Coghill[2], Ashwin Srinivasan[3], and Ross D. King[1]

[1] Department of Computer Science
University of Wales, Aberystwyth, United Kingdom
{smg,rdk}@aber.ac.uk
[2] Department of Computing Science
University of Aberdeen, Aberdeen, United Kingdom
gcoghill@csd.abdn.ac.uk
[3] Computing Laboratory
Oxford University, Oxford, United Kingdom
Ashwin.Srinivasan@comlab.ox.ac.uk

Abstract. We present a qualitative model-learning system, QOPH, developed for application to scientific discovery problems. QOPH learns the *structural* relations between a set of observed variables. It has been shown capable of learning models with intermediate (unmeasured) variables, and intermediate relations, under different levels of noise, and from qualitative or quantitative data. A biological application of QOPH is explored. An additional significant outcome of this work is the discovery and identification of kernel subsets of key states that must be present for model-learning to succeed.

1 Introduction

In this chapter, we present QOPH, a qualitative model-learning system. QOPH is developed for application to scientific discovery problems and learns models from observed data. The research behind QOPH is cross-disciplinary between machine learning, model based reasoning, and systems theory—where this type of learning is called *system identification* (Ljung, 1999).

The learning of models from data is a great challenge for machine learning methods since the search space is so large. The problem is also interesting because the data are *positive only*, i.e. when identifying a system from experimental data, nature only provides examples of states of the system, not examples the system can *not* be in. This hinders machine learning as there are no negative examples to restrict over-generalisation (Muggleton, 1996; Gold, 1967). Learning is also complicated when some of the variables required for a complete model are missing. To date, few methods have been shown capable of learning models containing these *intermediate variables*, and QOPH may be unique in that is can induce relations entirely between intermediate variables. The ability to induce

S. Džeroski and L. Todorovski (Eds.): Computational Discovery, LNAI 4660, pp. 248–272, 2007.

models with intermediate variables is essential for practical applications, as it can never be ensured that all relevant variables have been observed (measured).

The most common type of model in scientific applications is that of ordinary differential equations (ODEs). It would therefore be natural to use ODEs directly as the output modelling language, and this approach has indeed been tried successfully (Džeroski & Todorovski, 1995; Džeroski, 1992). However, qualitative inductive modelling provides an alternative to ODE learning, and is particularly useful when the learning data are noisy, missing or sparse. It contrast with the existing literature on methods for fitting parameters to ODE models where the structure is given (often by guesswork) in advance (e.g. (Ljung, 1999)), and we will not examine that problem here.

A comprehensive assessment of qualitative inductive modelling is presented, including: (i) a set of rigorous tests that demonstrate QOPH's ability to learn models from increasingly more realistic data, and (ii) a state-of-the-art review of the literature in this field, showing the main developments and themes. To this body of knowledge we add the discovery of *kernel subsets*—small groups of qualitative states that, when present, guarantee that modelling will succeed. We empirically demonstrate that:

- A small set of modelling constraints taken from machine learning and system science are sufficient to enable QOPH to learn models of systems science benchmark systems. Other systems have been investigated and their results are presented in (Garrett et al., 2001).
- Models can be successfully learnt from various sample qualitative states, and numerical models.
- Models can be successfully learnt in the presence of both large amounts of qualitative state noise, and numerical noise.
- Models can be learnt with (at least) four intermediate variables, and models may include intermediate relations.
- Models of complex, real biological systems can be learnt.
- Kernel subsets of states exist that, when given to QOPH, guarantee that a model will be learnt.

2 The Task of Discovering Qualitative Models

2.1 The Benefits of Qualitative Modelling

Since there exist ODE modelling tools, why develop a qualitative modelling system, or use a representation derived from qualitative reasoning (QR)? Some of the advantages of learning qualitative models *vs.* numerical models are:

Ease of understanding: QR models eliminate unnecessary detail and allow the user to focus on the essentials of the model and to extract quickly the required understanding of the system being modelled. Often what is of interest is what will happen if a variable increases, decreases or remains unchanged, not exactly how much that change is.

Finite state space: Unlike many real-valued numerical models, a qualitative model can only be in one of a limited number of states, and this set of states can be easily defined.

Increased quality assurance: All data gathered by physical measurements contain a degree of error due to the measuring process; on top of this is added human error. Modelling the data with QR overcomes many of these problems simply by considering the data's qualitative aspects alone. This means that models produced by QR can often be used with increased confidence.

Ease of induction: One of the main problems in numerical methods of system identification is parameter estimation. By its very nature qualitative reasoning does not have this problem because there is no need for the parameters. This reduces the workload of the inductive system by allowing it to concentrate on inducing just the structure of the model. Since qualitative data effectively summarises numerical data it should be possible to learn a QR model with less data than is required by a numerical modelling system.

2.2 Defining the Qualitative Formalism

There are several types of QR system (e.g. (Kuipers, 1994; Forbus, 1984)); we chose to use the representation and relations of QSIM. We do not use QSIM itself, which is a qualitative simulation engine, we merely use its formalism to represent models. The models produced by QOPH take the form of conjoined relations, each of which constrains the possible qualitative states that the model may legally take. We prefer the constraint-based formalism of QSIM to the process-based formalism of Forbus's QPT (Forbus, 1984) because a set of conjoined QSIM relations can be regarded as a qualitative differential equation (QDE), and this QDE can be directly related to an ODE of the same system. The two aspects of QSIM's formalism that we use are the form of representation of variables and the use of qualitative relations between those variables.

Each qualitative variable value consists of a $\langle qmag, qdir\rangle$ 2-tuple. Here, *qmag* is the qualitative magnitude of the variable, where $qmag \in \{\texttt{minf}...0, 0, 0...\texttt{inf})\}$, and *qdir* is the qualitative derivative, where $qdir \in \{\texttt{dec}, \texttt{std}, \texttt{inc}\}$. The possible values of the *qmag* set mean that the variable is in the interval $[-\infty, 0)$, is at the landmark '0', or is in the interval $(0, \infty]$. Infinity is considered to be a value. This set of possible states is known as a *quantity space*. In QSIM the resolution of this quantity space can be varied by adding other landmark points within an interval, but QOPH currently only uses the simple case just defined. The possible values of the *qdir* set mean that the variable is either decreasing (`dec`), steady (`std`) or increasing (`inc`). A qualitative variable can therefore only exist in one of nine possible states, which discretises and simplifies the state space to be searched by QOPH.

There are several kinds of constraint that can appear in a QSIM model, these include the usual algebraic operations, of addition, `ADD()`, subtraction, `SUB()`, multiplication, `MULT()`, and sign inversion, `MINUS()`. There is also a derivative predicate, `DERIV()`, stating that one variable is the derivative of another. Furthermore the relations $\mathtt{M}^-$ and $\mathtt{M}^+$ state that two variables, A and B may be functionally related in an unknown, non-linear but monotonic manner. If the variables are in a monotonically increasing relationship then `M+` is used, and if monotonically decreasing then `M−` is used. Thus, without knowing the exact numerical relation between A and B, `M+` and `M−` express useful information about that functional relation. This illustrates how the qualitative relations can also be used to express incomplete and imprecise knowledge. For more details about the QSIM formalism see (Kuipers, 1994; Kuipers, 1986).

2.3 The Task of Qualitative System Identification

Given this definition of the formalism used by QOPH, the task is to find a conjunction of qualitative relations that models the relationships between a set of measured variables, and to discover a number of putative, unmeasured intermediate variables that allow the model to be expressed in a complete form. There may be zero, one, or more intermediate variables, and where there are sufficient intermediate variables, QOPH should be able to discover *intermediate relations* exclusively between intermediate variables.

In order to minimise the search space, QOPH will use only the following relations: `ADD()`, `MINUS()`, `DERIV()` and `M+`. The `SUB()` and `M−` relations are redundant, given this set of relations, and the `MULT()` relation is not required for the benchmark systems described in Section 4.1, although it is used in Section 5.1.

3 The QOPH System

Since the target model will consist of conjoined relations, a machine learning paradigm was required that would naturally learn with logical clauses. Inductive logic programming (ILP) was therefore chosen. The use of ILP and a full description of QOPH's algorithm are now discussed, and an example is provided of QOPH's operation.

3.1 Inductive Logic Programming (ILP)

Given some observations (*evidence*), E, and background knowledge, B, the ILP task is to find a hypothesis, H, that, together with the background knowledge, explains the observations (Plotkin, 1971; Muggleton & Feng, 1990). It was natural to use ILP to learn clauses of conjoined qualitative relations (the hypothesis) that are in logical agreement with a set of qualitative data (the evidence), given a qualitative formalism (the background knowledge).

Many solutions to the ILP problem are possible, including the trivial solutions of 'E' (just restating the facts) or 'B → E' (the facts in the context of the qualitative formalism) (Muggleton, 1992). The problem is to find only those solutions that provide useful hypotheses. Note that although work in abduction, (Kakas et al., 1992; Flach & Kakas, 2000), states that solutions are restricted to ground facts; in ILP more general solutions are allowed (Plotkin, 1971; Muggleton &

Feng, 1990), even though there are still typically syntactic restrictions on what form solutions can take (Yamamoto, 1999).

For most scientific discovery problems it is clear that ILP is advantageous, as we wish to learn general theories; similarly ILP is a sensible choice for learning qualitative models, and we have previously applied machine learning and ILP to many scientific problems with great success (e.g. (King et al., 1992; Sternberg et al., 1994; Srinivasan et al., 1996; King & Srinivasan, 1997; Srinivasan & King, 1999)). QOPH builds on this foundation, and works in connection with either PROGOL (Muggleton, 1992; Muggleton, 1995) or, in this case, ALEPH (Srinivasan, 2000). ALEPH was used to process the clauses produced by the QOPH implementation, which was written separately.

QOPH's *background knowledge* of qualitative relations was implemented as a set of Prolog rules, not a (large) set of ground clauses, as in (Bratko et al., 1992). QOPH's *evidence* was given as a set of Prolog `state()` facts, that expressed the values of several qualitative variables at some point in time. An example state containing values for four variables might be,

```
state(0...inf/std, minf...0/dec, 0/std, 0...inf/std).
```

The states did not need to be temporally ordered since QOPH finds hypotheses that match the evidence set in any order by a covering algorithm (Fürnkranz, 1999). Given this information, QOPH induced hypotheses (models) that were in logical agreement with E and B.

3.2 The QOPH Algorithm

ILP, like much of learning, can be considered to be a search through a space of possible solutions (Mitchell, 1979). In this case the space consists of qualitative models, containing one or more relations, acting on two or more qualitative variables. Any relation, between any known variables, is a candidate for being part of a model. Moreover, relations may exist between known variables and an intermediate variable. Once an intermediate variable has been introduced it becomes available for use in other relations, and these other relations may in turn introduce other new intermediate variables. In this way it is eventually possible, where required, to posit relations that exclusively relate intermediate variables.

In general the model space is vast (see (Garrett et al., 2001) for an analysis of how large) and the QOPH algorithm searches this space by a best-first search, plus a number of pruning heuristics to minimise the search space. The full QOPH algorithm is shown in Table 1. More detail is now given.

The Search Algorithm. Table 1 progressively extends the *Clause* list from a single relation to a full model. A partial model is extended by appending in turn each of the relations in the list produced by `generateLegalRels`. This list, *LRList*, contains all possible relationships between all current variables (including intermediate variables introduced in previous calls to `generateLegalRels`), and an additional, new intermediate variable. The new list of variables is returned in $NewVariables$.

Table 1. The Qoph algorithm

```
qoph(Variables, RelPredicates, QualStates):
  extendClause(Variables, RelPredicates, QualStates, {})

extendClause(Variables, RelPredicates, QualStates, Clause):
  LRList, NewVariables ← generateLegalRels(RelPredicates, Variables)
  SRList ← sortRelsByCost(LRList)
  foreach Rel in SRList
    NewClause ← Clause + Rel
    if ( full(NewClause) and legal(NewClause) and
     accurate(NewClause, QualStates) ) then
      print NewClause
    else
      extendClause(NewVariables, RelPredicates, QualStates, NewClause)
    end if
  end foreach
```

Since each *NewClause* is created by appending *Rel* to *Clause*, the differences in cost are entirely due to the relations being conjoined, thus `sortRelsByCost` ensures that the *NewClauses* are extended in order of cost, which is simply the number of intermediate variables in the relation. This is a reasonable heuristic estimate for the size of the final clause because a larger number of variables generally requires a larger number of relations between those variables.

A partial model is declared a full model when it is consistent with all the following tests, otherwise a further extension is required:

`full`: This predicate ensures that the clause contains the same number of non-input variables (including intermediate variables) as relations; furthermore, it checks that the clause contains all the measured variables. The clause can not be a model of the qualitative states if either of these do not hold.

`accurate`: As with individual relations, *NewClause* should be logically in agreement with E to some degree of accuracy. This step is at every degree of clause extension because intermediate variables may only be properly constrained once the model is complete.

`legal`: The model must satisfy a (user definable) set of heuristics; the experiments described here require the following are true:

- `non-disjoint`. The model must not contain two or more disjoint models or model parts. The assumption is that if a set of measurements are being made within a particular context then the expectation is that they will emanate from the same system and not from two or more separate systems. Of course, one may be mistaken in this assumption; therefore models filtered under this heuristic could be cached and revisited if no suitable models have been found.
- `causal ordered`. The model must be causally ordered (Iwasaki & Simon, 1986) with an integral causality (Gawthrop & Smith, 1996). That is,

the causality runs through the algebraic constraints of the model from the magnitudes of the state variables to their derivatives; and from the derivatives to the magnitudes through a `DERIV()` constraint only. Since some systems can only be constructed in a non-causally ordered manner (Cellier, 1991), the models eliminated by this filter could be cached in case a suitable causally ordered model cannot be found.
- `non-hanging variables`. Endogenous variables that are only a member of one relation are named *hanging variables*. Such relations provide no useful constraint so the model can be complete if it contains such variables.

The Pruning Heuristics. For each call to `extendClauses`, *LRList* is filtered to remove relations that are invalid by ensuring the following predicates hold:

- `nonRedundant`. This prevents generally contradictory, repeating or equivalent relations from appearing, such as `add(A,B,C)` and `add(C,B,A)`.
- `languageOK`. This limits the number of times a relation can appear in a clause. For the experiments described in this chapter it was set to 3, meaning that there would be no more than three instances of any relation in a single clause.
- `dimensionalAnalysis`. This ensures that variables are typed and that only variables of the correct types occur together in qualitative relations; for example, adding variables that have units of flow and amount is meaningless (Bhaskhar & Nigam, 1990). This implements Say's suggestion, discussed in Section 6, that dimensional analysis should be central to the search procedure.

3.3 A Simple Example of QOPH's Operation

Referring to Table 1, consider the case where QOPH is attempting to learn a simple qualitative model of a tank of liquid with an input, an output and a level, and an unmeasured net flow into the tank,

```
DERIV(level,net),
M+(level,out),
ADD(net,out,in).
```

given a set of states for *level*, *in* and *out*, and a background knowledge of a set of qualitative relations, implemented by the predicates `DERIV()`, `ADD()`, M^+`()` and `MINUS()`. QOPH must introduce an intermediate variable, *net*, in some form, to complete the model.

Starting with *Clause* = [], QOPH will create all possible qualitative relations between *level*, *in* and *out* using the set of relations available. It will also posit an intermediate variable, $v1$ and create all possible relations between it and the other variables. For example this list might begin,

```
DERIV(level,in), DERIV(level,out), DERIV(level,v1),
DERIV(in,out), DERIV(in,v1), DERIV(out,v1), ADD(level,in,out),
ADD(level,in,v1), ADD(level,out,v1), ADD(in,out,v1), ...
```

In this list is `DERIV(`*level*`,`$v1$`)`, which (like all the relations above) is a partial model. It will be extended by adding each in turn of all the possible relations that can be created from the variables, *level*, *in*, *out*, $v1$ and a new intermediate variable, $v2$. One member of that set of relations will be `M+(level,out)` and thus there will be a partial model containing the first two of the three relations. The process continues a third time giving a partial model containing all three relations, which satisfies the `full`, `accurate` and `legal` predicates, and is presented as a model of the state data.

During this search process many of the possible options are pruned away. For example, types are attached to the variables, so that we have *l:level*, *f:in*, *f:out*, *x:v1*, where l means 'level', f means 'flow' and x means 'any type'. Given this information we can delete `DERIV(in,out)` as a possibility because two variables of 'flow' type can not be in a `DERIV()` relationship. Similarly for other relations in the list above. By a combination of best-first search, pruning heuristics and search completeness criteria, QOPH can find the required model far more quickly than would otherwise be the case.

4 Experiments with Benchmark Models

This section describes a set of experiments on benchmark systems, used to validate the QOPH method. Qualitative and numerical versions of the benchmarks were used; producing clean and noisy data of various sorts for model-learning. Numerical data was always converted to qualitative data before being input into QOPH. Fig. 1 shows how some of these elements are related.

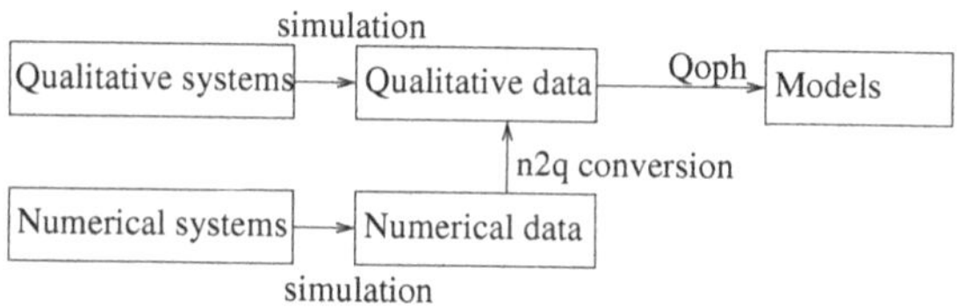

Fig. 1. An outline of the inputs and outputs of the QOPH system. The label 'n2q' stands for 'numerical to qualitative'.

4.1 The Benchmark Systems

Coupled tanks. The coupled tanks system is a U-shaped tube containing liquid, with an input and an output. The input, $inflow_A$, pours into the top of tank A and the output, $outflow_B$, pours out of the base of tank B (see Fig. 2). Note that $flow_{AB} \equiv outflow_A \equiv inflow_B$. The qualitative and ODE models are:

DERIV($level_A$, $netflow_A$),
DERIV($level_B$, $netflow_B$),
ADD($level_B$, $levelDiff$, $level_A$),
M^+($levelDiff$, $flow_{AB}$),
M^+($level_B$, $outflow_B$),
ADD($netflow_B$, $outflow_B$, $flow_{AB}$),
ADD($flow_{AB}$, $netflow_A$, $inflow_A$).

$$\frac{dlevel_A}{dt} = inflow - flow_{AB} \quad (1)$$

$$\frac{dlevel_B}{dt} = flow_{AB} - outflow \quad (2)$$

$$flow_{AB} = k \times (levelA - levelB)\,. \quad (3)$$

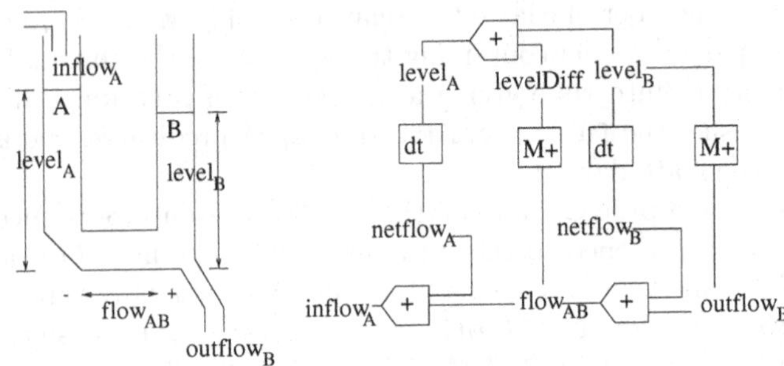

Fig. 2. The coupled tanks model: physical (left) and QSIM (right)

Here there are three intermediate variables, '$netflow_A$', '$netflow_B$', and '*levelDiff*'. For the system to be correctly learned these variables will have to be induced. Variable '$inflow_A$' (the input) is *exogenous* to the model and so appears only once. All other variables are *endogenous* (internal).

Mass Spring Damper. This system consists of a mass on a spring, the movement of which is damped and to which an external force is applied, in line with the direction of motion. Due to the damping, the oscillations do not continue indefinitely; due to the force, the rest point of the mass is displaced from the natural rest position (see Fig. 3). The mass M has displacement $disp_M$ from its rest position, and at any time, t, it is moving with velocity vel_M and accelerating at rate acc_M. The qualitative and ODE models are:

DERIV($disp_M$,vel_M),
DERIV(vel_M,acc_M),
M^+($disp_M$,H_1),
M^+(vel_M,H_2),
M^+(acc_M,H_3),
ADD(H_1,H_2,H_4),
ADD(H_3,H_4,$force$).

$$\frac{ddisp_A}{dt} \equiv vel_B \tag{4}$$

$$\frac{dvel_B}{dt} = k_3 \times (force - H_4) \tag{5}$$

$$H_4 = (k1 \times disp_A) + (k2 \times vel_B)\,. \tag{6}$$

Note the *four* intermediate variables, H_1, H_2, H_3 and H_4, as well as an intermediate relation ADD(H_1,H_2,H_4), which can only be learnt by first learning the intermediate variables it relates. The input force, $force$, is exogenous.

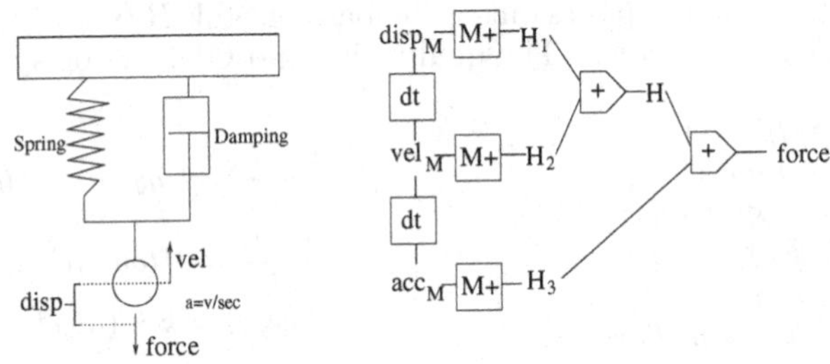

Fig. 3. The mass spring damper (a) physical; (b) QSIM

Qualitative simulations always result in a finite set of states for the model for a given input value' this is known as the model's *complete envisionment* (see Appendix A). The union of complete envisionments for all possible inputs is known as the *total envisionment.* The qualitative input data for QOPH are always a subset of states from one of these two sets.

Numerical simulations were constructed using the same relations as the qualitative models, plus a parameter for each monotonic relation, giving a linear relation between the two variables. This greatly simplifies the ODE of the spring model but still provides an interesting system. These parameters are only important insofar as the single qualitative behaviour they define after conversion from numerical data to qualitative state. Parameters were chosen such that the model approached a steady state during the time period of the test. A stiff ODE solver was used (i.e. one that had high resolution in the time dimension) which helped to reduce the amount of false oscillation about the steady state. Such oscillations can cause complications when the output is qualitised.

4.2 Clean Data Experiments

Ideally we would have wished to test QOPH's ability to learn models for all possible subsets of the total envisionment of a system, since this would detect any subsets of states that ensure successful model-learning. Unfortunately, if T is the number of states in the total envisionment of a particular model, then there are $(2^T - 1)$ possible non-empty subsets of those states. In the case of the coupled tanks it was only possible to perform the experiments on a single complete envisionment, with the *inflow* set to $\langle$`0,std`$\rangle$. In the case of the spring the state space is so large that even this is not possible. Even the smallest input partition would result in $2^{25} - 1$ experiments, and since it was also by far the slowest system to run this made it impossible to exhaustively examine all these state subset combinations. As a result the 31 states partitioned by $\langle$`(0,inf),std`$\rangle$ was randomly sampled 1000 times, and these 1000 sets of states formed the learning input.

Numerical data required conversion from real-valued data to qualitative data, since QOPH learns from qualitative states. This form of conversion is an important area of research in itself (e.g. (DeCoste, 1991; Cheung & Stephanopoulos, 1990a; Cheung & Stephanopoulos, 1990b; Bakshi & Stephanopoulos, 1994)). We employed a relatively simple but robust method. The n real-valued time series steps for a variable x were numerically differentiated by means of a central difference approach (Shoup, 1979) such that,

$$\left.\begin{aligned} \frac{dx_i}{dt} &= \frac{(x_i - x_{i-1}) + (x_{i+1} - x_i)}{2} \\ \frac{d^2x_i}{dt^2} &= (x_i - x_{i-1}) - (x_i - x_{i+1}) \end{aligned}\right\} i = 2 \cdots n-1\,,$$

then the first and second derivatives were smoothed by applying a Blackman filter (Blackman & Tukey, 1958) to their Fast Fourier Transforms (FFT) and taking the real part of the inverse FFT. Smoothing is required because noise is introduced by the process of quasi-differentiation and can not be avoided. Furthermore, temporal misalignments between two variables can occur during this process, introducing further errors.

In principle, having obtained values for x, $\frac{dx}{dt}$ and $\frac{d^2x}{dt^2}$ the numerical values would just be converted to a qualitative quantity space, $[x]$, such that:

$$\begin{aligned} x_i < 0 &\quad \Rightarrow \quad [x_i] = - \\ x_i = 0 &\quad \Rightarrow \quad [x_i] = 0 \\ x_i > 0 &\quad \Rightarrow \quad [x_i] = + , \end{aligned}$$

in practice, however, a small margin of signal error,

$$\begin{aligned} (x_t < 0) \wedge (\frac{3}{100} min(x_i)_{i=1}^{n} \leq x_t \leq 0), \Rightarrow x_t \mapsto 0 \\ (x_t > 0) \wedge (0 \leq x_t \leq \frac{3}{100} max(x_i)_{i=1}^{n}), \Rightarrow x_t \mapsto 0 \, , \end{aligned} \tag{7}$$

was allowed around the zero value since real values are highly unlikely to be exactly zero; this *zero envelope* allows us to assign near-zero real values to the qualitative zero landmark. Again there are errors introduced by this procedure since the signal may enter and leave the zero envelope any number of times, apparently jumping to and from zero. This may or may not be the desired behaviour: if the signal remains close to the edge of the envelope, small perturbations in the signal can cause errors in assigning the qualitative value.

For a qualitative variable $[x]$, the value of *qmag* (see p.3) is generated from x, and the value of the *qdir* is generated from the derivative of $[x]$. The magnitude and derivative of $[dx/dt]$, is obtained in a similar manner but generates its data from dx/dt and d^2x/dt^2 respectively. Using this approach it is possible to convert the real data to qualitative data in the QSIM representation.

Although learning from qualitative data is expected to be easier than learning from quantitative data (as discussed in Section 2), the conversion from quantitative data to qualitative data is noisy; therefore it was of interest to test whether the inductive mechanism was robust enough to learn the correct model from a single behaviour, under differing initial conditions, despite this conversion noise.

Four initial conditions were chosen for the two state variables in both benchmark systems: (0,0), (2,0), (0,3) and (2,3). The numerical models produced data from these initial conditions until a steady state was approached. The data produced were then converted to qualitative states in the manner just described, and these qualitative states formed QOPH's E data.

4.3 Noisy Data Experiments

In order to be useful, QOPH should be able to learn from noisy data. Noise was added to both qualitative and numeric data.

A noisy qualitative state is defined as a state that is not part of the complete envisionment but is of the same form, containing the same number and type of variables. This enlarges the E state space considerably, since as well as choosing subsets from the complete envisionment we now need to vary the amount of noise with those states, without making the number of experiments prohibitive. For both benchmarks, this was achieved by sampling the clean/noise state space.

Noise was added to the numerical data, and then converted to qualitative states, as follows. Clean numerical data was modulated with varying amounts of Gaussian noise. The degree of noise was described as a fraction of a noise vector, thus we applied 1/1000, 10/1000, 100/1000, and 1000/1000 of the noise to the signal. At 1000/1000, the noise was designed to be more powerful than the signal.

The combined signal and noise were denoised by again applying a Blackman filter to remove the high frequencies from the FFT domain, performing the inverse FFT and keeping the real part. This remaining signal was then converted to qualitative data (as above), giving an errant subset of the complete envisionment for each numerical simulation, and attempts were then made to learn the model from these data.

Since the conversion to qualitative states is itself a noisy process it is clear that deliberately adding noise to the signal prior to conversion would make it highly unlikely that QOPH would be given many clean qualitative states to learn from. Note, however, that this is due to the conversion process, not the learning algorithm, and numerical to qualitative conversion is not the focus of this work.

4.4 Experiments

Once the data were in qualitative form, they and the background knowledge were given to QOPH. QOPH then output zero, one or more models. A model was regarded as correct if it can be shown to be logically equivalent to a known, correct, *gold standard* clause. To simplify matters, we restricted analysis of the models to the first ten produced. If there was no model in this group that was logically equivalent to the gold standard model, then the test was regarded as a failure. In many cases the correct model would have been found if the search had been allowed to proceed beyond ten clausal models, but we wished to build a system that would generally find the correct model quickly.

If *some* of the clauses were found to be correct models, this was regarded as a partial success. The *reliability* of the experiment is the ratio of the number of correct models to the total number of models induced (upto the limit of ten models). This measure of reliability was used in the results of both the clean and noisy data experiments. If an experiment *only* produced models that were logically equivalent to the gold standard then this is said to *reliably* produce the correct model.

Since any given experiment will induce its models from a finite number of states, it is possible to plot the *average reliability* for all the experiments for a given number of states. For example, if there are ten states in a complete envisionment then 120 of the 1024 possible subsets of these states will contain three states, and each of these 120 subsets will have some measure of reliability; therefore it is possible to plot the average reliability for this, or any other, number of states.

For the noise experiments, there are three dimensions to be investigated: number of clean states, number of noise states and average reliability. However, the results will be presented as 2-D (the noise dimension is projected) and 3-D plots.

4.5 Results

Models Learnt from Qualitative Data

Coupled Tanks. The plots of the number of states against average reliability for the coupled tanks are shown in Fig. 4. Not surprisingly the noise experiments show a marked reduction in accuracy; this suggests that there are numerous false models of partial envisionments.

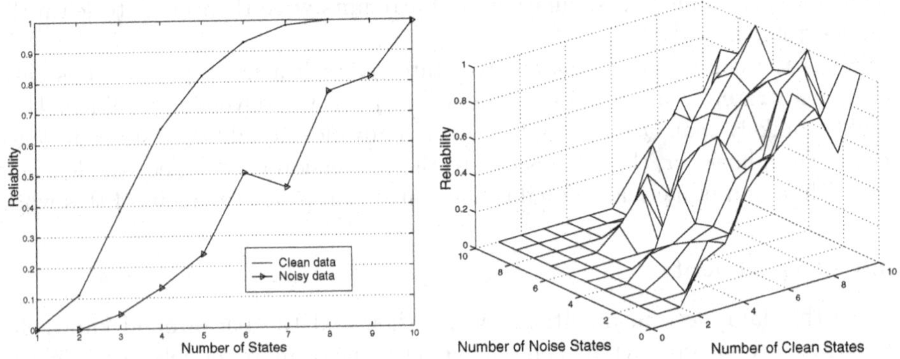

Fig. 4. Coupled tanks reliability graphs: comparative 2D plots (left); state vs. noise vs. reliability (right)

Analysis of the results also showed that there are a number of subsets of states must be present in other, larger subsets for a model to be correctly learnt. We term these subsets *kernel subsets.* The kernel subsets for the coupled tanks' clean data experiments are as follows:

```
[1,6],   [6,8],   [6,9]   (state 6 with state 1, 8 or 9)
[2,8],   [6,8],   [7,8]   (state 8 with state 2, 6 or 7)
[1,2,3], [1,2,4], [1,2,5] (states 1 and 2 with state 3, 4 or 5)
[1,3,7], [1,4,7], [1,5,7] (states 1 and 7 with state 3, 4 or 5)
[3,7,9], [4,7,9], [5,7,9] (states 2 and 9 with state 3, 4 or 5)
[2,3,9], [2,4,9], [2,5,9] (states 7 and 9 with state 3, 4 or 5).
```

The numbers represent the states that the coupled tanks may be in (refer to Appendix A). Note that there are five unique kernel subsets containing two states, and 12 kernel subsets containing 3 states. Comparing these kernel subsets to the envisionment graph (Fig. 5), seems to indicate that a selection of states from different behaviours (contiguous paths in the directed graph) is the key to reliably inducing the correct model.

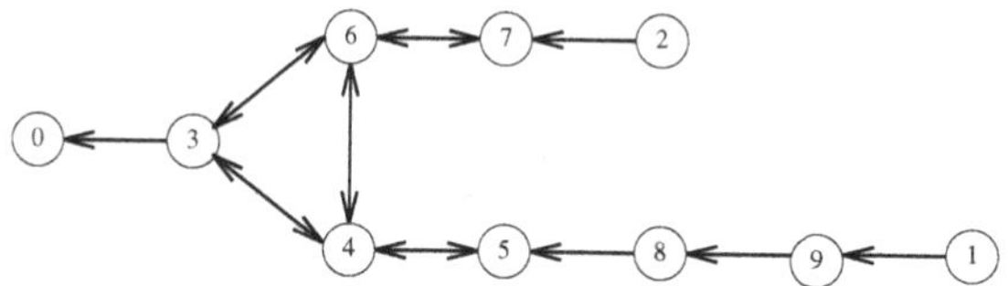

Fig. 5. The envisionment graph for the coupled tanks: *c.f.* Appendix A

The importance of this finding is that a knowledge of the kind of states that form kernel subsets might make it possible to learn models from very low numbers of states—in this case there were five cases where just two states out of ten would have been sufficient for totally reliable model-learning to take place.

Mass Spring Damper. The graph of number of states against average reliability is shown in Fig. 6. The graphs show that noise affects accuracy, though not by as much as the coupled tanks system. This might suggest that the model space for the spring contains fewer models of partial envisionments. Similar experiments on systems other than the coupled tanks and spring show a greater drop in accuracy than for the spring (Garrett et al., 2001), and in these cases QOPH would more easily find a number of errant models before finding the true model, and the run be regarded as a failure.

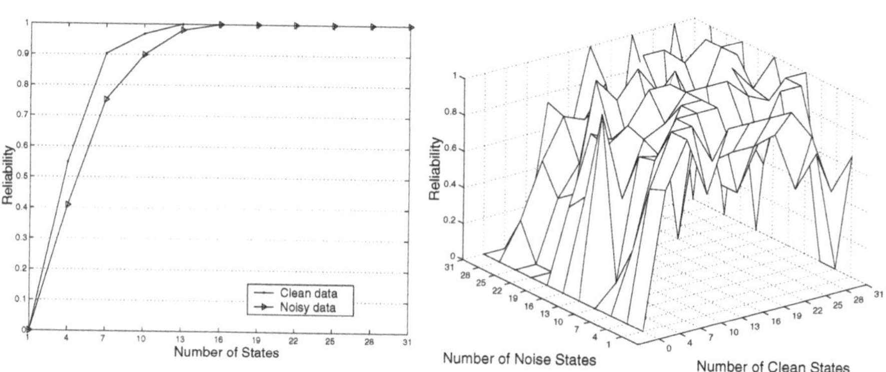

Fig. 6. Damped spring reliability graphs: comparative 2D plots (left); state vs. noise vs. reliability (right)

The 3-D graph does not have a very smooth surface due to the difficulty in performing enough experiments to cover the clean/noisy state space. Nevertheless, there appears to be a drop in reliability as the number of noise states increases to high values, as would be expected. The damped spring has a large number of possible clean data experiments (2^{31}) and this, coupled with the normal distribution of the experiments' state subsets over the number of states means it

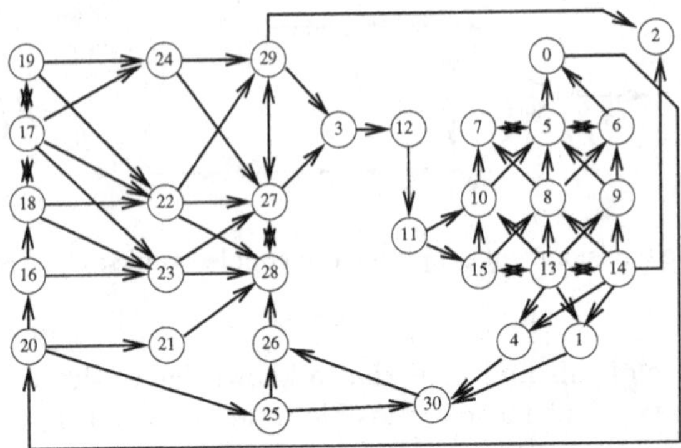

Fig. 7. The envisionment graph for the damped mass on a spring: *c.f.* Appendix A

is hard to perform enough experiments in the $N/2$ range of states to build a meaningful graph of the performance across all states, and makes it less likely that state subsets containing most of the complete envisionment are used in the experiments.

Kernel subset analysis was only meaningful when the experiments explored *all* subsets of a complete envisionment, unlike the spring experiments. Other examples of kernel subsets are presented in (Garrett et al., 2001).

Models Learnt from Numerical Data. The results from the numerical data experiments are presented in Fig. 8 as a summary of the results for each benchmark. At each level of noise the results of the four initial conditions are averaged, for each benchmark system.

The results show that it is possible to learn models from clean and noisy numerical data. Again the coupled tanks system provides more errant models, of the partial states that formed the basis for learning, than the spring system.

The drop in reliability is due to there being few states to learn from. The set of states, gleaned from quantitative to qualitative conversion, did not form a full behaviour for either the coupled tanks or the spring, which makes the ability to learn a model from them even more impressive. In all failure cases the kernel subsets were missing (see (Garrett et al., 2001) for details).

5 Qualitative Machine Learning in the Biological Domain

5.1 High-Level Components

So far we have limited the components of models to the basic qualitative relations, here we investigate the possibility of building models using predefined *collections* of these relations, to improve scalability. This is an example of the well-known AI principle of *chunking* (Laird et al., 1986). An example biological metabolic pathway will be modelled since they essentially contain only two types

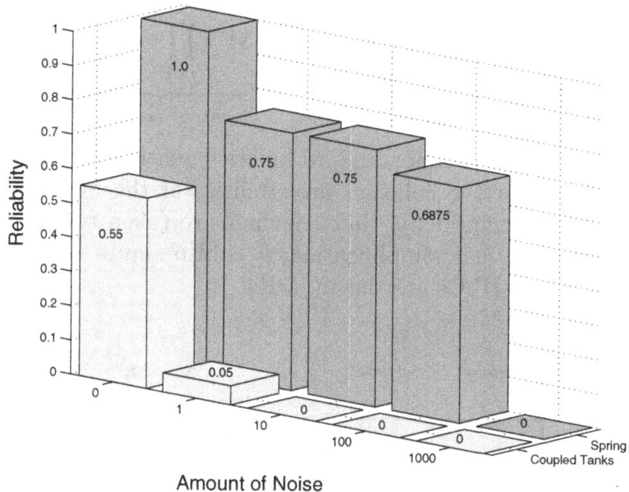

Fig. 8. Reliability of learning the correct model from numerical data vs. 1000ths of Gaussian noise for coupled tanks (front) and spring (back)

of molecule: metabolites and enzymes. We therefore designed two *high-level components* (HLCs), built from standard qualitative relations, to model metabolites and enzymes.

Concentrations of metabolites vary over time, as they are synthesised or utilised in enzymatically catalysed reactions, similar to a tank of liquid increasing and decreasing in level over time. Thus their concentration at time t is a function of their concentration at time $t-1$ and the amount that they are used or created. If a metabolite is catalysed by n enzymes, expressed as flow 'through' each enzyme, $Enzyme_i$, then its change in concentration can be expressed by,

$$\frac{dMetab}{dt} = Metab(t) + \sum_{i=0}^{n} Enzyme_i \,. \tag{8}$$

Enzymes are the other form of high-level component. Each enzyme is assumed to have one or two inputs (substrates) and one or two outputs (products). If there are two inputs or outputs these are considered to form an input or output complex, such that the amount of the complex is *proportional* to the amount of the inputs or outputs multiplied together. This qualitatively models the probability that both the inputs (or outputs) will collide with the enzyme with sufficient timeliness to be catalysed into the output complex (or input complex). The input complex is converted into the output complex which then disassociates into the output metabolites, and vice versa. The qualitative equation for the flow 'through' the enzyme component, given input and output metabolites, is therefore,

$$Enzyme = \underbrace{\mathbf{M}^{+}(\prod_{i=1}^{n} input_i)}_{input\ complex} - \underbrace{\mathbf{M}^{+}(\prod_{j=1}^{m} output_j)}_{output\ complex} . \tag{9}$$

Thus this is an abstraction of standard kinetic equations (Cleland, 1963) and is also an expression of the collision probabilities of the metabolites and enzyme. We assume, for simplicity, that enzymes are taken to exist in constant amounts. Although this is a simplification, it is also almost always assumed in ODE modelling. These HLCs are shown in Fig. 9.

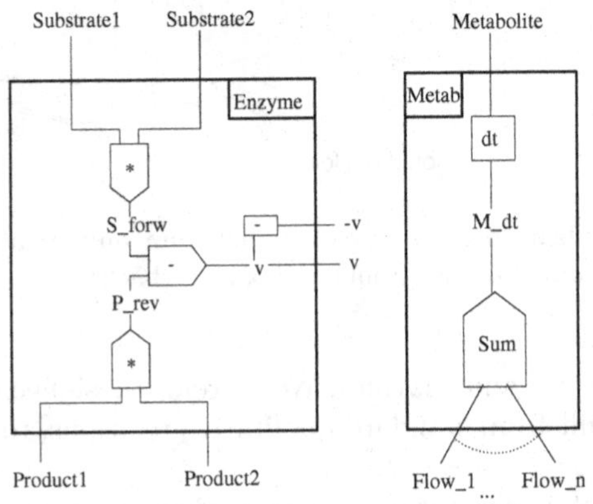

Fig. 9. High-level components (HLCs) for metabolic system modelling as described in the text: 'Enzyme' (left) and 'Metab' (right). V+ and V- are used as appropriate by the metabolite HLC to indicate consumption or production of the metabolite.

5.2 Glycolysis Model Learnt from Qualitative Data

The high-level qualitative components, just described, can now be used as background knowledge for model-learning, B, with qualitative data as evidence, E. The evidence was obtained by simulation of an existing model of part of the 'glycolysis' metabolic pathway[1] that used only the `METAB` (metabolite) and `ENZYME` (enzyme) relations.

The high level component model of part of the glycolysis pathway is shown in Fig. 10, and is introduced here[2]. This is the first report of a qualitative model of a metabolic pathway, and is potentially very useful for biologists since it improves on the predictive ability of their current models of such pathways. A Prolog version of this model, using the HLCs is,

[1] Glycolysis is the process is converting glucose into pyruvate. Here only the first half of the conversion process is used.

[2] A justification for the structure of the model is given in (Garrett et al., 2001).

```
glyc_top:-
    enzyme([Glc,ATP], [F16BP,ADP],HKInvFlow,HKFlow),
    metab(l:F16BP, [f:HKFlow, f:AldInvFlow]), % F16BP DEFINED

    enzyme([ADP,ADP], [ATP,AMP],PFlow, PInvFlow),
    metab(l:ATP, [f:HKInvFlow, f:PFlow]),       % ATP DEFINED
    metab(l:ADP, [f:HKFlow, f:PInvFlow]),       % ADP DEFINED
    metab(l:AMP, [f:HKFlow]),                   % AMP DEFINED

    enzyme([l:F16BP], [l:DHAP,l:G3P], f:AldInvFlow, f:AldFlow),
    enzyme(l:DHAP, l:G3P, f:TrioseInvFlow, f:TrioseFlow),
    metab(l:DHAP, f:DHdt, [f:AldFlow, f:TrioseInvFlow]),
    metab(l:G3P, f:G3dt, [f:AldFlow, f:TrioseFlow]).
```

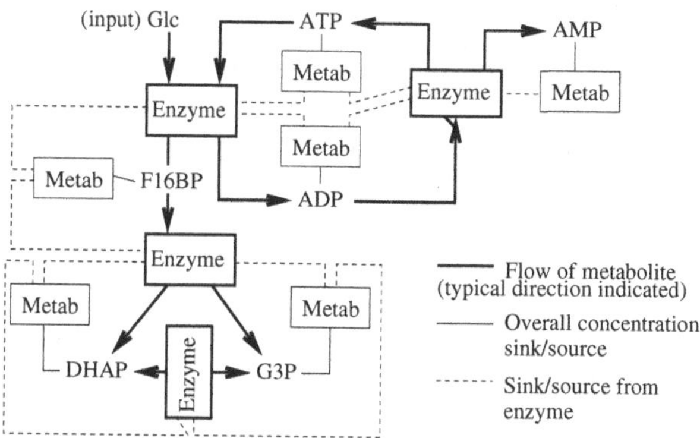

Fig. 10. The glycolysis metabolic pathway, learnt from high-level components

Given this model, its states can be simulated to produce its total envisionment. This resulted in 77 states, for all possible inputs. This envisionment was then used as the data for learning the model, with the HLCs as background knowledge. Using just this information, with no knowledge about chemical interactions between the molecules in glycolysis, it was possible to learn the top part of the model; however, the size of model represented the computational limit of QOPH's abilities at the time of the experiments. Real biological data were not obtained and qualitised due to experimental difficulties, but this does not affect the demonstration here since we have already shown that it is possible to learn from discretised numerical data.

Learning a model of this sort represents a major step forward because the learned system consisted of 36 qualitative relations, and this assumes that `SUM` and `PRODUCT` are to be considered as atomic rather than being implemented as a series of `ADD` or `MULT` relations. It was calculated that introducing HLCs has more than doubled the complexity of the models that can be learned, where complexity is measured by the number of QSIM relations, as well as making the resulting models easier to read.

6 Related Work

Coiera and Hau. One of the earliest reports on inductive learning of qualitative models was Coiera's GENMODEL, the results of which first appeared in 1989 (Coiera, 1989a; Coiera, 1989b). GENMODEL has an effective strategy for using positive-only data (identified by Hau & Coiera (Hau & Coiera, 1993) as form of relative least general generalisation (RLGG) (Plotkin, 1971)), and it can also make use of negative examples if they are available. However, GENMODEL can not introduce or allow intermediate variables, and the results produced are usually *overconstrained*; a feature of many systems for learning qualitative models. This occurs because GENMODEL works by filtering the set of all possible relations and, as implemented, it was not usually possible to remove all superfluous relationships. QOPH takes a constructive approach which avoids this problem.

Hau updated GENMODEL (Hau & Coiera, 1993) and showed that QSIM models are efficiently PAC learnable from the given data. The work was also notable for pointing out that dimensional analysis (Bhaskhar & Nigam, 1990) can be considered as a form of directed negative example generation that uses background knowledge about the structure of the domain. Hau also demonstrated GENMODEL working on qualitative data generated from real-valued experimental data in an impressive manner, but this was limited by the need for all the variables to be known from the outset, which is unrealistic in most domains.

MISC method (Kraan et al., 1991) consists of three stages: quantitative to qualitative data conversion; generating and testing all possible constraints (relations), and building models from the constraints. Although the generation and testing of *all* constraints is computationally intensive, this is mitigated to some extent by performing dimensional analysis on the variables (Bhaskhar & Nigam, 1990). On this basis, certain relations can *a priori* be regarded as impossible, as illustrated in Sections 3.2 and 3.3.

If stage two of the method is not properly defined then stage three has to test all possible subsets of constraints, which can lead to an exponential explosion of possible subsets that all have to be tested. This was only limited by the heuristics that a model must contain at least one derivative relation, is not disjoint and constrains all given variables.

The main limitations of MISC were that the models produced tended to be overconstrained, particularly when less than complete variable information was given, such as the types involved. MISC relied heavily on a consistent set of behaviours and a "complete description" (i.e. no intermediate variables). Ramachandran's follow up to MISC, MISC-RT (Ramachandran et al., 1994), addressed the problem of model induction over multiple operating regions. However, it did not appear to add to the general algorithm.

Richards' work, after MISC, addresses the issue of inducing QSIM models of systems, and was able to find intermediate variables, given positive-only data (i.e. there were no negative qualitative behaviours) (Richards et al., 1992; Richards & Mooney, 1995).

The method described in (Bratko et al., 1992) used Muggleton and Feng's GOLEM (ILP) program (Muggleton & Feng, 1990) along with QSIM to produce a model of the U-tube system (well-known in systems theory, e.g. (Say & Kuru, 1996; Garrett et al., 2001)). This work was important since it was the first to show that intermediate variables (of a sort) could be introduced into the final model. Their system "borrowed" (as they put it) an unused variable to use as an intermediate variable. However, the Bratko et al. (1992) method also has a number of awkward requirements that made its application a rather limited solution:

- it required the use of negative examples because GOLEM could not induce from positive-only evidence. To provide these, six hand-generated negative examples were generated. Hau & Coiera (Hau & Coiera, 1993) point out that badly chosen negative examples cause an inappropriate clause to be formed;
- due to scalability problems (Muggleton, private communication) they were not able to extend their work to more complex systems;
- the model found by their system was shown to be logically equivalent to the standard U-tube model, but it was not *physically* equivalent from a systems theory point of view, and was overconstrained;
- since GOLEM could only represent background knowledge extensionally (i.e. it could not have programs in the background knowledge), Bratko *et al.* had to represent QSIM as a list of thousands of ground Prolog facts.

Say and Kuru. (Say & Kuru, 1996) reviewed the work in the field and presented an interesting approach (QSI). This starts with correlations, and iteratively introduces new variables, building a model and comparing the output of that model with the known states until a satisfactory model is found. Say and Kuru characterised this approach as one of "diminishing oscillation" as it aims to home in on the correct model. QSI works for positive-only data and can introduce intermediate variables (termed "deep variables").

Given the variables (which Say and Kuru, after Kuipers, call parameters) and relations (which they call constraints) QSI searches the induction lattice, or model space. QSI starts with no information about the types or dimensions of the variables, nor with any background information about the nature of those types, such as "amounts cannot be negative." This appears unnecessarily general as in the target application areas such information is usually readily available. Say and Kuru's idea of a model was slightly different from that reported here: they considered a single $\mathtt{M}^+$ relation between two variables to be a *complete* model of a given system (the U-tube) (Say & Kuru, 1996). They had to take this approach because, unlike QOPH, their system cannot initially form models that contain relations relating intermediate variables. Say and Kuru used redundancy checks and hanging variable tests (see Section 3.2), however the final model presented in their paper was overconstrained. Furthermore, it is unclear how QSI can choose between a number of candidate models that satisfy all the filtering criteria, although in their conclusions they mention the need for QSI's output to be pruned of "coincidental and non-interesting constraints."

Džeroski and Todorovski. Džeroski and Todoroski (Džeroski & Todorovski, 1995) describe QMN, which learns QSIM models directly from quantitative data, as well as LAGRANGE, which learns quantitative models (Džeroski, 1992). Both introduce intermediate variables by means of the multiplication operator, whereas LAGRAMGE (Todorovski & Džeroski, 1997), which can also learn quantitative models, uses background functions to introduce variables. Only QMN is fully relevant here, however the means of introducing variables in the other systems is relevant. An older review of qualitative model-learning can be found in Lavrač and Džeroski's book (Lavrač & Džeroski, 1994).

7 Discussion and Conclusion

We have explored the robustness of QOPH under a variety of conditions—clean qualitative data, clean numerical data, noisy qualitative data and noisy numerical data—as well as demonstrating learning using high level components. An examination of these results allows us to draw a number of conclusions here.

7.1 Contributions to Machine Learning of Qualitative Models

The first general point to note is that, for both benchmarks, QOPH had to find intermediate variables, and their relationships to known variables, for the correct model to be learned. This is true of other systems too (Garrett et al., 2001). Furthermore, QOPH is able to introduce intermediate *relations* where required.

Analysis of the clean data experiments showed that it was always possible to find reliably the correct model, given the complete envisionment of the coupled tanks and spring. As expected there was a gradual degradation in reliability as the number of states presented as data was reduced.

The discovery of kernel subsets suggests that it may be possible to minimise the amount of state data required, making model-learning faster. It is also of value theoretically and could lead to more efficient learning algorithms.

The noise experiments showed that it is possible to learn models reliably, despite a degree of noise. This was demonstrated in both in the qualitative and numerical cases. As the noise level increases the search space explodes and the number of models suggested also increases rapidly; however, to some degree QOPH is able to learn from a partial, errant envisionment, and from states converted from numerical data.

Finally, we have shown it is possible to learn qualitative models of biological systems, using high level components. Being able to qualitatively model such a system is itself useful to biologists, since it gives a degree of mathematical formality without requiring a full numerical model. However, being able to routinely learn models from data, in the manner demonstrated by QOPH, would be extremely valuable, particularly in bioinformatics.

7.2 Future Work

Having validated the model induction method on noisy qualitative and numerical data, and demonstrated its ability to learn complex systems, the next step is to explore how successful it can be at modelling real experimental data. We are developing QOPH to learn models from bioinformatic data.

Different search algorithms and heuristics could be explored for further speed benefits. For example, the use of a genetic programming (Koza, 1992) to search the model space may be faster, although there are other implementational difficulties with this approach. If causal ordering is always to be used (i.e. variables are only introduced where they can be fully constrained) then it would be more efficient to make it part of the model construction process. However, since causal ordering is not always possible, our current approach is more flexible.

An important question, raised by Džeroski (personal communication), is whether intermediate, unmeasured variables can be expressed in terms of the measured ones. Introducing variables that can not be expressed in this manner is much more difficult, since it would ultimately also allow unknown state variables to be learned. Nevertheless, given the algorithm above (Section 3.2), it seems likely that QOPH *can* introduce unmeasured, exogenous variables if it is required to complete a model, however this has yet to be tested.

Finally, we intend to apply use of high-level components and QOPH to other biological problems.

Acknowledgements: This work is supported by BBSRC/EPSRC grant BIO10479. The authors would like to thank Stephen Oliver and Douglas Kell for some useful discussions on the biological aspects of this chapter.

References

Bakshi, B.R., Stephanopoulos, G.: Representation of process trends – Part 3: multiscale extraction of trends from process data. Computers and Chemical Engineering 18, 267–302 (1994)

Bhaskhar, R., Nigam, A.: Qualitative physics using dimensional analysis. Artificial Intelligence 45, 73–111 (1990)

Blackman, R.B., Tukey, J.W.: The measurement of power spectra. John Wiley and Sons, New York (1958)

Bratko, I., Muggleton, S., Varšek, A.: Learning qualitative models of dynamic systems. In: Muggleton, S. (ed.) Inductive logic programming, pp. 437–452. Academic Press, San Diego, CA (1992)

Cellier, F.E.: Continuous system modelling. Springer, Berlin (1991)

Cheung, J.T.-Y., Stephanopoulos, G.: Representation of process trends – Part 1: a formal representation framework. Computers and Chemical Engineering 14, 495–510 (1990a)

Cheung, J.T.-Y., Stephanopoulos, G.: Representation of process trends – Part 2: the problem of scale and qualitative scaling. Computers and Chemical Engineering 14, 511–539 (1990b)

Cleland, W.W.: The kinetics of enzyme-catalysed reactions with two or more substrates and products: 1. nomenclature and rate equations. Biochimica et Biophysica Acta 67, 104–137 (1963)

Coiera, E.W.: Generating qualitative models from example behaviours (Technical Report 8901). University of New South Wales, Deptartment of Computer Science (1989a)

Coiera, E.W.: Learning qualitative models from example behaviours. In: Proceedings of the Third Workshop on Qualitative Physics, Stanford, CA, pp. 45–51 (1989)

DeCoste, D.: Dynamic across-time measurement interpretation. Artificial Intelligence 51, 273–341 (1991)

Džeroski, S.: Learning qualitative models with inductive logic programming. Informatica 16, 30–41 (1992)

Džeroski, S., Todorovski, L.: Discovering dynamics: from inductive logic programming to machine discovery. Journal of Intelligenty Information Systems 4, 89–108 (1995)

Flach, P.A., Kakas, A.C.: Abduction and induction: Essays on their relation and integration. Kluwer Academic Publishers, Amsterdam, The Netherlands (2000)

Forbus, K.D.: Qualitative process theory. Artificial Intelligence 24, 169–204 (1984)

Fürnkranz, J.: Separate-and-conquor rule learning. Artificial Intelligence Review 13, 3–54 (1999)

Garrett, S.M., Coghill, G.M., King, R.D., Srinivasan, A.: On learning qualitative models of qualitative and real-valued data (Technical Report UWA-DCS-01-037). University of Wales, Aberystwyth (2001)

Gawthrop, P.J., Smith, L.P.S.: Metamodelling: Bond graphs and dynamic systems. Prentice Hall, Hemel Hempstead, Herts, England (1996)

Gold, E.M.: Language identification in the limit. Information and Control 10, 447–474 (1967)

Hau, D.T., Coiera, E.W.: Learning qualitative models of dynamic systems. Machine Learning 26, 177–211 (1993)

Iwasaki, Y., Simon, H.A.: Causality in device behavior. Artificial Intelligence (See also de Kleer and Brown's rebuttal and Iwasaki and Simon's reply to their rebuttal in the same volume of this journal) 29, 3–32 (1986)

Kakas, A.C., Kowalski, R.A., Toni, F.: Abductive logic programming. Journal of Logic and Computation 2, 719–770 (1992)

King, R.D., Muggleton, S., Lewis, R.A., Sternberg, M.J.E.: Drug design by machine learning — the use of inductive logic programming to model the structure-activity-relationships of trimethoprim analogs binding to dihydrofolate-reductase. In: Proceedings of the National Academy of Sciences of the USA, vol. 89, pp. 11322–11326 (1992)

King, R.D., Srinivasan, A.: The discovery of indicator variables for QSAR using inductive logic programming. Journal of Computer-Aided Molecular Design 11, 571–580 (1997)

Koza, J.R.: Genetic programming. MIT Press, Cambridge, MA (1992)

Kraan, I.C., Richards, B.L., Kuipers, B.J.: Automatic abduction of qualitative models. In: Proceedings of the Fifth International Workshop on Qualitative Reasoning about Physical Systems, Austin, TX, pp. 295–301 (1991)

Kuipers, B.: Qualitative simulation. Artificial Intelligence 29, 289–338 (1986)

Kuipers, B.: Qualitative reasoning. MIT Press, Cambridge, MA (1994)

Laird, J.E., Rosenbloom, P.S., Newell, A.: Chunking in SOAR: The anatomy of a general learning mechanism. Machine Learning 1, 11–46 (1986)

Lavrač, N., Džeroski, S.: Inductive logic programming: Techniques and applications. Ellis-Horwood, New York (1994)

Ljung, L.: System identification: Theory for the user, 2nd edn. PRT Prentice Hall, Upper Saddle River, NJ (1999)

Mitchell, T.M.: Version spaces: An approach to concept learning. Doctoral dissertation, Stanford University, CA (1979)

Muggleton, S.: Inductive logic programming. Academic Press, London (1992)

Muggleton, S.: Inverse entailment and PROGOL. New Generation Computing 13, 245–286 (1995)

Muggleton, S.: Learning from positive data. In: Proceedings of the Sixth International Workshop on Inductive Logic Programming, Stockholm, Sweden, pp. 358–376 (1996)

Muggleton, S., Feng, C.: Efficient induction of logic programs. In: Proceedings of the First Conference on Algorithmic Learning Theory, Tokyo, Japan, pp. 368–381 (1990)

Plotkin, G.: Automatic methods of inductive inference. Doctoral dissertation, Edinburgh University (1971)

Ramachandran, S., Mooney, R.J., Kuipers, B.J.: Learning qualitative models for systems with multiple operating regions. In: Working Papers of the Eighth International Workshop on Qualitative Reasoning about Physical Systems, Nara, Japan, pp. 212–223 (1994)

Richards, B.L., Kraan, I., Kuipers, B.J.: Automatic abduction of qualitative models. In: Proceedings of the Tenth National Conference on Artificial Intelligence, San Jose, CA, pp. 723–728 (1992)

Richards, B.L., Mooney, R.J.: Automated refinement of first-order horn-clause domain theories. Machine Learning 19, 95–131 (1995)

Say, A.C.C., Kuru, S.: Qualitative system indentification: deriving structure from behavior. Artificial Intelligence 83, 75–141 (1996)

Shoup, T.E.: A practical guide to computer methods for engineers. Prentice-Hall, Englewood Cliffs, NJ (1979)

Srinivasan, A.: Aleph web site (2000), `http://web.comlab.ox.ac.uk/oucl/research/areas/mach-learn/Aleph/aleph_toc.html`

Srinivasan, A., King, R.D.: Feature construction with inductive logic programming: A study of quantitative predictions of biological activity aided by structural attributes. Data Mining and Knowledge Discovery 3, 37–57 (1999)

Srinivasan, A., Muggleton, S.H., Sternberg, M.J.E., King, R.D.: Theories for mutagenicity: A study in first-order and feature-based induction. Artificial Intelligence 85, 277–299 (1996)

Sternberg, M.J.E., King, R.D., Lewis, R.A., Muggleton, S.: Application of machine learning to structural molecular biology. Philosophical Transactions of the Royal Society of London Series B - Biological Sciences 344, 365–371 (1994)

Todorovski, L., Džeroski, S.: Declarative bias in equation discovery. In: Proceedings of the Fourteenth International Conference on Machine Learning, Nashville, TN, pp. 376–384 (1997)

Yamamoto, A.: Revising the logical foundations of inductive logic programming systems with ground reduced programs. New Generation Computing 17, 119–127 (1999)

A Complete Envisionments Used

```
COUPLED TANKS Input = 0/std
 StateNo  State Description
 -----------------------------------------------------
 0  state(l:0/std,l:0/std,f:0/std,f:0/std,f:0/std).
 1  state(l:0/inc,l:0...inf/dec,f:0/std,f:minf...0/inc,f:0...inf/dec).
 2  state(l:0...inf/dec,l:0/inc,f:0/std,f:0...inf/dec,f:0/inc).
 3  state(l:0...inf/dec,l:0...inf/dec,f:0/std,f:0...inf/dec,f:0...inf/dec).
 4  state(l:0...inf/dec,l:0...inf/dec,f:0/std,f:0...inf/std,f:0...inf/dec).
 5  state(l:0...inf/dec,l:0...inf/dec,f:0/std,f:0...inf/inc,f:0...inf/dec).
 6  state(l:0...inf/dec,l:0...inf/std,f:0/std,f:0...inf/dec,f:0...inf/std).
 7  state(l:0...inf/dec,l:0...inf/inc,f:0/std,f:0...inf/dec,f:0...inf/inc).
 8  state(l:0...inf/std,l:0...inf/dec,f:0/std,f:0/inc,f:0...inf/dec).
 9  state(l:0...inf/inc,l:0...inf/dec,f:0/std,f:minf...0/inc,f:0...inf/dec).
```

```
MASS-SPRING DAMPER Input = 0...inf/std
 StateNo  State Description
 -----------------------------------------------------
 0  state(f:0...inf/std,d:minf...0/std,v:0/inc,f:0...inf/dec).
 1  state(f:0...inf/std,d:0/std,v:0/inc,f:0...inf/dec).
 2  state(f:0...inf/std,d:0...inf/std,v:0/std,f:0/std).
 3  state(f:0...inf/std,d:0...inf/std,v:0/dec,f:minf...0/inc).
 4  state(f:0...inf/std,d:0...inf/std,v:0/inc,f:0...inf/dec).
 5  state(f:0...inf/std,d:minf...0/dec,v:minf...0/inc,f:0...inf/std).
 6  state(f:0...inf/std,d:minf...0/dec,v:minf...0/inc,f:0...inf/dec).
 7  state(f:0...inf/std,d:minf...0/dec,v:minf...0/inc,f:0...inf/inc).
 8  state(f:0...inf/std,d:0/dec,v:minf...0/inc,f:0...inf/std).
 9  state(f:0...inf/std,d:0/dec,v:minf...0/inc,f:0...inf/dec).
 10  state(f:0...inf/std,d:0/dec,v:minf...0/inc,f:0...inf/inc).
 11  state(f:0...inf/std,d:0...inf/dec,v:minf...0/std,f:0/inc).
 12  state(f:0...inf/std,d:0...inf/dec,v:minf...0/dec,f:minf...0/inc).
 13  state(f:0...inf/std,d:0...inf/dec,v:minf...0/inc,f:0...inf/std).
 14  state(f:0...inf/std,d:0...inf/dec,v:minf...0/inc,f:0...inf/dec).
 15  state(f:0...inf/std,d:0...inf/dec,v:minf...0/inc,f:0...inf/inc).
 16  state(f:0...inf/std,d:minf...0/inc,v:0...inf/std,f:0/dec).
 17  state(f:0...inf/std,d:minf...0/inc,v:0...inf/dec,f:minf...0/std).
 18  state(f:0...inf/std,d:minf...0/inc,v:0...inf/dec,f:minf...0/dec).
 19  state(f:0...inf/std,d:minf...0/inc,v:0...inf/dec,f:minf...0/inc).
 20  state(f:0...inf/std,d:minf...0/inc,v:0...inf/inc,f:0...inf/dec).
 21  state(f:0...inf/std,d:0/inc,v:0...inf/std,f:0/dec).
 22  state(f:0...inf/std,d:0/inc,v:0...inf/dec,f:minf...0/std).
 23  state(f:0...inf/std,d:0/inc,v:0...inf/dec,f:minf...0/dec).
 24  state(f:0...inf/std,d:0/inc,v:0...inf/dec,f:minf...0/inc).
 25  state(f:0...inf/std,d:0/inc,v:0...inf/inc,f:0...inf/dec).
 26  state(f:0...inf/std,d:0...inf/inc,v:0...inf/std,f:0/dec).
 27  state(f:0...inf/std,d:0...inf/inc,v:0...inf/dec,f:minf...0/std).
 28  state(f:0...inf/std,d:0...inf/inc,v:0...inf/dec,f:minf...0/dec).
 29  state(f:0...inf/std,d:0...inf/inc,v:0...inf/dec,f:minf...0/inc).
 30  state(f:0...inf/std,d:0...inf/inc,v:0...inf/inc,f:0...inf/dec).
```

Logic and the Automatic Acquisition of Scientific Knowledge: An Application to Functional Genomics

Ross D. King[1], Andreas Karwath[2], Amanda Clare[1], and Luc Dehaspe[3]

[1] Department of Computer Science,
University of Wales, Aberystwyth, U.K.
[2] Albert-Ludwigs Universität, Institut für Informatik, Georges-Köhler-Allee 079,
D-79110 Freiburg, Germany
[3] PharmaDM,
Heverlee, Belgium

"If we trace out what we behold and experience through the language of logic, we are doing science" "The grand aim of all science [is] to cover the greatest number of empirical facts by logical deduction from the smallest number of hypotheses or axioms"
A. Einstein

Abstract. This paper is a manifesto aimed at computer scientists interested in developing and applying scientific discovery methods. It argues that: science is experiencing an unprecedented "explosion" in the amount of available data; traditional data analysis methods cannot deal with this increased quantity of data; there is an urgent need to automate the process of refining scientific data into scientific knowledge; inductive logic programming (ILP) is a data analysis framework well suited for this task; and exciting new scientific discoveries can be achieved using ILP scientific discovery methods. We describe an example of using ILP to analyse a large and complex bioinformatic database that has produced unexpected and interesting scientific results in functional genomics. We then point a possible way forward to integrating machine learning with scientific databases to form intelligent databases.

1 Introduction

1.1 The Need for Automation of Reasoning in Science

Perhaps the most characteristic feature of science at the start of the third millennium is the "data explosion" (Reichart, 1999). From particle-physics to ecology, from neurology to astronomy, almost all the experimental sciences are experiencing an unprecedented increase in the amount and complexity of available data. Scientific databases with Terabytes (10^{12} bytes) of data are now common place, and soon Petabyte (10^{15} bytes) database will be so. In some fields plans are already underway to deal with Exabytes (10^{18} bytes) and Yottabytes (10^{21} bytes) of data. To take a number of high profile cases:

S. Džeroski and L. Todorovski (Eds.): Computational Discovery, LNAI 4660, pp. 273–289, 2007.

- In physics - the new CERN Large Hadron Collider (LHC) is expected to generate ~100 Petabytes of data over its 15 year lifetime.
- In geology - the US Geological survey is photographing the entire US from the air at high resolution. This will form a ~12 Terabyte database.
- In biology - transcriptome (microarray) data from gene knockouts of the human genome is expected to generate ~1 Petabyte of raw data - this data is also extremely complicated, as it has semantic links to many other biological databases.

These databases are made possible by increasingly sophisticated instrumentation, and by ever more powerful information technology. They represent a great achievement for science, and within them lies a wealth of new scientific knowledge. However, full analysis of such databases presents enormous difficulties, and it would seem probable that traditional approaches to scientific data analysis will not be sufficient. The reasons for this are:

- Traditional approaches did not emphasise the efficiency of access to the data, and many analysis methods have implicitly assumed that all the data can be held in main memory.
- Lack of data has been traditionally considered the main problem, not overabundance, hence the emphasis on small sample statistics.
- Traditional methods have focused on answering single questions, they have been "hypothesis driven". The formation and analysis of large databases argues for a more "data driven" analysis of data - pioneering work in this area such as Tukey (1977) was given the then derogatory name "data mining".
- Human decision making has been tightly coupled to the analysis process.
- Humans have been tightly coupled to the experiment planning process, this is becoming impossible in some areas, e.g. the control of molecules by lasers where a million separately designed experiments can be done in a second (Rabitz, *et al.* 2000).

Therefore, new more automated data analysis methods are essential to refine the vast quantities of scientific data into communicable scientific knowledge.

1.2 Scientific Discovery

The branch of Artificial Intelligence devoted to developing algorithms for acquiring scientific knowledge is known as "scientific discovery". Work has proceeded in this area for over thirty years and much has been achieved (e.g. Buchanan *et al.*, 1969; Langley *et al.*, 1987; Sleeman *et al.*, 1989; Gordon *et al.*, 1995; King *et al.*, 1996; Valdes-Perez, 1999). However, the field has not made a significant impact on science generally, and no significant scientific discovery is directly attributable to an artificial intelligence program. By this we mean that no scientific discovery program has made a contribution to science that would be worthy of a major scientific prize. In fact, to the best of our knowledge, no scientific discovery program has made a contribution worthy of co-authorship in one of the main three general journals (Nature, Science, and Proceedings of the National Academy of Science, USA - PNAS). Work on applying Progol (Muggleton, 1995) to drug design has been published in PNAS

(King *et al.*, 1992; *King et al.*, 1996), but this was mostly on the basis of the methodology, not the discoveries on their own.

Why then do scientific discovery programs not rival scientists in the way that the chess playing programs rival chess Grand masters? We believe the answer is that chess is an isolated world where, despite the almost limitless complications possible, everything that is relevant to the game is described in the rules. In contrast, scientific problems cannot be so easily isolated from general reasoning. To be successful scientific discovery programs do not need to know about everything - "Cabbages and Kings" (Boden, 1977); but they do need access to large amounts of background knowledge about the scientific field they are applied to. Humans still have the decisive edge over scientific discovery programs because scientific knowledge is difficult to encode, and we do not have a good theory of relevance (although some interesting theoretical work has been done on "relevant logics", e.g. *Plato.* As the focus of science moves towards problems involving very large and complex datasets, the balance of advantage must shift towards machine intelligence.

1.3 Inductive Logic Programming / Relational Data Mining

The issue of how best to represent knowledge for machine inference has been a hotly disputed one in AI. We believe that logic should be the basis for representing knowledge in scientific discovery programs as it has the best understood semantics and inference mechanisms (Russell & Norvig, 1995).

The use of propositional logic based methods currently dominate the field of data analysis (Mitchell, 1997). There are a number of reasons for this:

- The idea of representing data as a table with the examples as rows, and the attributes as columns is intuitively easily to understand - so much so that it can be hard to convince some domain scientists that there is an alternative.
- Propositional methods are highly effective when they have access to good attributes.
- Propositional methods can be highly efficient.

Despite these advantages we consider propositional logic to be too weak to be generally used for scientific discovery problems: it cannot express much of the background knowledge we require about scientific problems, nor can it represent scientific theories in the concise way scientists expect. If first-order predicate logic (FOPL) is used these disadvantages are alleviated: while the advantages of logic of a well-understood semantics and inference mechanism are retained. FOPL allows the representation of data and hypotheses in a perhaps even more intuitive way than propositional logic, and it has been successfully applied to scientific discovery problems (see below).

The main disadvantage of FOPL is that inferences are less efficient than with propositional logic; e.g. to learn (inductively infer) a FOPL hypothesis normally involves a search through a larger search space than for a propositional hypothesis. For datasets with ~$10^{3\text{-}6}$ examples this greater computation cost of using FOPL is not important, but to deal with the very large datasets of the 21st century, computational cost is important. Despite this, we consider the great advantage FOPL has over propositional logic in representing knowledge to outweigh the greater computational cost in learning. Greater efficiency in FOPL can be obtained by parallelisation (e.g. Fujita, *et al.* 1996) and/or sampling (Srinivasan, 2001).

Higher-order, temporal logics, etc. have more powerful representative abilities than FOPL; they could for example be used to recognise scientific analogies. The theory and practise of learning using higher order logics has generally lagged behind that of FOPL - however, for some interesting current work in this area e.g. see Flach *et al.*, 1998, and Bowers *et al.*, 2000. For learning, higher-order logics are currently even less efficient than FOPL. In our opinion, this inefficiency makes such methods, especially when large databases are involved, currently too computationally expensive to be used. To conclude, we consider FOPL methods to be currently the most appropriate as they strike the best balance between computational tractability and expressive power.

Machine learning and data mining methods that employ FOPL to represent background knowledge and theories come from the field of Inductive Logic Programming (ILP) (Muggleton, 1990; Muggleton, 1992; Lavrac & Dzeroski, 1994) or Relational Data Mining (RDB) (Dzeroski & Lavrac, 2001). The different terms reflect contrasting original emphases. In ILP there has been a strong interest in inducing computer programs, whereas in RDB the emphasis has been more on analysing databases with multiple tables. For convenience in this paper will refer to them both as ILP. In ILP, background knowledge (B), examples (E), and hypotheses (H) are represented as logic programs. It is first assumed that there is a requirement for an inductive hypothesis, i.e. $B \nvDash E$ (prior necessity). The core ILP problem is to find hypotheses H such that $B \wedge H \vDash E$. Matters are complicated slightly by the fact that evidence in ILP is usually divided into two types: E^+ data that is consistent with the conjoined hypothesis and background information, and E^- data that is not consistent with them. Therefore H is required to meet the criterion: $B \wedge H \vDash E^+$, and $B \wedge H \nvDash E^-$. It is usual that H is also restricted by necessarily meeting other requirements such as: being non-trivial (not e.g. E, or $B \rightarrow E$), parsimonious, etc.

In practical applications there is the additional problem that there is usually noise present (i.e. some of the logical statements are inconsistent with reality), this means that the above conditions need to be relaxed such that the best hypothesis maximises its coverage of E^+ and minimises its coverage of E^-. These two values give a cost ratio that represents the divergence from the ideal. The problem of learning H is normally designed as a search problem through the space of models meeting the above criteria (Mitchell, 1982).

ILP has shown its value in many scientific problems: in chemistry (e.g. King, *et al.*, 1992; King, *et al.*, 1996; Dehaspe, *et al.*, 1998; Kramer *et al.*, 1998; Finn *et al.*, 1998; Srinivasan & King, 1999; Dzeroski *et al.*, 1999) and molecular biology (Muggleton, *et al.*, 1992; King, *et al.*, 1994; Sternberg *et al.*, 1994; Turcotte *et al.*, 2001) where it has found solutions not accessible to standard statistical, neural network, or genetic algorithms. It is not well understood how to objectively measure whether one form of theory is more comprehensible than another, and the number of symbols used is often used as a proxy for this. In terms of comprehensibility the theories produced by ILP appear competitive with those generated by other methods. "Sternlish" an automatic translation of induced Prolog rules describing proteins for the protein structure expert Mike Sternberg is a notable example of this.

1.4 Summary of the Case for the Increasing Use of ILP in Scientific Discovery

In science there is an explosion in the amount of available data, with terabytes of data already common, and petabytes expected soon. These vast new databases cannot be analysed using traditional approaches, and novel, more automated data analysis methods are essential to refine the vast quantities of scientific data into communicable scientific knowledge. The branch of Artificial Intelligence devoted to developing algorithms for acquiring scientific knowledge is known as "scientific discovery". We believe that logic should be the basis for representing knowledge in scientific discovery programs, and that traditional propositional logic based scientific discovery methods are too weak for this task. Therefore, first-order predicate logic (FOPL) based methods are more suitable. Machine learning and data mining methods that employ FOPL or more expressive logics are known as Inductive Logic Programming (ILP) or Relational Data Mining (RDB).

2 An Example Application of ILP Scientific Discovery to Functional Genomics

In this section we describe an example application of ILP to an important problem in functional genomics which illustrates how ILP in combination with propositional learning methods can be used to glean scientific knowledge from a large database. Full details of work on this application can be found in: King *et al.*, 2000a; King *et al.*, 2000b; King *et al.*, 2001.

2.1 Scientific Background

Molecular Biology is currently experiencing a data explosion. The genomes of over 60 micro-organisms have now been completely sequenced (e.g. Blattner, *et al.*, 1997; Cole, *et al.*, 1998; Goffeau, *et al.*, 1996), as have those of the multicellular organisms *Caenorhabditis elegans* (C. elegans Sequencing Consortium, 1998), *Drosophilia melanogaster* (Adams *et al.*, (2000), Arabidopsis (Arabidopsis genome initiative, 2000). The first draft of the sequencing of the ~$3x10^9$ bases of the human genome is now also complete (HGP) (International Human Genome Sequencing Consortium, 2001; Venter *et al.*, 2001).

The data from these sequencing projects is revolutionising biology. Perhaps the most important discovery from the sequenced genomes is that the functions of only 40-60% of the predicted genes are known with any confidence. The new science of *functional genomics* (Hieter & Boguski, 1997; Bussey, 1997; Bork *et al.*, 1998; Brent, 1999; Kell & King, 2000) is dedicated to determining the function of the genes of unassigned function, and to further detailing the function of genes with purported function.

A pressing need in functional genomics is better ways of accurately predicting a protein's function from its sequence. Currently this is most usually done by using sequence similarity methods to find a similar (homologous) protein in the database that has a labelled function (Pearson & Lipman, 1988; Altschul, *et al.,* 1997). The function of the new sequence is then inferred to be the same as the homologous protein as it has been conserved over evolution. This is a kind of nearest-neighbour type inference in sequence space (Mitchell, 1997).

2.2 Methodology

We have developed a complementary approach to sequence similarity methods for predicting a protein's function from its sequence. This approach is based on learning symbolic rules to predict a protein's functional class based on using a set of proteins of known function. To test this approach we selected the *Mycobacterium tuberculosis* and *Escherichia coli* genomes.

The *M. tuberculosis* bacterium is the causative agent of tuberculosis, which kills around two million people each year. Concern about the growing tuberculosis epidemic has led the World Health Organisation to declare a global emergency. The genome of *M. tuberculosis* has been sequenced and 3,924 proteins identified (ORFs) (Cole, *et al.,* 1998). A database of these proteins and their identified functions can be found at *TB*.

E. coli has probably the best characterised genome, and is the "model" bacteria. It has an estimated 4,289 identified proteins (Blattner *et al.*, 1997). These proteins have also, where possible, been assigned function, see *GenProtEC*. This database uses a different functional schema than that for *M. tuberculosis.*

For both organisms the assignments of function were organised in a strict hierarchy, where each higher level in the tree is more general than the level below it, and the leaf nodes are the individual functions of proteins - which is typical of current genome annotation. A subsection of the function hierarchy for *M. tuberculosis* is shown in Figure 1. A typical protein in the genome is L-fuculose phosphate aldolase (Rv0727c fucA). Its top-level class assignment is "Small-molecule metabolism", its second-level class is "Degradation", and its third-level class is "Carbon compounds". Note that there are errors in the annotation of function (Brenner, 1999), and proteins may have more than one function from which one is arbitrarily chosen, which adds "noise" to the assignments. The organisation of functions into classes allows generalisation of the sequences over these classes using machine learning.

To facilitate induction of a mapping between protein sequence and function we first formed a Datalog database (Ullman, 1988) containing all the data we could find on the protein sequences. The most commonly used technique in bioinformatics to gain information about a sequence is to run a sequence similarity search, and this was used as the starting point in forming descriptions. The basic data structure in the Datalog database is the result of a PSI-BLAST sequence similarity search (Altschul, *et al.,* 1997). For each protein in the genome we formed an expressive description based on:

the frequency of individual and pairs of residues in the protein; the phylogeny ("family tree") of the organism from which each homologous protein was obtained (SWISS-PROT (Bairoch A. & Apweiler (2000) a standard protein database was used); SWISS-PROT protein keywords from homologous proteins; the length and molecular weight of the protein; and its predicted secondary structure (Ouali & King, 2000). The details of these predicates, and examples are given in King *et al.*, 2000b and King *et al.*, 2001. In total 5,895,649 Datalog facts were generated for *M. tuberculosis* and 10,097,865 for *E. coli*. Such databases are clearly too large to analyse manually.

This data requires ILP for analysis for the following reasons:

- The use of homology search produces a 1:m relationship between yeast proteins and their homologs, and each homolog has in turn many attributes. As there are ~100,000 proteins which could be homologous, and each homolog could have ~10,000 attributes, a propositional representation of the problem would require $\sim 10^9$ attributes.
- The representation of protein secondary structure is inherently relational, as it involves the sequential relationship of different types of secondary structure.
- Some background knowledge is relational, such as the phylogenic tree extracted from SWISS-PROT.
- Some background knowledge, such as keywords, could only be very inefficiently represented using propositions.

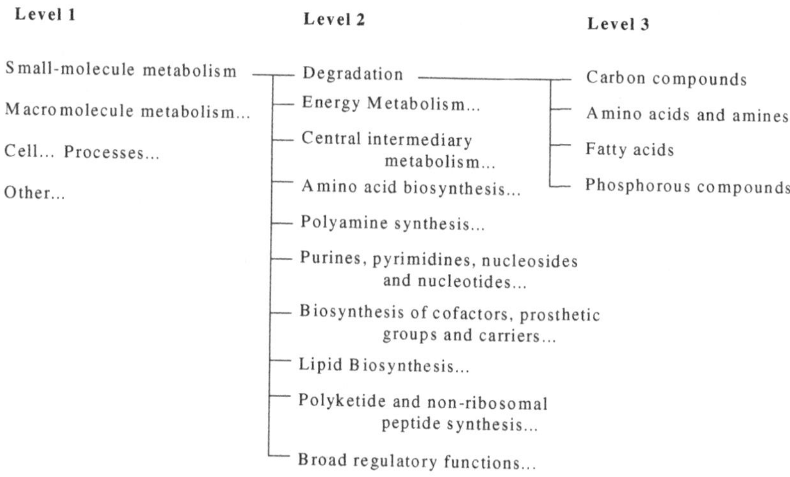

Fig. 1. An example subset of the gene functional hierarchy in M. tuberculosis. The gene L-fuculose phosphate aldolase is in the Level 3 class "carbon compounds". This example illustrates only three out of the four possible classification levels.

To mine our databases we used a hybrid combination of ILP and propositional tree learning to analyse the data (Figure 2). The ILP data mining program Warmr

(Dehaspe, *et al.*, 1998) was first used to identify frequent patterns (conjunctive queries) in the databases. Warmr is a general purpose data mining algorithm that can discover knowledge in structured data. It can learn patterns reflecting one-to-many and many-to-many relationships over several tables. No standard data mining program can do this, as they are restricted to simple associations in single tables. Warmr uses the efficient levelwise method known from the Apriori algorithm (Fayyad, *et al.*, 1996). This allows it to be used on very large databases. The Warmr levelwise search algorithm (Mannila & Toivonen, 1997) is based on a breadth-first search of the pattern space. This space is ordered by the generality of patterns. The levelwise method searches this space one level at a time, starting from the most general patterns. The method iterates between candidate generation and candidate evaluation phases: in *candidate generation*, the lattice structure is used for pruning non-frequent patterns from the next level; in the *candidate evaluation* phase, frequencies of candidates are computed with respect to the database. Pruning is based on the anti-monotonicity of specificity with respect to frequency – if a pattern is not frequent then none of its specialisations can be frequent. The application of Warmr can be considered as a way of identifying the most important structure in a database. For example, in the *M. tuberculosis* database Warmr discovered ~18,000 frequent queries, and these frequent patterns were converted into 18,000 Boolean (indicator) attributes for propositional rule learning, where an attribute gets value 1 for a specific gene if the corresponding query succeeds for that gene, and 0 otherwise. In this way we essentially propositionalised the data, see e.g. Kramer *et al.* (2001). A typical example frequent query is "two β-strands are predicted at i, i+2, the first >3, and the second >1" see Figure 3.

The propositional machine learning algorithms C4.5 and C5 (Quinlan, 1993) were used to induce rules that predict function from these attributes. Good rules were selected on a validation set, and the unbiased accuracy of these rules estimated on a test set. Rules were selected to balance accuracy with unidentified gene coverage. The prediction rules were then applied to genes that have not been assigned a function to predict their functions. Note we did not aim for a general model of the relationship between sequence and function, and we were satisfied with finding good rules to cover part of the space.

In the learning methodology (Figure 2) we used 2/3 of the data to learn frequent patterns. The remaining third was set aside as a final test set. The machine learning algorithm C4.5 and C5 was used to learn rules that predict function from the descriptional attributes using 4/9 of the data (2/3 of 2/3). Good rules were selected on the validation data - the remaining 2/9 of the dataset. The unbiased accuracy of these rules estimated on the 1/3 test set. The selection criteria for good rules were that on the validation data they covered at least two correct examples, had an accuracy of at least 50%, and an estimated deviation of ≥ 1.64. These parameters were based on investigation of the actual biological predictions in the validation data, but are admittedly somewhat arbitrary. This process was only carried out once because of computational resource limitations.

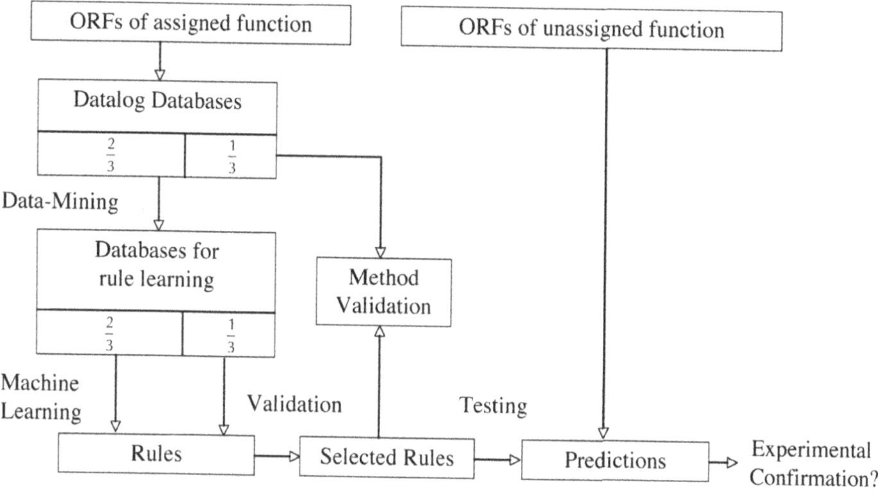

Fig. 2. Flow chart of the experimental methodology

2.3 Results

It was possible to find good rules that predict function from sequence at all levels of the functional hierarchies in both organisms, as shown in Table 1. The number of rules found are those selected on the validation set. A key biological result of the work is the ability of predict functional class even in the absence of homology, as traditional sequence similarity based methods for predicting function are based on homology. A rule predicts more than one homology class, if there is more than one sequence similarity cluster in the correct test predictions. A rule predicts a new homology class, if there is a sequence similarity cluster in the test predictions that has no members in the training data. Average test accuracy is the accuracy of the predictions on the test proteins of assigned function (if conflicts occurred, the prediction with the highest *a priori* probability was chosen). Default test accuracy is the accuracy that could be achieved by always selecting the most populous class. "New functions assigned" is the number of proteins of unassigned function predicted. It would have been better to use cross-validation or some similar resampling method (Mitchell, 1997) to estimate the variance in these values, however this would have been computationally infeasible because of the large size of the databases. The test accuracy estimates may be too pessimistic, as proteins may have more than one functional class but only one of these is considered correct (see examples in the text). However, it is also possible to argue that estimates are too optimistic, as the proteins of unassigned function come from a different distribution from that used to train the rules. Only by empirically testing the prediction rules can the true accuracy of the rules be determined.

Table 1. A summary of the results of learning to predict prtein functional class from sequence in *M. tuberculosis* and *E. coli*

	M. tuberculosis				E. coli		
	Level 1	Level 2	Level 3	Level 4	Level 1	Level 2	Level 3
Number of rules found	25	30	20	3	13	13	13
Rules predicting more than one homology class	19	18	8	1	9	10	3
Rules predicting a new homology class	14	15	1	0	9	5	3
Average test accuracy	62%	65%	62%	76%	75%	69%	61%
Default test accuracy	48%	14%	6%	2%	40%	21%	6%
New functions assigned	886 (58%)	507 (33%)	60 (4%)	19 (1%)	353 (16%)	267 (12%)	135 (6%)

If The amino-acid dipeptide trp-pro does **not** occur in the ORF **and**
Two β-strands are predicted at i, i+2, the first >3, and the second >1 **and**
A homologous protein was found in a kinetoplastida sp. with the keyword "transmembrane" **and**
A homologous protein was **not** found in an Epsilon subdivision sp. with the keyword "inner_membrane"
Then its functional class is "Energy Metabolism carbon"

Fig. 3. An example learnt rule. Note its relational nature. This rule had an accuracy of 50% on the test set (6/12), the default accuracy was (9.8%). The probability of this happening by chance is estimated at ~$5x10^{-4}$). The rule covers a set of phosphatases and oxidoreductive enzymes. These appear not all to be homologous. The rule requires a transmembrane homologue in the kinetoplastida (e.g. trypanosoma) but not in the bacterial epsilon subdivision (e.g. helicobacter). The fact that the epsilon subdivision bacteria are generally microaerophilic may be significant. The requirement for no trp-pro dipeptides may be connected with a transmembrane structural restriction (both have five-membered rings).

The test accuracy of the induced rules is far higher than possible by chance. Of the genes originally of unassigned function class, the rules predicted ~65% in both organisms to have a function at one or more levels of the hierarchy with good accuracy.

The rule learning data, the rules, and the predictions, are given at *FUNCTION*. We illustrate the value of the rules by describing the *E. coli* rule seq+sim+str-20 shown in Figure 3.

Perhaps the most important scientific result of this work was the unexpected discovery that it was possible to predict a protein's function in the absence of homology to a protein of known function. To demonstrate this we carried out all-against-all sequence similarity searches using PSI-BLAST for those proteins correctly predicted by each rule. If all the proteins could be linked together by liberal sequence similarity scores (PSI-BLAST e-values < 10) then the proteins were considered homologous. It was found that many of the predictive rules were more general than possible using sequence homology. Rules were found correctly to predict the function of sets of proteins that are not homologous to each other in the test set, and the function of proteins that are not homologous to any in the training data (Table 1). We speculate that such rules are caused by convergent evolution causing forcing proteins with similar function to resemble each other, or horizontal evolution has transferred functionally related groups of protein into the organisms.

2.4 Summary of Bioinformatic Data Mining

A key problem in current biology is to understand the relationship between protein sequence and function. We have applied a hybrid ILP data mining approach to this problem. The idea was to collect as much information as possible that could be computed purely from protein sequences in a genome, and then to use this data to empirically learn mappings between sequence and known function. Because of the relational nature of the data, to form this mapping it was necessary to use a combination of ILP/Relational Data Mining and propositional learning techniques. We applied our approach to the *M. tuberculosis*, and *E. coli* genomes with good success. For both genomes it was possible to learn accurate rules at all levels of the functional hierarchy. These rules predict the functional class of a large percentage of the proteins of currently unknown function, and provide insight into the relative roles of homology, convergent evolution, and gene swapping in the evolution of the complement of protein functions in cells.

3 Intelligent Databases for Functional Genomics

We are working on extending our work on *M. tuberculosis* and *E. coli* by developing an intelligent database (ID) for its functional genomics data. The ID will be designed to add value to the underlying data. The rough design of the ID is given in Figure 4. The database will:

- Store the large amount of new data being generated on the genomes, e.g. from transcriptome (DeRisi *et al.*, 1997; Brown & Botstein, 1999; Alizadeh *et al.*, 2000), proteome (Humphery-Smith *et al.*, 1997; Blackstock & Weir, 199), and metablome (Oliver & Baganz, 1998; Gilbert *et al.*, 1999) experiments.

- Enable standard database queries of the data.
- Incorporate background molecular biology knowledge.
- Enable deductive inferences involving the data and the background knowledge.
- Enable inductive inferences involving the data and the background knowledge.

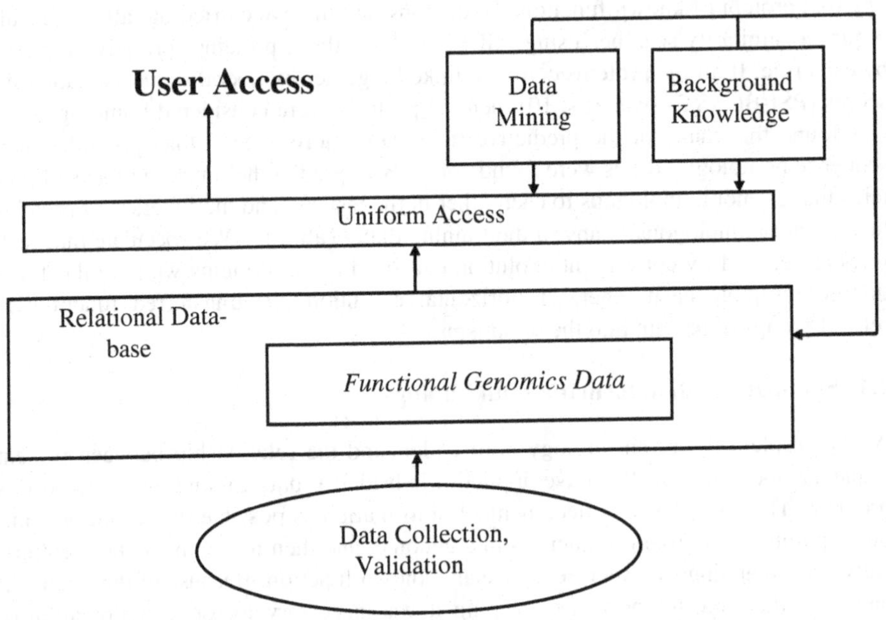

Fig. 4. Design of the proposed Intelligent Database

To populate the database, data will be collected through our existing research links to the functional genomics projects, and from primary bioinformatic databases. The basic data will be stored in a standard relational database management system. The background biological knowledge necessary for deduction will be stored together with the deduction engine in one connected module. One important element that is missing from the use of logic programs to represent scientific knowledge is a way of representing uncertainty (Russell & Norvig, 1995). How best to do this is unclear, but we favour Bayesian probabilistic methods as they have the best theoretical grounding (Jaynes, 1994). Probability theory is often considered to be grounded on propositional logic (Jaynes, 1994), and until recently little work has been done combining logic programs with probabilities. However, there has been a recent upsurge of research in this area (e.g. Muggleton, 2001; Kersting & DeRaedt, 200; Cussens, 2001).

Induction will be carried out using conventional (propositional) data mining and Inductive Logic Programming (ILP). The ILP data mining approaches will be connected with the background knowledge and deductive engine. A small section of the planned structure of data and background knowledge of ID is shown in Figure 5. The results of the induction will be stored with the basic data in the relational database, as in an inductive database (Mannila, 1997). The ID will be made accessible to the scientific community using a variety of standard methods, including Web Browsers, ODBC, XML, CORBA, etc.

The aim of the Intelligent Database is to directly integrate scientific databases and scientific discovery tools to provide scientists with a tool that can help automate science. We believe that such Intelligent Databases are essential to analyse the large amounts of new scientific data that is being generated and to refine this data into scientific knowledge.

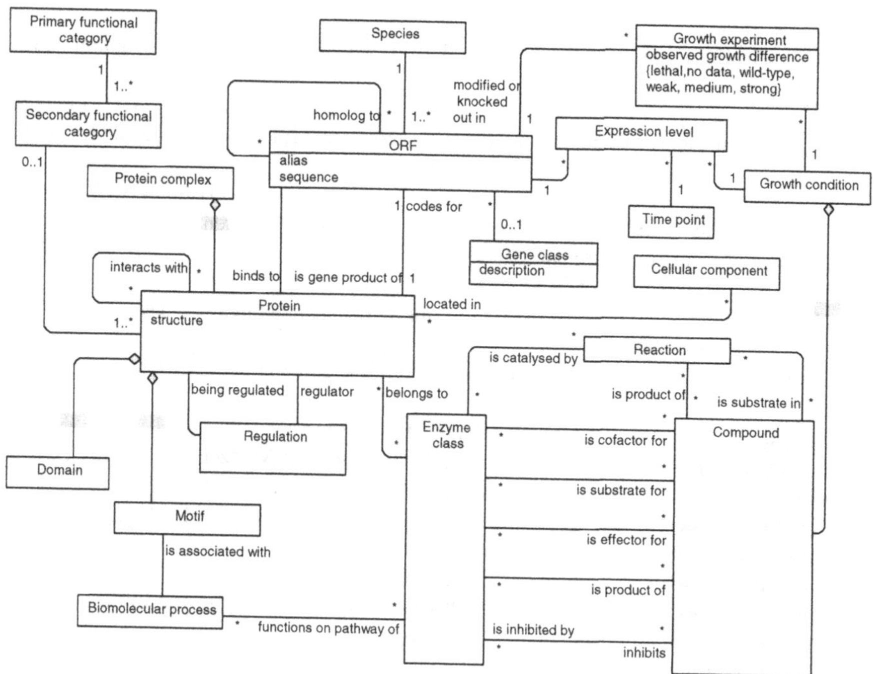

Fig. 5. A relational/UML model of part of the background knowledge in the proposed intelligent database

Acknowledgement: We would like to thank Ugis Sarkans and Nigel Hardy for their help in this paper.

References

Adams, et al.: The genome sequence of Drosophilia Melanogaster. Science 287, 2185–2195 (2000)

Alizadeh, A., et al.: Distinct types of diffuse large B-cell lymphoma identified by gene expression profiling. Nature 403, 503–511 (2000)

Altschul, S.F., Madden, T.L., Schaffer, A.A, Zhang, J., Zhang, Z., Miller, W., Lipman, D.J.: Gapped BLAST and PSI-BLAST: a new generation of protein database search programs. Nucl. Acid Res. 25, 3389–3402 (1997)

The Arabidopsis genome initiative. Analysis of the genome sequence of the flowering plant Arabidopsis thaliana. Nature 408, 796–815 (2000)

Bairoch, A., Apweiler, R.: The SWISS-PROT protein sequence database and its supplement. TrEMBL Nucleic Acids Research 28, 45–48 (2000)

Blackstock, W.P., Weir, M.P.: Proteomics: quantitative and physical mapping of cellular proteins. Tibtech 17, 121–127 (1999)

Blattner, F.R., et al.: The complete genome sequence of Escherichia coli K-12. Science 277, 1453–1461 (1997)

Boden, M.: Artificial intelligence and natural man. The Harvester Press, Brighton, Sussex (1977)

Bork, P., Dandekar, T., Diaz-Lazcoz, Y., Eisenhaber, F., Huynen, M., Yuan, Y.P.: Predicting function: From genes to genomes and back. Journal of Molecular Biology 283, 707–725 (1998)

Bowers, A.F., Giraud-Carrier, C., Lloyd, J.W.: Classification of Individuals with Complex Structure. In: Proceedings of the Seventeenth International Conference on Machine Learning, pp. 81–88. Morgan Kaufmann, San Francisco (2000)

Brenner, E.: Errors in gene annotation. Trends in Genetics 15, 132–133 (1999)

Brent, R.: Functional genomics: Learning to think about gene expression data. Current Biology 9, 338–R341 (1999)

Brown, P.O., Botstein, D.: Exploring the new world of the genome with DNA microarrays. Nature Genetics 21, 33–37 (1999)

Buchanan, B.G., Sutherland, G.L., Feigenbaum, E.A.: Heuristic DENDRAL: A program for generating explanatory hypotheses in organic chemistry. In: Meltzer, B., Michie, D. (eds.) Machine Intelligence 4, pp. 209–254 Edinburgh University Press (1969)

Bussey, H.: 1997 ushers in an era of yeast functional genomics. Yeast 13, 1501–1503 (1997)

C. elegans Sequencing Consortium: Genome sequence of the nematode C. elegans: A platform for investigating biology. Science 282, 2012–2018 (1998)

Cole, S.T., et al.: Deciphering the biology of Mycobacterium tuberculosis from the complete genome sequence. Nature 393, 537–544 (1998)

Cussens, J.: Parameter estimation in stochastic logic programs. Machine Learning 44, 245–271 (2001)

Dehaspe, L., Toivonen, H., King, R.D.: Finding frequent substructures in chemical compounds. In: The Fourth International Conference on Knowledge Discovery and Data Mining, pp. 30–36. AAAI Press, Menlo Park, CA (1998)

DeRisi, J.L., Iyer, V.R., Brown, P.O.: Exploring the metabolic and genetic control of gene expression on a genomic scale. Science 278, 680–686 (1997)

Dzeroski, S., Blockeel, H., Kompare, B., Kramer, S., Pfahringer, B., Van Laer, W.: Experiments in Predicting Biodegradability. In: Džeroski, S., Flach, P.A. (eds.) Inductive Logic Programming. LNCS (LNAI), vol. 1634, pp. 80–91. Springer, Heidelberg (1999)

Dzeroski, S., Lavrac, N.: Relational Data Mining. Springer, Heidelberg (2001)

Fayyad, U., Piatetsky-Shapiro, G., Smyth, P., Uthurusamy, R.: Advances in Knowledge Discovery and Data Mining. AAAI/MIT Press, Boston (1996)

Finn, P., Muggleton, S., Page, D., Srinivasan, A.: Pharmacophore discovery using the inductive logic programming system Progol. Machine Learning 30, 241–271 (1998)

Flach, P.A., Giraud-Carrier, C., Llyoyd, J.W.: Strongly typed inductive concept learning. In: Page, D.L. (ed.) Inductive Logic Programming. LNCS, vol. 1446, pp. 185–194. Springer, Heidelberg (1998)

Fujita, H., Yagi, N., Ozaki, T., Furukawa, K.: A new design and implementation of Progol by bottom-up computation. In: Inductive Logic Programming. LNCS, vol. 1314, pp. 163–174. Springer, Heidelberg (1997)

FUNCTION – http://www.aber.ac.uk/compsci/Research/bio/ProteinFunction

GenProtEC –http://genprotec.mbl.edu

Gilbert, R.J., Johnson, H.E., Winson, M.K., Rowland, J.J., Goodacre, R., Smith, A.R., Hall, M.A., Kell, D.B.: Genetic programming as an analytical tool for metabolome data. In: Langdon, W.B., Poli, R., Nodin P., Fogarty, T. (eds.): Late-breaking papers of EuroGP-99, Software Engineering, CWI, pp. 23–33 (1999)

Goffeau, A., et al.: Life with 6000 genes. Science 274, 546–567 (1996)

Gordon, A., Sleeman, D., Edwards, P.: Informal Qualitative Models: A Systematic Approach to their Generation. In: Valdes-Perez, R. (ed.) Proceedings of AAAI 1995 Spring Symposium on Systematic Methods of Scientific Discovery, pp. 18–22. AAAI Press, Stanford (1995)

HGP – http://www.sanger.ac.uk/HGP

Hieter, P., Boguski, N.: Functional genomics: it's all how you read it. Science 278, 601–602 (1997)

Humphery-Smith, I., Cordwell, S.J., Blackstock, W.P.: Proteome research: complementarity and limitations with respect to the RNA and DNA worlds. Electrophoresis 18, 1217–1242 (1997)

International human genome sequencing consortium: Initial Sequencing and analysis of the human genome. Nature 409, 860–921 (2001)

Kell, D., King, R.D.: On the optimization of classes for the assignment of unidentified reading frames in functional genomics programmes: the need for machine learning. Trends in Biotechnology 18, 93–98 (2000)

Kersting, K., DeRaedt, L.: Bayesian Logic Programs. Linkoping Electronic Articles in Computer and Information Science. 5(034) (2001)

King, R.D., Muggleton, S., Lewis, R.A., Sternberg, M.J.E.: Drug design by machine learning - the use of inductive logic programming to model the structure-activity-relationships of trimethoprim analogs binding to dihydrofolate-reductase. Proceedings of the National Academy of Sciences of the USA 89, 11322–11326 (1992)

King, R.D., Clark, D.A., Shirazi, J., Sternberg, M.J.E.: On the use of machine learning to identify topological rules in the packing of beta-strands. Protein Engineering 7, 1295–1303 (1994)

King, R.D., Muggleton, S.H., Srinivasan, A., Sternberg, M.J.E.: Structure-activity relationships derived by machine learning: The use of atoms and their bond connectivities to predict mutagenicity by inductive logic programming. Proceedings of the National Academy of Sciences of the USA 93, 438–442 (1996)

King, R.D., Karwath, A., Clare, A., Dehapse, L.: Genome scale prediction of protein functional class from sequence using data mining. In: Ramakrishnan, R., Stolfo, S., Bayardo, R., Parsa, I. (eds.) The Sixth ACM SIGKDD International Conference on Knowledge Discovery and Data Mining. The Association for Computing Machinery, New York, USA, pp. 384–389 (2000a)

King, R.D., Karwath, A., Clare, A., Dehapse, L.: Accurate prediction of protein class in the M. tuberculosis and E. coli genomes using data mining. Yeast (Comparative and Functional Genomics) 17, 283–293 (2000b)

King, R.D., Karwath, A., Clare, A., Dehapse, L.: The utility of different representations of protein sequence for predicting functional class. Bioinformatics 17, 445–454 (2001)

Kramer, S., De Raedt, L., Helma, C.: Molecular feature mining in HIV Data. In: Proceedings of the Seventh ACM SIGKDD International Conference on Knowledge Discovery and Data Mining, pp. 136–143 (2001)

Kramer, S., Lavrac, N., Flach, P.: Propositionalization approaches to relational data mining. In: Dzeroski, S., Lavrac, N. (eds.) Relational Data Mining, Springer, Heidelberg (2001)

Jaynes, E.T.: Probability theory: The logic of Science (1994), http://omega.albany.edu:8008/JaynesBook.html

Langley, P., Simon, H.A., Bradshaw, G.L., Zytkow, J.M.: Scientific Discovery: Computational Explorations of the Creative Process. MIT Press, Cambridge, MA (1987)

Lavrac, N., Dzeroski, S.: Inductive logic programming: techniques and applications. Ellis Horwood, Chichester (1994)

Mannila, H.: Inductive database and condensed representations for data mining. In: Maluszynski, J. (ed.) Proceedings of the International Logic Programming Symposium, pp. 21–30. MIT Press, Cambridge (1997)

Mannila, H., Toivonen, H.: Levelwise search and borders of theories in knowledge discovery. Data Mining and Knowledge Discovery 1, 241–258 (1997)

Mitchell, T.M.: Generalization as search. Artificial Intelligence 18, 203–226 (1982)

Mitchell, T.M.: Machine Learning. McGraw-Hill, London (1997)

Muggleton, S.H.: Inductive Logic Programming. New Generation Computing 8, 295–318 (1990)

Muggleton, S.H.: Inductive Logic Programming. Academic Press, London (1992)

Muggleton, S.: Inverse Entailment and Progol. New Generation Computing Journal 13, 245–286 (1995)

Muggleton, S., King, R.D., Sternberg, M.J.E.: Protein secondary structure prediction using logic-based machine learning. Protein Engineering 5, 647–657 (1992)

Muggleton, S.: Learning Stochastic Logic Programs. Linkoping Electronic Articles in Computer and Information Science 5(041) (2001)

Oliver, S.G., Baganz, F.: The yeast genome: systematic analysis of DNA sequence and biological function. In: Copping, L.G., Dixon, G.K., Livingstone, D.J. (eds.) Genomics: commercial opportunities from a scientific revolution, Bios, pp. 37–51, Oxford (1998)

Ouali, M., King, R.D.: Cascaded multiple classifiers for secondary structure prediction. Protein Science 9, 1162–1176 (2000)

Pearson, W.R., Lipman, D.J.: Improved tools for biological sequence comparison. Proceedings of the National Academy of Sciences of the USA 85, 2444–2448 (1988)

Plato – http://plato.stanford.edu/entries/logic-relevance

Quinlan, R.: C4.5: Programs for machine learning. Morgan Kaufmann, San Mateo (1993)

Rabitz, H., de Vivie-Riedle, R., Motzkus, M., Kompa, K.: Whither the Future of Controlling Quantum Phenomena? Science 288, 824–828 (2000)

Reichardt, T.: It's sink or swim as a tidal wave of data approaches. Nature 399, 517–520 (1999)

Russel, S.J., Norvig, P.: Artificial Intelligence: A modern approach. Prentice Hall, Englewood Cliffs (1995)

Sleeman, D.H., Stacy, M.K., Edwards, P., Gray, N.A.B.: An Architecture for Theory-Driven Scientific Discovery. In: Morik, K. (ed.) Proceedings of the Fourth European Working Session on Learning, pp. 11–23, Pitman, London (1989)

Srinivasan, A., King, R.D.: Feature construction with Inductive Logic Programming: A study of quantitative predictions of biological activity aided by structural attributes. Data Mining and Knowledge Discovery 3, 37–57 (1999)

Srinivasan, A.: A study of two probabilistic methods for searching large spaces with ILP. Data Mining and Knowledge Discovery 3, 95–123 (2001)

Sternberg, M.J.E., King, R.D., Lewis, R.A., Muggleton, S.: Application of machine learning to structural molecular biology. Philosophical Transactions of the Royal Society of London Series B- Biological Sciences 344, 365–371 (1994)

TB - http:/www.sanger.ac.uk/Projects/M_tuberculosis/gene_list_full.shtm

Tukey, J.W.: Exploratory Data Analysis. Addison-Wesley, London (1977)

Turcotte, M., Muggleton, S.H., Sternberg, M.J.E.: The effect of relational background knowledge on learning of protein three-dimensional fold signatures. Machine Learning 12, 81–96 (2001)

Ullman, J.D.: Principles of databases and knowledge-base systems, vol. 1. Computer Science Press, Rockville, MD (1988)

Valdes-Perez, R.E.: Discovery tools for science applications. Communications of the ACM 42, 37–41 (1999)

Venter, J.C., et al.: The sequence of the human genome. Science 291, 1304–1351 (2001)

Drug Discovery as an Example of Literature-Based Discovery

Marc Weeber*

Lister Hill National Center for Biomedical Communications
National Library of Medicine, Bethesda, Maryland, USA
marc@weeber.net

Abstract. Since Swanson's introduction of literature-based discovery in 1986, new hypotheses have been generated by connecting disconnected scientific literatures. In this paper, we present the general discovery model and show how it can be used for drug discovery research. We have developed a discovery support tool that employs Natural Language Processing techniques to extract biomedical concepts from Medline titles and abstracts. Using semantic knowledge, the user, typically a biomedical scientist, can efficiently filter out irrelevant information. This chapter provides an algorithmic description of the system and presents a potential drug discovery. We conclude by discussing the current and future status of literature-based discovery in the biomedical research domain.

1 Introduction

The amount of scientific knowledge has grown immensely during the past century. Science expands constantly because scientists continue to be curious about the world that surrounds them. If a scientist has found something new, he immediately wonders what its implications are, and tries to formulate new hypotheses that he subsequently tests, which leads to new insights and discoveries. The fact that Nobel prizes, the most prestigious appraisals for scientists, are awarded to people who make breakthrough scientific discoveries, shows that discovery is at the heart of science.

The study of *discovery in science*, characterized by Valdés-Pérez as the "generation of novel, interesting, plausible, and intelligible knowledge about the objects of study" (Valdés-Pérez, 1999), is an interesting one. Questions arise as to what the prerequisites are for discovery in terms of existing knowledge and data gathering. How does a scientist recognize patterns in data and how does he define generalizations or even laws? Also, once new facts have been discovered, how does he disseminate and communicate these to other researchers, and how do his colleagues react and integrate this new knowledge?

Research into artificial intelligence has tried to analyze and mimic these processes. Some computer systems are able to simulate the discoveries of natural

* The author can be contacted at the Dept. of Medical Informatics, Erasmus University Rotterdam, Dr. Molewaterplein 50, 3015 GE Rotterdam, The Netherlands.

S. Džeroski and L. Todorovski (Eds.): Computational Discovery, LNAI 4660, pp. 290–306, 2007.

laws based on a database of observations, see (Simon et al., 1997) for a short overview. Also, computer systems have been developed that assist the human scientist in the scientific discovery process. Both Valdés-Pérez and Langley discuss a wide variety of systems such as MECHEM (catalytic chemistry), ARROWSMITH (biomedicine), GRAFFITI (graph theory), DAVICCAND (mettalurgy), and MPD/KINSHIP (anthropological linguistics) that have successfully been used to assist in the creation of new scientific knowledge (Valdés-Pérez, 1999; Langley, 2000).

One of the characteristics of increasing scientific knowledge is that individual scientists have to interpret vast amounts of existing knowledge and acquire specialist skills before they are able to contribute to their scientific domain by discovering new knowledge. Additionally, keeping abreast of the latest developments in order to integrate newly created knowledge with his own research is not a trivial task for a scientist. Simon et al. (1997) state that scientific publications, as a public blackboard, is the principal instrument for the cumulation and coordination of scientific knowledge. Swanson has shown that it is possible to use scientific publications to generate new knowledge in the context of literature-based discovery.

This chapter describes our reseach in literature-based discovery (Weeber, 2001). Our goals are three-fold. First, we integrate Swanson's generic discovery model (Swanson, 1986) with Vos's drug discovery model (Vos, 1991). Second, we use advanced natural language processing (NLP) to efficiently analyze the scientific literature, and third, we develop a tool that may assist researchers in their scientific discovery process. In this paper we will discuss the discovery models, NLP techniques, and the tool in a case study on discovering new applications for the forty year-old drug thalidomide.

2 Models of Discovery

Since 1986, Swanson and his colleague Smalheiser have continuously made discoveries in biomedicine by connecting disconnected knowledge structures, see (Smalheiser & Swanson, 1998) for an overview. The premise of their approach is that there are two bodies, or structures of scientific knowledge that do not communicate. However, part of the knowledge of one such a domain may complement the knowledge of the other one.

Suppose that one scientific community knows that B is one of the characteristics of disease C. Another scientific group (discipline, or knowledge structure) has found that substance A affects B. Discovery in this case is making the implicit link AC through the B-connection. Figure 1 depicts this situation, see also (Swanson & Smalheiser, 1997).

Vos's model of discovery uses the concept of drug profiles interacting with disease profiles. A profile of a particular drug consists of all the effects it has in the human body. Some of them are intended, or *wished for*, i.e. the drug has specifically been developed with these characteristics in mind, others are not wished for. Vos calls all effects the *operational functional characteristics* of a drug.

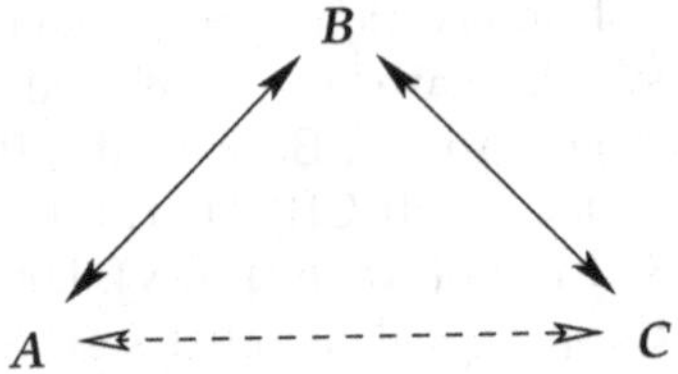

Fig. 1. Swanson's ABC model of discovery. *A*, *B*, and *C* are general concepts. The relationships *AB* and *BC* are known and reported in the literature. The implicit relationship *AC* is a putative new discovery.

Standard drug development involves the optimization of the wished for characteristics together with a minimization of the negative operational functional characteristics, or adverse effects. However, the not wished for characteristics can be viewed positively in a different context (Vos, 1991; Rikken & Vos, 1995; Rikken, 1998).

A well-known example is the anti-hypertensive drug minoxidil. Some patients developed extra hair growth as a not wished for result. Women, for instance, may value this negatively, especially if it concerns facial hair growth. In the different context of baldness, stimulation of hair growth is beneficial. Interestingly, the manufacturers of minoxidil did register male pattern baldness as a new indication for minoxidil. Consequently, hair growth became a new wished for characteristic.

A disease profile consists of a cluster of relevant signs and symptoms, or in other words, the characteristics of the disease. Vos defines the process of drug discovery as the rapprochement of the drug and the disease with respect to their profiles. The more characteristics are relevant to both, the more promising the drug is for treating the disease (Vos, 1991).

Figure 2 shows how Vos's model can be considered as a specification of Swanson's general model in a drug discovery context. The characteristics of the profiles in Vos's model are the intermediate *B*s in Swanson's model. The profile for drug *A*, for instance, may include the therapeutic characteristic (*B*) of "reduction of oxygen demand" whereas "increase of oxygen demand" may be a characteristic of disease *C* (Vos, 1991). Or, patients with Raynaud's disease (*C*) have the characteristic of elevated blood viscosity (*B*). One of the characteristics of dietary fish oil (*A*) is blood viscosity reduction (Swanson, 1986).

3 Discovery Space

There are two approaches to discovery that we have named as *open* and *closed* (Weeber et al., 2001). The closed discovery starts with known *A* and *C*. This may be an observed association, or an already generated hypothesis. The discovery in this situation concerns finding novel *B*s that may explain the observation. In the model, the letters *A*, *B*, and *C* refer to general scientific concepts that researches use.

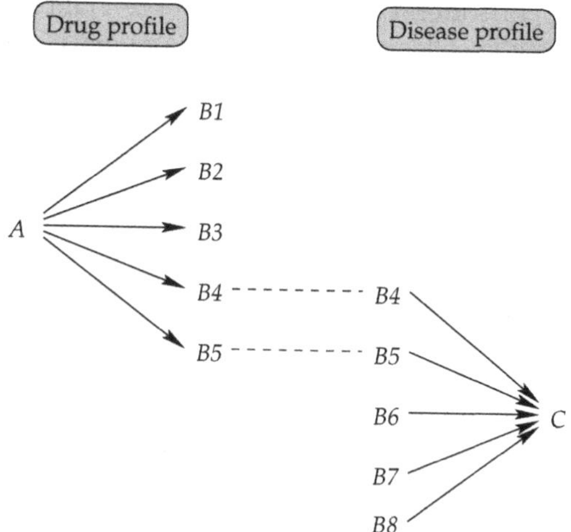

Fig. 2. Vos's and Swanson's model of discovery combined. The linking of a disease profile to a drug profile may be used to find the therapeutic application (disease) C for the drug A through pathways (therapeutic characteristics) $B4$ and $B5$.

The open discovery process starts in the knowledge structure in which the scientist takes part (A). The first step is to find potential B-connections. These will likely be found within his domain. The crucial step, however, is from B to C which is most likely outside the scientist's scope, and might therefore be in any point of the knowledge space of science. Or even outside that space.

We can illustrate this with the similarity of a person's social life. In a continuously growing world population (total science), our main character (A) knows an increasing but limited number of persons (B). Keeping up to date with his social structure is not a trivial task for A. Knowing the social structure (C) of any B-person included in his own structure is impossible. Our main character will not know all his friends' friends.

A closed discovery process starts with an initial hypothesis that A has some association to C. The nature of this association is unknown or not fully understood. The goal of the closed approach is first to unveil new possible explanations for an AC association, and second to provide an evaluation of the strength of the association. The likely outcome of the closed approach is to either strengthen or to reject the AC association.

Similar to Swanson, we define discovery in biomedicine as connecting disconnected structures (or disciplines or domains) of biomedical scientific knowledge in biomedicine. Note that just any science can be selected, the discovery model holds true for any discovery space. The literature of the selected discipline, biomedicine in our case, is the most comprehensive and accessible format of scientific knowledge in which experimental results, facts, theories, models, and hypotheses are reported. Discovery by connecting different structures implies connecting different (collections of) scientific texts. We therefore pursue *literature-based discovery.*

A system that supports literature-based discovery should have the potential of exploring the complete knowledge space. Because we have selected biomedicine as our scientific discipline, we use MEDLINE, the most comprehensive biomedical bibliographical database with over 11,000,000 citations as the representation of the *knowledge universe* in which discoveries may be made. Each citation consists of at least a title. In many cases, an abstract is available as well. Also, other bibliographic information is included, such as authors, journal, date of publication, and keywords (called Medical Subject Headings, or MeSH). Using PubMed (`http://www.pubmed.gov`), the online interface to MEDLINE, and using NLP techniques, we have developed a discovery support tool called *Literaby* to explore this vast space.

The definition of the discovery space allows us to specify the model letters *A*, *B*, and *C* used in the model. Swanson uses MEDLINE titles, therefore, the model letters are (combinations of) title words. As our implementation of the model is concept based, the letters are represented by biomedical concepts (see next section).

Literaby implements both the open and the closed approach to discovery. In the open discovery, it first analyzes the literature of the starting point: *A*. Selecting interesting terms, the literature on these *B*-terms is downloaded and analyzed to find the final *C*-term. In the closed discovery, both the literatures on *A* and *C* are downloaded and analyzed to search for interesting overlapping *B*-terms to strengthen (or reject) the initial *AC*-hypothesis. In most cases, an open discovery concerns *generating* a hypothesis that is *evaluated* in a closed process.

4 Text Analysis

Swanson's first discovery of the probable therapeutic effects of fish oil on patients with Raynaud's disease (Swanson, 1986) was a coincidence (Swanson, personal communication). He was asked to study the literature on the Inuit diet. Fish is a main ingredient of this diet, and the effects of fish oil on the the cardiovascular system in Inuit has been studied. Reduced blood viscosity and blood platelet aggregation, and certain vasoreactive characteristics were observed in Inuit.

In another context, Swanson had been studying the literature on Raynaud's disease. From this literature he had learned that patients with this disease have a relatively high blood viscosity and increased platelet aggregation function. Also, they were characterized by certain vasoreactive phenomena.

Combining the knowledge from two contexts, he hypothesized that the active ingredients of fish oil, omega-3 fatty acids, may help Raynaud's patients. With this hypothesis in mind, he studied the literatures both on fish oil and on Raynaud's disease to find out that there was no overlap at that time (1986). Using the model of disconnected bodies of biomedical knowledge, he published a second hypothesis that magnesium insufficiency is involved in migraine. No one had pointed this out in the literature, while Swanson found eleven indirect connections in the literature (Swanson, 1988).

The first two discoveries were done by extensive manual searching in literature databases and reading many titles and abstracts of scientific publications. Since 1988, Swanson has used computational text analysis tools to assist him in studying the literature. These tools have evolved into a discovery support tool called ARROWSMITH (Swanson & Smalheiser, 1997).

In ARROWSMITH, the user can upload a file of Medline titles on A and on C (an implementation of the closed approach). The tool provides a list of overlapping Bs. Additionally, the context of the Bs can be viewed in a juxtaposed (AB next to BC-sentences). The list of B-terms is potentially very long, and filtering is needed. The current analytic approach is to use an extensive stop list, a list of words such as determiners and adverbs that are considered non-relevant. Also, words with a too general biomedical meaning are included in this list. The stop list has mainly been compiled during rediscovering his first discoveries, incorporating expert knowledge from users.

Gordon and Lindsay used a more principled analytic approach based on word frequency (lexical) statistics used in Information Retrieval (IR) research (Gordon & Lindsay, 1996; Lindsay & Gordon, 1999). In addition to MEDLINE titles, they use the abstracts of citations as well. Gordon and Lindsay emply the statistics to find a rank-ordered list of potentially relevant words. They use the most highly ranked words to walk through their open discovery approach.

Gordon and Lindsay are able to replicate Swanson's first two discoveries. Gordon and Lindsay (1996) use specific measures and provide a likely explanation why these techniques work in the Raynaud–fish oil case. However, when applied to the migraine–magnesium case, the same statistics fail and different ones had to be used (Lindsay & Gordon, 1999). Therefore, there still does not exist a unifying, principled lexical statistical approach.

Our approach to the analysis of titles and abstracts of scientific publications is to use advanced NLP techniques to identify biomedical concepts in text. The Unified Medical Language System (UMLS)® (Lindberg et al., 1993) provides the largest biomedical thesaurus to date: the Metathesaurus®. The Metathesaurus provides a uniform, integrated distribution format from over 60 biomedical source vocabularies and classifications, and links many different names for the same concept. Over 700,000 biomedical concepts are represented with over 1,500,000 text strings.

The use of concepts has several advantages. First, different textual representations, i.e., spelling variants, synonyms, derivations, and inflections are all linked to one concept. For instance, IL-12, IL12, interleukin 12, CLMF, cytotoxic lymphocyte maturation factor(s), and natural killer cell stimulatory factor(s) refer to the same concept: *Interleukin-12*. Second, many biomedical ideas or concepts are expressed by more than one word. Finding meaningful multi-word terms in text is non-trivial in NLP. Different word statistical strategies may be employed (Weeber et al., 2000b), and results always include noise. By using concepts, we select only existing, biologically meaningful, ones. We employ the MetaMap program (Rindflesch & Aronson, 1994; Aronson, 1996; Aronson, 2001) to find UMLS concepts in natural language text.

The most important reason to use concepts, however, is the availability of the UMLS semantic classification scheme. Each concept has been assigned to one or more semantic categories. There is a total of 134 categories including "Disease or Syndrome", "Gene or Genome", "Amino Acid, Peptide, or Protein". The concept *Thalidomide*, for instance, has been assigned the semantic types "Organic Chemical", "Pharmacologic Substance", and "Hazardous or Poisonous Substance". At different stages of the discovery process, we can select only certain semantic types to filter the output of the text analysis. For instance, if we are looking for diseases in text, we select only the semantic type "Disease or Syndrome" which will result in a list of disease concepts extracted from natural language sentences. Figure 4 on page 300 provides a part of the interface to semantic filter in our discovery system. We provide a more extensive overview of the our text analysis techniques in (Weeber et al., 2001).

Hristovski et al. (2001) also use a concept-based approach. They use MeSH keywords added to Medline citations. The UMLS provides co-occurrence tables of major MeSH keywords. Using these frequency data, Hristovski et al. (2001) apply association rules to compute the most interesting associations. In an interactive interface, the user, typically a biomedical scientist, can quickly assess these associations. In this book, Hristovski and his collegues show how their system can be used.

5 Literaby

Literaby, our current, web-based, discovery support tool has evolved from our first tool called the *DAD*-system (Weeber et al., 2000a). The acronym *DAD* expands to Disease – Adverse Drug Reaction – Drug, or the other way around. It represents our interest in drug discovery. The new version, Literaby, shows that our analytic approach can be generalized. Other changes are that the query generation phase is now fully automated, and the interface for presenting the bibliographic evidence has been overhauled.

The underpinnings, however, have not been changed. This section provides an algorithmic overview of the semi-automated discovery process. The high level

Table 1. High level description of the Literaby system

GIVEN: Current version of Medline, Metathesaurus
INPUT: text string A

1. *open* discovery phase: generating a hypothesis
   a. From A to B using the Algorithm 1 (automatic)
   b. User selects most promising B-concepts based on computer output and literature
   c. From B to C using Algorithm 2 (automatic)
   d. User selects most promising C-concepts:these are the putative new discoveries
2. *closed* discovey phase: evaluating a hypothesis (Algorithm 2)

OUTPUT: Set of concepts C that have likely new connections to A through B

Table 2. Literaby Algorithm 1: one step in the *open* process

GIVEN: Current version of Medline, Metathesaurus
INPUT: text string A

1. Find set of concepts SA for text string A in Metathesaurus
2. Find textual variants SA-VAR for SA using MetaMap
3. Find titles and abstracts A-CIT in Medline query composed of SA-VAR
4. Find ALL-B concepts in titles and abstracts of A-CIT using MetaMap
5. Select set of concepts SEN-B from ALL-B that co-occur in sentences with SA
6. User forms a semantic filter FILT-B by subsetting the 134 semantic categories
7. Apply semantic filter FILT-B to SEN-B to retrieve set of concepts SB

OUTPUT: Set of concepts SB

Table 3. Literaby Algorithm 2: *closed* process

GIVEN: Current version of Medline, Metathesaurus
INPUT: text string A, text string C

1. Find set of concepts SA for A, SC for C in Metathesaurus
2. Find textual variants SA-VAR for SA and SC-VAR for SC using MetaMap
3. Find citations A-CIT and C-CIT in Medline using SA-VAR and SC-VAR, respectively
4. Find ALL-AB concepts in titles and abstracts of A-CIT using MetaMap
5. Find ALL-BC concepts in titles and abstracts of C-CIT using MetaMap
6. Select set of concepts SUB-AB from ALL-AB that co-occur in sentences with SA
7. Select set of concepts SUB-BC from ALL-BC that co-occur in sentences with SC
8. Select potential concepts POT-B such that SUB-AB = SUB-BC
9. User forms a semantic filter FILT-B by subsetting the 134 semantic categories
10. Apply semantic filter FILT-B to POT-B to retrieve set of concepts SB

OUTPUT: Set of concepts SB

description is presented Table 1. The user starts with an *open* discovery process. He starts with his term of interest, for instance, a drug. The system then tries to find concepts that are in some way related to this drug using algorithm 1 (Table 2).

Algorithm 1 maps the initial query string to biomedical concepts, and "back-translates" this concept to all synonyms and textual varians that are available in the natural language text of Medline. Literaby formulates the query to PubMed, the online version of Medline and retrieves the citations that matched the query. Literaby also takes care of finding concepts in these citations through MetaMap, and finally assists the user in selecting a semantic filter (or use a predefined one).

The output of Algorithm 1 is a set of concepts that have some (user defined semantic) relationship with the starting text string. In case of a drug, typical

B-concepts are processes that are a likely biologic actions of this drug. The user selects a few most promising ones, basing his selections on expert knowledge, and assessing the bibliographic evidence provided by Literaby.

The next step is a replay of Algorithm 1, but now with the B-concepts as input, and likely C-concepts as output. Typical C-concepts in the case of drug discovery are diseases or pathological processes. Again using the available bibliographic information, the user selects the most interesting C-concepts to start the second phase of the discovery process, the CLOSED discovery using Algorithm 2 (Table 3).

With this algorithm, the user tries to evaluate the generated hypotheses in the open discovery process. The main idea is that the more relations (Bs) there are between A and C, the more plausible the association AC is. In the next section, we illustrate the use of Literaby to assist scientists by following the discovery of new potential therapeutic applications for the drug thalidomide (Weeber et al., 2003).

6 Literaby and Thalidomide

Between 1959 and 1961, thalidomide was a popular over the counter sedative. Devastating teratogenic effects led to withdrawal from the market only a few years after its introduction. In recent years, however, interest in thalidomide has intensified based on its reported anti-inflammatory and immunomodulatory properties. In 1998, the FDA approved thalidomide for the indication of erythema nodosum leprosum, an inflammatory manifestation of leprosy. Additionally, thalidomide seems to have beneficial effects on ulcers and wasting associated with HIV infection.

The first step (A in the discovery model) is to identify concepts in the UMLS that are related to thalidomide. Entering the string `thalidomide` results in a list of 33 concepts that map to this string. Figure 3 depicts part of this list.

By using the hierarchy of the thesaurus we not only find the concept *Thalidomide*, which is the generic name of the drug, but also the brand names, which are children concepts in the thesaurus, and the chemical description of the compound. The user has the option to (de)select these concepts, and then proceeds. Employing MetaMap, Literaby maps the concepts back to their textual variants to automatically generate and execute a query to PubMed. For instance, the text string `thalidomide` maps to the concept *Thalidomide*. The UMLS provides us the drug brand name, among other "thalidomid", "supidimide" and "sedoval". These brand names are included in the PubMed query.

The resulting citations are downloaded and analyzed to extract concepts from the titles and abstracts, if available. After this step, the user is involved again; the B-concepts have to be selected. For this, the user's expert knowledge is needed. In this case, we collaborated with an immunologist, because the newly registered application involves the immune system. We hypothesized that we might find new therapeutic applications through thalidomide's apparently successful immunologic pathway modulation. Literaby presents the semantic filter

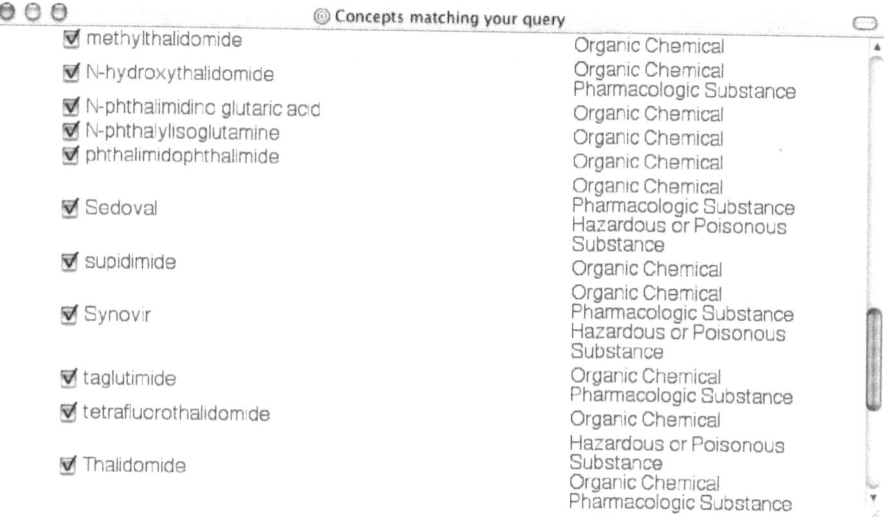

Fig. 3. The search string `thalidomide` in the UMLS Metathesaurus resulted in 33 concepts, for instance, different chemical names for the substance, but also brand names for the drug

to the user, where he can choose from a the list of the 134 categories or semantic types (Fig. 4).

At this stage, we only select the semantic type of "Immunologic Factor", and Literaby returns a list of 93 immunologic factors that co-occur in sentences mentioning a textual representation of the concept *Thalidomide*. Figure 5 shows the twelve most frequent ones. The domain expert selected *Interleukin-12* (IL-12) as the B-concept of potential interest. Clicking on the button before the concept, the user may see the sentences in which this B-concept co-occurs with thalidomide. For *Interleukin-12*, we observe sentences such as:

- Inhibition of **IL-12** production by *thalidomide.*
- *Thalidomide* potentially suppressed the production of **IL-12** by PBMC [...].
- *Thalidomide*-induced inhibition of **IL-12** production [...].

Indeed, it appears that thalidomide's inhibitory effects on IL-12, together with the reported stimulatory effect on IL-10 production, seems to be the mechanism of how thalidomide favors the differentiation of T-helper 0 (Th0) immune system cells into T-helper 2 (Th2) cells by blocking differentiation of Th1 cells. Our hypothetical model of action (Weeber et al., 2003) suggests that patients with, in particular, auto-immune diseases may benefit from thalidomide treatment.

Using *Interleukin-12* as the selected B-concept, we downloaded all citations from PubMed that include (variants of) IL-12 in either title or abstract. The resulting citations were MetaMapped to UMLS concepts, and Literaby provides the user again with the semantic filter. At this stage, we looked for C-concepts,

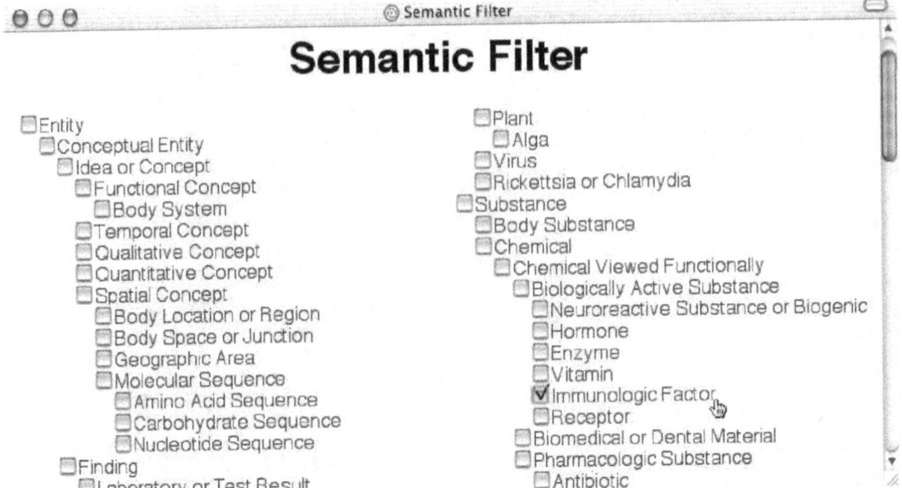

Fig. 4. The semantic filter of the discovery support tool Literaby. There are 134 semantic types that the user may select.

disease concepts in our case. We selected the semantic type "Disease or Syndrome", which resulted in a list of 420 diseases that co-occur with IL-12. After a partly automated filtering process, (see Weeber et al. 2003), we studied the sentences that related IL-12 to the reduced set of diseases. Examples are:

- **IL-12** [...] expression in mononuclear cells in response to acetylcholine receptor is augmented in *myasthenia gravis.*
- Possible involvement of **IL-12** expression by Epstein-Barr virus in *Sjögren syndrome.*
- *Acute pancreatitis* patients had serum concentrations of total **IL-12, IL-12p40**, and IL-6 significantly higher ($p < 0.05$) than those of the healthy subjects.
- Expression of B7-1, B7-2, and **IL-12** in anti-Fas antibody-induced *pulmonary fibrosis* in mice.

The previous sentences indicate that IL-12 is overexpressed in these diseases. Studying the sentences, their complete abstracts, and sometimes even the online full text papers, we hypothesized for twelve diseases that thalidomide might be a useful therapy through its inhibitory effects of IL-12. These twelve hypotheses were the starting point of twelve closed discovery processes Literaby downloaded and analyzed each of these C-literatures. The discovery process consisted of finding (a lack of) overlapping immunologic B-concepts to strengthen (or reject) the initial hypotheses.

Using chronic hepatitis C (CHC) as an example, the semantic filter, again set to "Immunologic Factor", provided us with a list of 60 immunologic factors, presented in a similar way as Fig. 5. We find additional citations in the CHC literature that IL-12 is augmented in patients with this disease. Figure 6 provides

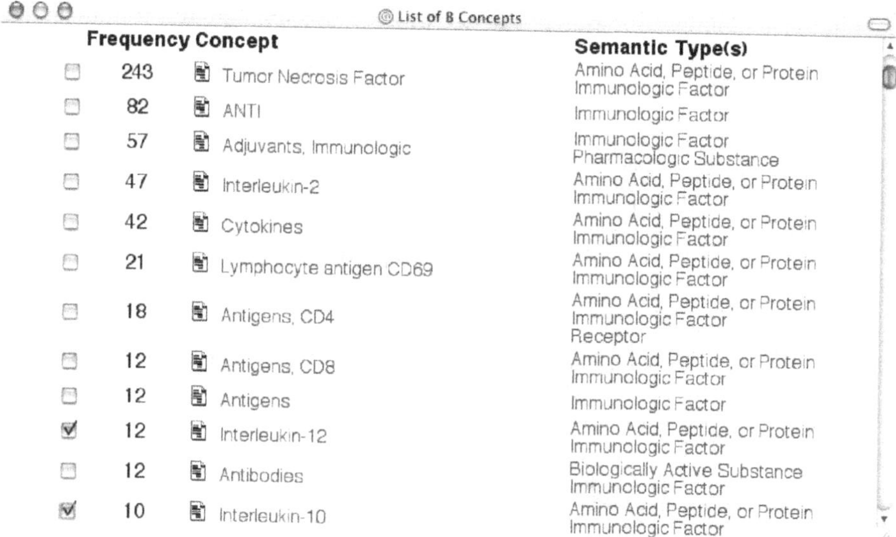

List of B Concepts

	Frequency	Concept	Semantic Type(s)
☐	243	Tumor Necrosis Factor	Amino Acid, Peptide, or Protein Immunologic Factor
☐	82	ANTI	Immunologic Factor
☐	57	Adjuvants, Immunologic	Immunologic Factor Pharmacologic Substance
☐	47	Interleukin-2	Amino Acid, Peptide, or Protein Immunologic Factor
☐	42	Cytokines	Amino Acid, Peptide, or Protein Immunologic Factor
☐	21	Lymphocyte antigen CD69	Amino Acid, Peptide, or Protein Immunologic Factor
☐	18	Antigens, CD4	Amino Acid, Peptide, or Protein Immunologic Factor Receptor
☐	12	Antigens, CD8	Amino Acid, Peptide, or Protein Immunologic Factor
☐	12	Antigens	Immunologic Factor
☑	12	Interleukin-12	Amino Acid, Peptide, or Protein Immunologic Factor
☐	12	Antibodies	Biologically Active Substance Immunologic Factor
☑	10	Interleukin-10	Amino Acid, Peptide, or Protein Immunologic Factor

Fig. 5. Top of the list of immunologic factors that co-occur in sentences with the *Thalidomide*

the interface in which the bibliographical information on thalidomide–IL-12 and IL-12–CHC is juxtaposed. In one overview the user can assess the AB and BC-information to infer the hypothesis AC.

In addition to IL-12, we also find the concept *Tumor Necrosis Factor* (TNFα). It is widely known that thalidomide inhibits TNFα through mRNA degradation. It turns out that CHC is characterized by increased levels of TNFα. Thus, we have strengthened our initial hypothesis that thalidomide may be used in CHC by elucidating an additional potential pathway.

In the closed discovery processes, we were able to find strong bibliographical evidence that supports the hypotheses that thalidomide may be a therapeutic drug for helicobacter pylori-induced gastritis, acute pancreatitis, chronic hepatitis C, and myasthenia gravis. For the latter three serious diseases, there is no known cure or therapy. The bibliographical findings merit experimental and clinical studies that should provide information on the cost/benefit trade-off of effects and side effects of thalidomide in these diseases.

7 Discussion

In the presented example, the discovery was made by human scientists supported by a tool for analyzing huge amounts of text. We do not regard, or pursue, literature-based discovery as an automatic process. The reason for this is that expert knowledge is indispensable in studying the output of the support system, not only to filter out non-interesting information but also to assess potentially contradicting information.

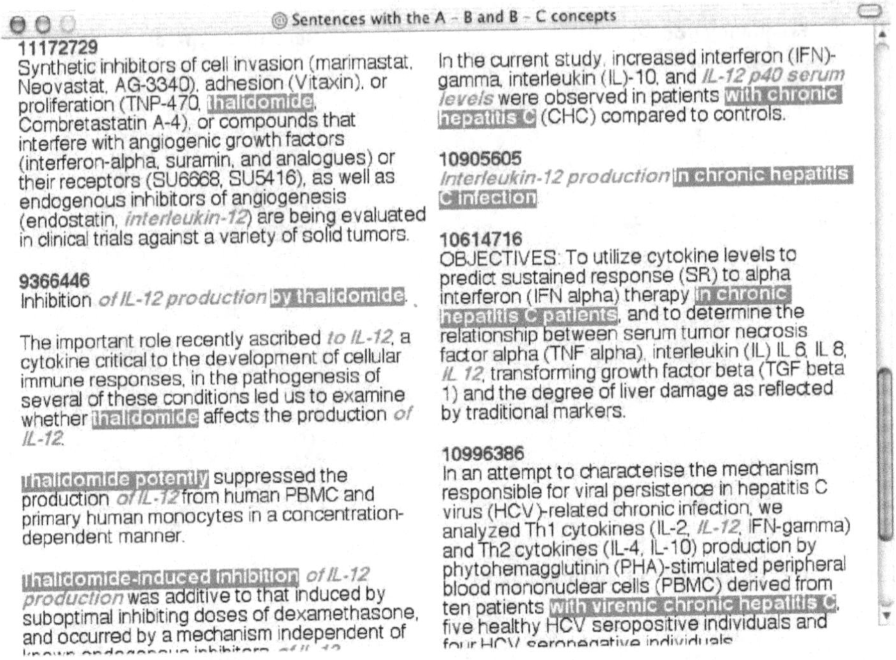

Sentences with the A - B and B - C concepts

11172729
Synthetic inhibitors of cell invasion (marimastat, Neovastat, AG-3340), adhesion (Vitaxin), or proliferation (TNP-470, thalidomide, Combretastatin A-4), or compounds that interfere with angiogenic growth factors (interferon-alpha, suramin, and analogues) or their receptors (SU6668, SU5416), as well as endogenous inhibitors of angiogenesis (endostatin, *interleukin-12*) are being evaluated in clinical trials against a variety of solid tumors.

9366446
Inhibition *of IL-12 production* by thalidomide.

The important role recently ascribed *to IL-12*, a cytokine critical to the development of cellular immune responses, in the pathogenesis of several of these conditions led us to examine whether thalidomide affects the production *of IL-12*.

Thalidomide potently suppressed the production *of IL-12* from human PBMC and primary human monocytes in a concentration-dependent manner.

Thalidomide-induced inhibition *of IL-12 production* was additive to that induced by suboptimal inhibiting doses of dexamethasone, and occurred by a mechanism independent of

In the current study, increased interferon (IFN)-gamma, interleukin (IL)-10, and *IL-12 p40 serum levels* were observed in patients with chronic hepatitis C (CHC) compared to controls.

10905605
Interleukin-12 production in chronic hepatitis C infection.

10614716
OBJECTIVES: To utilize cytokine levels to predict sustained response (SR) to alpha interferon (IFN alpha) therapy in chronic hepatitis C patients, and to determine the relationship between serum tumor necrosis factor alpha (TNF alpha), interleukin (IL) IL 6, IL 8, *IL 12*, transforming growth factor beta (TGF beta 1) and the degree of liver damage as reflected by traditional markers.

10996386
In an attempt to characterise the mechanism responsible for viral persistence in hepatitis C virus (HCV)-related chronic infection, we analyzed Th1 cytokines (IL-2, *IL-12*, IFN-gamma) and Th2 cytokines (IL-4, IL-10) production by phytohemagglutinin (PHA)-stimulated peripheral blood mononuclear cells (PBMC) derived from ten patients with viremic chronic hepatitis C, five healthy HCV seropositive individuals and

Fig. 6. Bibliographic information that suggests that chronic hepatitis C may benefit from thalidomide through IL-12 inhibition. The left column shows sentences in which A (thalidomide) and B(interleukin 12) concepts co-occur, the right column shows the relevant sentences for B and C(chronic hepatitis C).

For instance, there is one MEDLINE citation that co-mentions thalidomide and myasthenia gravis and it claims that thalidomide is not effective in Lewis rats with myasthenia gravis. This information potentially refutes our hypothesis that thalidomide may be benefical for patients with this disease. However, the expert provided the knowledge that Lewis rats have an altered immune system. Conclusions based on these experiments may therefore not be transferred to a human context. We think it impossible to model such domain knowledge in a discovery system. Even if it is possible to model knowledge to such detailed extent, one has to consider that the model should comprise the total biomedical knowledge available, as this is the knowledge space in which literature-based discovery takes place.

The second reason why we do not pursue automated discovery is that it will result in just another database, in this case one of hypotheses. How to make a decision as to what hypothesis to test experimentally? Again, human experts are needed to decide. Some bibliographically well founded hypotheses may not be interesting to test. For instance, since thalidomide has some severe side effects, a clinical application may only be interesting in severe diseases or diseases for which there is no treatment at all. We concur with Smalheiser (2002) who views a literature-based discovery approach not as a replacement but as an added value to current hypothesis driven experimental research. Smalheiser envisions a research environment in which informatics tools make hypothesis-driven research more efficient and productive.

There is some scepticism towards literature-based discovery and its potential for scientific research. Results are considered too obvious and once a hypothesis is proposed, people might say "it's logical" or "of course", and the hypothesis may have activated existing knowledge that was already available in one person. We have also encountered remarks such as "but then you can also hypothesize that..." originally intended to downplay the discovery, but actually resulting in yet another plausible hypothesis. This can be seen as a kind of activation of dormant knowledge in the mind of a scientist.

We can counter these criticisms with two facts. First, Swanson and his colleague Smalheiser have made eight literature-based discoveries that have been published in relevant, peer-reviewed, scientific journals. Swanson's first two discoveries have even been corroborated experimentally and clinically. A paper describing the new potential uses of thalidomide resulting from literature-based discovery is currently under review for a biomedical journal.

Second, no one has denied the premise of the model, i.e., that there are disconnected structures in science that may benefit from connection. This is shown by the relative ease with which we have discovered new hypothetical applications for the controversial and well-known drug thalidomide. This is not surprising, because biomedical scientists work in widely varying and highly specialized disciplines and contexts.

For instance, we observe a distinction between *in vivo*, or clinical research in humans, *in vitro*, preclinical research in laboratory and animal experiments, and *in silico*, computer-based research. The transfer of knowledge from one domain to the other is non-trivial. The research interests and goals of both domains are very different. Also, educational background of the scientists diverges largely, being clinical (medicine), experimental (biology, pharmacy, biochemistry), or computational (computer science, mathematics), respectively.

Current literature-based discoveries have mainly been made in biomedicine. Both Swanson and Spasser (Spasser, 1997) have noted that the biomedical bibliography is particularly suited for this because of the explicit titles that often state the main outcome of the research, for instance:

- Inhibition of IL-12 production by thalidomide.
- Thalidomide treatment in chronic constrictive neuropathy decreases endoneurial TNFα, increases IL-10 and has long-term effects on spinal cord dorsal horn met-enkephalin.
- Inhibition of TNFα synthesis with thalidomide for prevention of acute exacerbations and altering the natural history of multiple sclerosis.

However, not only titles are interesting. In the thalidomide case, there are only two titles mentioning IL-12 together with the drug. There were ten more sentences in MEDLINE abstracts that provided additional useful information. Of course, using abstracts also introduces more noise, but the employed filtering techniques were able to suppress this. More importantly, Cory showed that literature-based discovery is possible in humanities, a scientific discipline that is not famous for its explicit titles (Cory, 1997).

This suggests that the presented approach to generating scientific hypotheses is valid for science in general. As long as there are comprehensive bibliographic databases, reported knowledge can be combined to generate new, hypothetical knowledge. Additionally, it would be interesting to combine databases from different disciplines. Biomedicine may profit from more chemically and biologically oriented databases, such as Biological and Chemical Abstracts. Even wider gaps between disciplines may result in interesting new insights.

Research in literature-based discovery has been acknowledged as important in information and library sciences, but unfortunately, it has received little attention in biomedicine. In seems that the disconnection between biomedicine and information science prevents further developments and use of the ideas of Swanson (Spasser, 1997). Recently, however, a substantial National Institutes of Health grant has been awarded to Dr. Smalheiser (University of Illinois at Chicago) in the context of The Human Brain Project and neuroinformatics (Smalheiser, personal communication, see also `http://arrowsmith.psych.uic.edu`). The goal of this project is to use informatics tools to optimize communication between neuroscientists and to connect individual research projects, data, and results. Researchers in five neuroscience laboratories will use a further developed version of ARROWSMITH to generate new hypotheses that they will test experimentally. This research is the first step in transferring literature-based discovery support tools from the computer and information science lab into the wet lab.

Acknowledgments: Over the past years, the presented research has benefited from the input of many people. I would like to thank Rein Vos, Lolkje de Jong - van den Berg, Henny Klein, all at the University of Groningen, and Don Swanson for their many contributions and discussions and Grietje Molema as the domain expert in the thalidomide discovery. I am grateful to Alan Aronson and Jim Mork for discussions and access to the National Library of Medicine's natural language processing tools.

This research was supported in part by an appointment to the National Library of Medicine Research Participation Program administered by the Oak Ridge Institute for Science and Education through an interagency agreement between the U.S. Department of Energy and the National Library of Medicine.

References

Aronson, A.R.: The effect of textual variation on concept based information retrieval. In: Proceedings of the Annual Symposium of American Medical Informatics Association AMIA-96, Nashville, TN, pp. 373–377 (1996)

Aronson, A.R.: Effective mapping of biomedical text to the UMLS metathesaurus: The MetaMap program. In: Proceedings of the Annual Symposium of American Medical Informatics Association AMIA-01, Washington, DC, pp. 17–21 (2001)

Cory, K.A.: Discovering hidden analogies in an online humanities database. Computers and the Humanities 31, 1–12 (1997)

Gordon, M.D., Lindsay, R.K.: Toward discovery support systems: A replication, re-examination, and extension of Swanson's work on literature-based discovery of a connection between Raynaud's and fish oil. Journal of the American Society for Information Science 47, 116–128 (1996)

Hristovski, D., Stare, J., Peterlin, B., Dzeroski, S.: Supporting discovery in medicine by association rule mining in medline and umls. In: Proceedings of the Tenth World Congress on Medical Informatics, London, UK, pp. 1344–1348 (2001)

Langley, P.: The computational support of scientific discovery. International Journal of Human-Computer Studies 53, 393–410 (2000)

Lindberg, D.A.B., Humphreys, B.L., McCray, A.T.: The unified medical language system. Methods of Information in Medicine 32, 281–291 (1993)

Lindsay, R.K., Gordon, M.D.: Literature-based discovery by lexical statistics. Journal of the American Society for Information Science 50, 574–587 (1999)

Rikken, F.: Adverse drug reactions in a different context. Doctoral dissertation, University of Groningen, The Netherlands (1998)

Rikken, F., Vos, R.: How adverse drug reactions can play a role in innovative drug research. Pharmacy World & Science 17, 195–200 (1995)

Rindflesch, T.C., Aronson, A.R.: Ambiguity resolution while mapping free text to the UMLS Metathesaurus. In: Proceedings of the Annual Symposium of American Medical Informatics Association AMIA-94, San Francisco, CA, pp. 240–244 (1994)

Simon, H.A., Valdés-Pérez, R.E., Sleeman, D.H.: Scientific discovery and simplicity of method. Artificial Intelligence 91, 177–181 (1997)

Smalheiser, N.R.: Informatics and hypothesis-driven research. EMBO Reports 3, 702 (2002)

Smalheiser, N.R., Swanson, D.R.: Using ARROWSMITH: A computer-assisted approach to formulating and assessing scientific hypotheses. Computer Methods and Programs in Biomedicine 57, 149–153 (1998)

Spasser, M.A.: The enacted fate of undiscovered public knowledge. Journal of the American Society for Information Science 48, 707–717 (1997)

Swanson, D.R.: Fish oil, Raynaud's syndrome, and undiscovered public knowledge. Perspectives in Biology and Medicine 30, 7–18 (1986)

Swanson, D.R.: Migraine and magnesium: Eleven neglected connections. Perspectives in Biology and Medicine 31, 526–557 (1988)

Swanson, D.R., Smalheiser, N.R.: An interactive system for finding complementary literatures: A stimulus to scientific discovery. Artificial Intelligence 91, 183–203 (1997)

Valdés-Pérez, R.E.: Principles of human computer collaboration for knowledge discovery in science. Artificial Intelligence 107, 335–346 (1999)

Vos, R.: Drugs looking for diseases. Kluwer Academic Publishers, The Netherlands, Dordrecht (1991)

Weeber, M.: Literature-based discovery in biomedicine. Doctoral dissertation, University of Groningen, The Netherlands (2001)

Weeber, M., Klein, H., Aronson, A.R., Mork, J.G., de Jong-van den Berg, L.T.W., Vos, R.: Text-based discovery in biomedicine: The architecture of the DAD-system. In: Proceedings of the Annual Symposium of American Medical Informatics Association AMIA-00, Los Angeles, CA, pp. 903–907 (2000a)

Weeber, M., Vos, R., Baayen, R.H.: Extracting the lowest-frequency words: Pitfalls and possibilities. Computational Linguistics 26, 301–317 (2000b)

Weeber, M., Vos, R., Klein, H., de Jong-van den Berg, L.T.W.: Using concepts in literature-based discovery: Simulating Swanson's Raynaud – fish oil and migraine – magnesium discoveries. Journal of the American Society for Information Science and Technology 52, 548–557 (2001)

Weeber, M., Vos, R., Klein, H., de Jong-van den Berg, L.T.W., Molema, G.: Generating hypotheses by discovering implicit associations in the literature. a case report of a search for new potential therapeutic uses for thalidomide. Journal of the American Medical Informatics Association 10, 254–262 (2003)

Literature Based Discovery Support System and Its Application to Disease Gene Identification

Dimitar Hristovski[1], Borut Peterlin[2], Sašo Džeroski[3], and Janez Stare[1]

[1] IBMI, Medical Faculty; Vrazov trg 2/2, 1105 Ljubljana, Slovenia
dimitar.hristovski@mf.uni-lj.si,
janez.stare@mf.uni-lj.si
[2] Department of Human Genetics, Clinical Center Ljubljana
Zaloška, 1000 Ljubljana, Slovenia
borut.peterlin@guest.arnes.si
[3] Department of Knowledge Technologies, Jozef StefanInstitute
Jamova 39, 1000 Ljubljana, Slovenia
saso.dzeroski@ijs.si

Abstract. We present an interactive discovery support system, which for a given starting concept of interest, discovers new, potentially meaningful relations with other concepts that have not been published in the medical literature before. The known relations between the medical concepts come from the MEDLINE bibliographic database and the UMLS (Unified Medical Language System). We use association rules to mine for the relationships between medical concepts. Then we demonstrate a successful application of the system for predicting a gene candidate for a disease, a fact recently confirmed via the positional cloning approach. We conclude that the discovery support system we developed is a useful tool, complementary to the already existing bioinformatic tools in the field of human genetics. The system described in this chapter is available at http://www.mf.uni-lj.si/bitola/

1 Introduction

With the rapidly growing body of scientific knowledge and increasing over-specialization, it is likely that the scientific work of one research group might solve an important problem that arises in the work of another group. Yet, the two groups might not be aware of the work of each other. However, a great deal of knowledge is recorded, at least in a secondary form, in bibliographic databases such as MEDLINE for the field of biomedicine. Also very important for current biomedical research are various specialized molecular biology databases. In the present context, these vast databases provide both an opportunity and a need for developing advanced methods and tools for computer supported knowledge discovery.

The main points addressed in this paper are: 1) Is it possible to discover new, potentially meaningful relations (knowledge) between medical concepts by searching and analyzing the documents from a bibliographic database such as MEDLINE? 2) To what degree can the discovery process be automated? and 3) Can this process be used for candidate gene discovery for a human disease? To deal with these issues, we

S. Džeroski and L. Todorovski (Eds.): Computational Discovery, LNAI 4660, pp. 307–326, 2007.

have developed an interactive discovery support system based on association rule mining of the MEDLINE bibliographic database. Its intended use is as a generator for research ideas that should be then investigated by traditional scientific methods.

The remainder of this chapter is organized as follows. Section 2 gives some background, in particular about literature-based discovery and Swanson's ARROWSMITH system [3]. Section 3 describes the materials and methods used within our interactive discovery support system, which include MEDLINE, MeSH, UMLS, and association rule discovery. The system itself is described in Section 4, where we first list the goals and basic premises, present the discovery algorithm and discuss the various facets of the user interface and the practical implementation. The evaluation of the system, which includes its use to discover a candidate gene for a specific disease, is presented in Section 5. We conclude with a discussion and outline some directions for further work in Section 6.

2 Background

The idea of using a bibliographic database for generating new medical discoveries that need to be later verified by traditional follow-up studies was proposed by Swanson [1]. He managed to make seven medical discoveries just by searching the MEDLINE database with some smart strategies and by analyzing the resulting bibliographic records. These discoveries were later confirmed and published in relevant medical journals. Swanson's discovery support process is based on the concepts of complementary literatures and noninteractive literatures. If one set of articles (XY) reports an interesting relation between concepts X and Y, and a different set of articles (YZ) reports a relation between Y and Z, but nothing has been published concerning a possible link between X and Z, then XY and YZ are called complementary literatures. Generally, XY and YZ are complementary if a potentially new relation can be inferred by considering them together that cannot be inferred from either of them separately. For example, X might be a disease, Y a physiological function associated with X, and Z a substance or drug which induces or regulates the physiological function Y. If the readers and authors of one body of literature are not acquainted with another, as might often be the case with two different specialties, then the two literatures are noninteractive. By combining the concepts of complementary and noninteractive literatures, Swanson developed the concept of undiscovered public knowledge, meaning that although the literatures XY and YZ represent publicly available knowledge, the potentially new relation between X and Z remains undiscovered and is a valuable source of new discoveries.

The first published example of a discovery Swanson made was about Raynaud's disease and fish oil [1]. Articles on Raynaud's disease (X) and articles on eicosapentaenoic acid (Z) when considered together indicated that dietary fish oil rich in eicosapentaenoic acid might be beneficial for treating Raynaud patients. One Y concept was, for example, blood viscosity. The line of reasoning used was: dietary eicosapentaenoic acid (Z) can decrease blood viscosity (Y), which has been reported in patients with Raynaud's disease (X).

Another notable discovery made by Swanson was about the relation between migraine (X) and magnesium deficiency (Z) [2]. It was discovered that magnesium deficiency was the cause of certain physiological effects (Y), which were associated with migraine.

In the beginning, Swanson performed the discovery process manually by searching the MEDLINE database. Later he added software support for some of the stages of the process. His current system is called ARROWSMITH and is described in detail in [3]. Swanson's discovery methodology contains two steps that are usually done sequentially, but each one can be done independently as well.

The goal of the first step is, for a given starting concept X, to find potentially new relations to concepts Z that are unknown at the beginning. The user starts by searching MEDLINE for all the articles about a starting concept of interest (X). The articles found are then uploaded into ARROWSMITH. Then the titles of the articles are analyzed and a list of all words and phrases is made. This list is, of course, very large, and a stop word list is used to reduce it. Another way to alter the list is by manual user editing. The remaining words and phrases are considered concepts (Y) that are somehow related to the starting concept (X). Now a set of search strategies is generated in order to search MEDLINE for each of the Y concepts. The search results are uploaded into the system and again a list of words and phrases appearing in the titles is produced. This is the set of concepts Z related to Y. Those Z concepts for which there are articles in MEDLINE containing both X and Z are eliminated. The remaining concepts in Z represent possible candidates of novel relations between X, Y and Z, where Y is some intermediary concept linking X to Z.

The goal of the second step of Swanson's discovery methodology is, for given concepts X and Z, to find intermediate concepts Y through which X and Z are related. It is possible that more than one Y leads from X to Z and ARROWSMITH orders the Z concepts by decreasing number of Y connections. So, in ARROWSMITH, the frequency of words or phrases in article titles is used as a measure of relational strength between medical concepts. The publicly available version of ARROWSMITH supports only the second step of the discovery methodology.

Some of the Swanson's discoveries were repeated with different methods by Gordon and Lindsay [4], and by Weeber [5]. Weeber also discovered several hypothetical new therapeutical applications of existing drugs. For more details about Swanson's approach and its comparison with the others, see the chapter by Weeber (this volume).

Our system is based on Swanson's ideas, but there are however, several notable novelties in our approach. Instead of using title words as a representation of the meaning of the MEDLINE documents, we use MeSH (Medical Subject Headnings) descriptors. We use association rules as a measure of relationship between medical concepts while Swanson uses word frequencies. We have built a large association rule base by pre-calculating and storing the association rules in a database management system. This allows us to build a truly interactive discovery support system with a fast response to user queries.

3 Materials and Methods

This section describes the materials and methods used within our interactive discovery support system. These include the MEDLINE bibliographic database, MeSH, which comprises the controlled vocabulary and thesaurus used to index the

MEDLINE articles, and UMLS, a set of knowledge sources for linking diverse medical vocabularies. Finally, it introduces association rule discovery.

Title:
Improving the convenience of home-based interferon beta-1a therapy for multiple sclerosis.
Authors:
Lesaux J, Jadback G, Harraghy CE.
Abstract:
Subcutaneous interferon beta-1a (Rebif) therapy has been recognized as a significant advance in the treatment of relapsing-remitting multiple sclerosis (MS).
...
MeSH Terms:

- Adjuvants, Immunologic/therapeutic use*
- Adjuvants, Immunologic/administration & dosage
- Adult
- Home Nursing/methods*
- Human
- Injections, Subcutaneous
- Interferon-beta/therapeutic use*
- Interferon-beta/administration & dosage
- Middle Age
- Multiple Sclerosis, Relapsing-Remitting/nursing*
- Multiple Sclerosis, Relapsing-Remitting/drug therapy*
- Ontario
- Patient Compliance
- Patient Education
- Self Administration

....

Fig. 1. An example of a MEDLINE record. Only the title, authors, a part of the abstract and the MeSH terms (descriptors) fields are shown.

MEDLINE

The MEDLINE database is a product of the US National Library of Medicine (NLM). Because of its coverage and free accessibility, MEDLINE is the most important bibliographic database in the field of biomedicine. It contains bibliographic citations and author abstracts from over 4,600 biomedical journals. Each citation is associated with a set of MeSH terms that describe the content of the item (Figure 1). In 2001, the database comprised over 12 million records dating back to 1966 [6]. MEDLINE is available for free searching on many websites of government and health agencies. One of the most popular is the NLM Web based product PubMed. There are also about 80 commercial products that provide access to MEDLINE.

In our system, we use MEDLINE as the source of the known relations between biomedical concepts. We extract these relations and store them in a knowledge base. The discovery algorithm then operates on this knowledge base as described later.

Medical Subject Headings (MeSH)

MeSH comprises NLM's controlled vocabulary and thesaurus used for indexing articles and for searching MeSH-indexed databases, including MEDLINE. It contains biomedical subject headings (descriptors), subheadings, and supplementary chemical terms. MeSH terms provide a consistent way to retrieve information that may use different terminology for the same concepts. MeSH organizes its descriptors in a hierarchical structure that permits searching at various levels of specificity from narrower to broader. This structure also provides an effective way for searchers to browse MeSH in order to find appropriate descriptors. A retrieval query is formed using MeSH terms to find items on a desired topic. Similarly, indexers normally use the most specific descriptors available to describe the subject content of an article. Problems with MeSH indexing may arise when not all important terminology in a field is covered: when using descriptors, new concepts may be worded in a way that nonexpert users cannot readily identify. Problems may also arise because of inconsistency of human indexing [7]. In 2001, MeSH included more than 21,000 descriptors, over 132,000 Supplementary Concept Records (formerly Supplementary Chemical Records), and over 300,000 synonyms and related terms [8].

In our system, MeSH represents the set of biomedical concepts we are dealing with. In the first phase, we extract known relations between these concepts and in the second, we try to discover new relations between them.

Unified Medical Language System (UMLS)

Providing improved access to search terms as well as databases is the goal of the Unified Medical Language System (UMLS) project that NLM began in 1986. The UMLS project was undertaken in order to provide a mechanism for linking diverse medical vocabularies as well as sources of information; namely, the proliferation of disparate vocabularies, none of which was compatible with any other, was recognized as a significant impediment to the development of integrated applications. The project develops "Knowledge Sources" that can be used by a wide variety of applications programs to overcome retrieval problems caused by differences in terminology and the scattering of relevant information across many databases. There are now three UMLS Knowledge Sources: the Metathesaurus, Semantic Network, and SPECIALIST Lexicon [9,10].

The Metathesaurus provides a uniform, integrated distribution format for terms from about 60 biomedical vocabularies and classifications used in patient records, administrative health data, bibliographic and full-text databases and expert systems.

The Semantic Network contains information about the types or categories (e.g., "Disease or Syndrome," "Virus") to which all concepts in the Metathesaurus have been assigned and the permissible relationships among these types (e.g., "Virus" causes "Disease or Syndrome"). The Semantic Network, through its 134 semantic types, provides a consistent categorization of all concepts represented in the UMLS Metathesaurus.

The Lexicon contains syntactic information for many terms, component words, and English words, including verbs, which do not necessarily appear in the Metathesaurus.

UMLS research has made progress on some of the many research issues associated with interpretation of user queries, mapping between the language of different information sources, and medical indexing and retrieval techniques. Much of the serious investigation and prototype system development involving links between patient data and knowledge-based information sources has been performed using UMLS components [10].

In our system, we use UMLS information regarding the semantic types of the biomedical concepts and their co-occurrence in MEDLINE records.

Association Rules

Association rules [11] were originally developed with the purpose of market-basket analysis, where it is of interest to find patterns of the form X→Y, with the intuitive meaning "baskets that contain X tend to contain Y". A basket corresponds to a single visit of a customer to a store and is called a transaction, while individual products in the basket are called items. The approach is general enough to apply to bibliographic databases, where transactions are documents and items are words or descriptors used for indexing the documents. Association rules here have the form Word1→Word2 or Descriptor1→Descriptor2. One example of an association rule would be: Disease X (Multiple sclerosis)→Symptom Y (Optic neuritis). Another example of an association rule would be: Disease X (Multiple sclerosis)→Treatment Y (Interferon-beta).

The task of association analysis is typically performed in two steps. First, all frequent itemsets are found, where an itemset is frequent if it appears in at least a given percentage *s* (called support) of all transactions. Next, association rules are found of the form X→Y, where X and Y are frequent itemsets and the confidence of the rule (the percentage of transactions containing X that also contain Y) passes a threshold *c*.

4 System Description

This section describes our interactive discovery support system for the field of medicine (named BITOLA). We first state the goal of this development and the basic premises behind it, and then proceed with an outline of the discovery of the algorithm. We then describe in detail the different facets of the user interface, which include searching for a starting concept X, finding the related concepts Y, finding the related concepts Z, and searching and browsing related MEDLINE records. We conclude with a brief summary of the implementation details.

Goal and Basic Premises

The system we developed is an interactive discovery support system for the field of medicine and is intended to be used as a generator of new, potentially meaningful relations between a starting, known concept of interest and other concepts.

The MEDLINE database is used heavily by biomedical researchers. Traditionally it has been used to check what is new in the literature on a particular topic of interest or to check if a medical discovery has already been published. In the latter case, the researcher has already made the discovery or at least has a general discovery idea, and just wants to check if someone else has already published that discovery. In addition, the various information retrieval systems used for searching MEDLINE are geared towards the task of searching for documents about a topic well known in advance. In contrast to the traditional use of MEDLINE, where it is used to check if a discovery is new or not, our system actively helps in the discovery process by generating potentially new discoveries and research ideas by analyzing the MEDLINE database.

When building our discovery support system, we started with the same basic ideas as Swanson. However, the methods we use for discovering new relations between concepts are different as well as the software implementation of these methods.

We use the major MeSH descriptors assigned to a MEDLINE record as a representation of the contents of the article. Some of the MeSH descriptors are designated as major (followed by an asterisk in the MEDLINE record). Major descriptors are those that form the main topic of the article. See Figure 1 for an example of a MEDLINE record and MeSH descriptors.

We use association rules [11] between pairs of medical concepts as a method to determine which concepts are related to a given starting concept. In our system an association rule of the form

$X \rightarrow Y$ *(confidence, support)*

means that in *confidence* percent of articles containing X, Y is also present and that there are *support* number of articles containing both X and Y. In other words, we take concept co-occurrence as an indication of a relation between concepts. If X is a disease, for example, then some possible relations might be: *has-symptom*, *is-caused-by*, *is-treated-with-drug* and so on. The system does not try to find out the kind of relation currently. This cannot be done directly by using the MeSH descriptors assigned to an article because there is no explicit information about the relation between the descriptors stored in the MEDLINE record. However, the system prepares a query on demand and retrieves the MEDLINE records containing both X and Y. By reading the titles and abstracts of these records, the user can determine the nature of the relation.

Discovery Algorithm

We calculated all the associations between the major MeSH descriptors for two subsets of MEDLINE citations. We did this regardless of the confidence and support values and for two MEDLINE time segments: 1990-1995 and 1996-1999. The calculated associations are stored in a database management system: there are currently more than 11.000.000 associations in the rule base. The calculation of the association rules was much simplified by the use of the data contained in the UMLS, especially the co-occurrence files. Actually, for these calculations it was not necessary to access the full MEDLINE records at all.

The large association rule base is a foundation upon which the algorithm for discovering new relations between concepts proceeds as described in Table 1. The main idea is to first find all the concepts Y related to the starting concept X (e.g., if X is a disease then Y can be pathological functions, symptoms, ...). Then all the concepts Z related to Y are found (e.g., if Y is a pathological function, Z can be a chemical regulating that function). As a last step, we check if X and Z appear together in the medical literature. If they do not appear together, we have discovered a potentially new relation between X and Z. The user of the system should then evaluate the proposed (X, Z) pairs and select among them those that deserve further investigation. It should be stressed that in general, it is possible to have more then one intermediate concept Y on the path from X to Z, and it is also possible to get from X to Z through different paths.

Table 1. The algorithm for discovering new relations between medical concepts

1. Let X be a given starting concept of interest.
2. Find all concepts Y such that there is an association rule X→Y.
3. Find all concepts Z such that there is an association rule Y→Z.
4. Eliminate those Z for which an association X→Z already exists.
5. The remaining Z concepts are candidates for a new relation between X and Z.

Because in MEDLINE each X concept can be associated with many Y concepts, each of which can be associated to many Z concepts, the possible number of X→Z combinations can be extremely large. In order to deal with this combinatorial problem, the algorithm incorporates *filtering (limiting)* and *ordering* capabilities. By filtering, we try to limit the number of X→Y or Y→Z associations and to minimize the number of accidental associations.

The default filtering that cannot be relaxed is that only the associations between major MeSH headings are considered by the system. The other filtering possibilities are optional and can be interactively enforced by the user of the system.

The related concepts can be limited by the semantic type to which they belong. Each MeSH descriptor belongs to one or more semantic types. For example, if the starting concept X is a disease (semantic type *disease or syndrome*) then the user can request that Y concepts are of semantic type *pathologic function* and that Z concepts are of semantic type *pharmacologic substance*. Consequently, the system will only consider chains of associations of the form: *disease or syndrome→pathologic function→pharmacologic substance*. The information about the semantic types to which a concept belongs is drawn from the Semantic Network component of the UMLS. The last possibility for limiting the number of related concepts is by setting thresholds on the support and confidence measures of the association rules in steps 2 and 3 of the algorithm. In fact, all of the filtering options can be interactively set individually or several of them can be selected in combination.

Because the usefulness of the system relies to a large degree on human judgment, special attention is paid to the order in which the candidate related concepts are presented to the user of the system. Thus, the goal of the ordering is to present best candidates first to make human review as easy as possible. Currently the default ordering is by decreasing association rule confidence, but it is also possible to order by support or semantic type.

User Interface

During a discovery session, the user has to browse and evaluate many potential discovery candidates. Therefore, user-friendliness and short response time were our most important goals when designing the user interface of the discovery support system (Figure 2). Additionally, the user needs a standard web browser to access MEDLINE through the Entrez system, through which it is also possible to search the GenBank, SwissProt, OMIM and other databases. We will now describe the elements of the user interface, its use in the discovery process and its integration with the Entrez system.

Searching for a Starting Concept X. The user initiates a discovery session by searching for a starting concept X, which is usually from his own research area. The query is performed with a *query-by-form* method using the same screen form for browsing data and for searching. The user can specify a full or partial name in the *Str* field. When specifying a partial name, the % (percent) sign is a wild-card character and is a replacement for any character string, including the empty string. After entering the search criteria, the user presses the *Query* button to retrieve the results, which are shown in the *Starting Concept* frame and the semantic types it belongs to are shown in the frame to the right.

When specifying a partial name, it is possible that more then one concept matches the search criteria. In that case, the system shows the first matching concept. However, by using the navigation buttons (<<, <, >,>>), it is possible to see the other concepts as well.

The fields above the concept name show the frequency of occurrence of the concept in MEDLINE when used as a major MeSH descriptor for two time intervals (1990-1995) and (1996-1999).

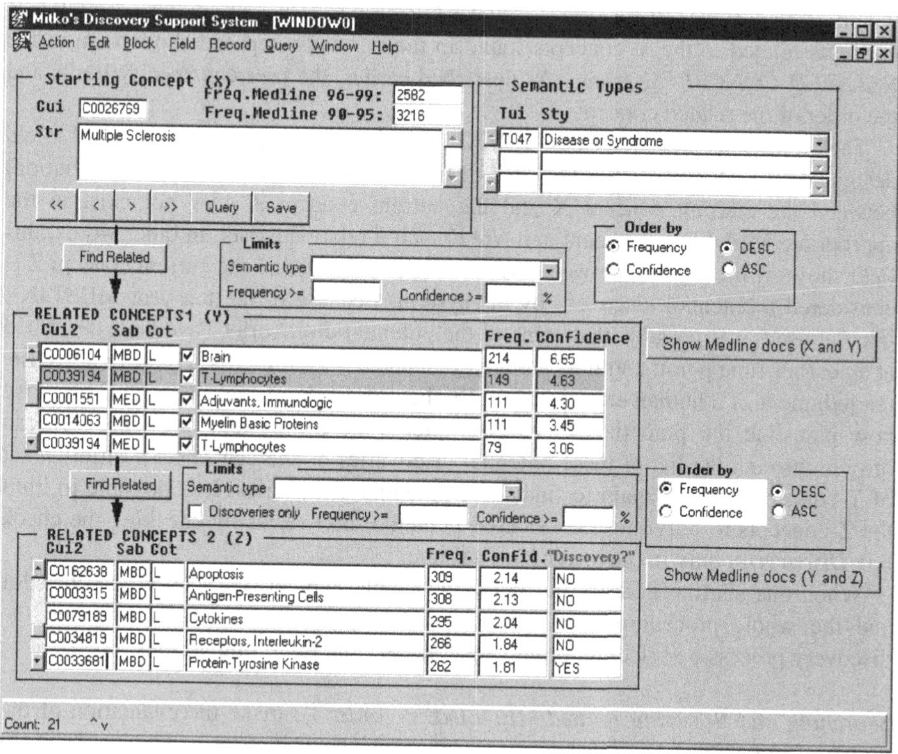

Fig. 2. The user interface of the interactive discovery support system

Finding the Related Concepts Y. The concepts Y related to the starting concept are found by pressing the *Find Related* button that is under the *Starting Concept* frame. Before finding related concepts, the user can specify limits such as the semantic type of the related concepts or the minimal confidence and frequency (support) of the association rules. This is done in the upper *Limits* frame. In addition, the order in which the related concepts are presented can be specified in the upper *Order by* frame. It can be descending or ascending by confidence or frequency (support). When the user wants to try a new limiting and ordering combination, he/she has to press the *Find Related* button again.

The related concepts are presented in the *Related Concepts1* frame. The *Sab* (Source abbreviation) column shows the source of relationship (MBD - MEDLINE 1990-1995, and MED - MEDLINE 1996-1999). Apart from the related concept names, the frequency of co-occurrence in the particular segment is shown as well as the confidence of the association rule between the starting concept and the related concept. The user can browse through the list of related concepts and select those that need further investigation. This is done by selecting the check box to the left of the related concept name.

Finding the Related Concepts Z. Now the user can press the *Find Related* button, which is under the *RELATED CONCEPTS1* frame. This action will find the Z concepts related to the Y concepts found in the previous step and show them in the *RELATED CONCEPTS2* frame. As described earlier, the user can specify limits and the order of the related concepts.

The frame *RELATED CONCEPTS2* contains an important additional field designated as *"Discovery?"*. The value of this field is YES if a relation (association) between the starting concept X and the current concept Z does not exist in the appropriate MEDLINE segment and *NO* if such a relation exists. In other words, this field shows if the relation between the starting concept X and the current concept Z is considered a potential discovery by the system. Because of the ten year MEDLINE interval used, it is possible that some of the potential discoveries have been described at an earlier time point, and thus are not good candidates for a discovery. In any case, the judgment of a human expert (hopefully the user of the system) is needed to verify how plausible the potential discoveries offered by the system are. The user can browse through the list of these potential discoveries or she can select a different set of Y concepts and try again to find some potential discoveries. It is possible to limit the Z concepts to only those considered potential discoveries by checking the check box *Discoveries only* in the lower *Limits* frame.

When one starting concept has been dealt with, another one can be searched for and the whole procedure can be repeated. The user can interactively guide the discovery process by selecting promising concepts and by setting various limits.

Searching and Browsing related MEDLINE records. To make the evaluation of the proposed potential discoveries easier for the user, the system provides the ability to search and display the MEDLINE records related to the concept currently under investigation. The user can read these records, decide which relations deserve further attention and guide the discovery process accordingly.

The search and display of MEDLINE records is accomplished by pressing either the *Show MEDLINE docs (X and Y)* or the *Show MEDLINE docs (Y and Z)* buttons. The first one searches for MEDLINE records containing both the starting concept X and the current concept Y. Similarly, the second button searches for records containing both the current Y and Z. When the user presses one of these buttons, the discovery system does the following: 1) prepares an appropriate search request, 2) starts a standard web browser, 3) connects to the search system Entrez, and 4) runs the prepared search request. The user can then browse through the resulting records and change the search request if necessary. With some basic knowledge of the Entrez system, it is also possible to display the related proteins and nucleotide sequences.

Implementation

The association rule base and the other necessary tables needed in the discovery process are stored in an Oracle relational database management system on a UNIX server. We have developed three versions of the end user program. The first one was developed using Oracle Forms6i. It communicates with the database server in a client/server manner over a TCP/IP network, which can also be the Internet. We use this version only in-house because it requires software installation on the user's computer. To make the system widely available, we developed the second and third versions. They are both Web based and require only a Web browser on the user side. The second version, which has functionality identical to the first, is a three-tier application with the user tier being implemented in Java. The third version is a CGI-BIN style application and does not require Java support on the user's side.

5 Evaluation and Results

We evaluated the system in three different ways: first, by checking the medical meaning of the relations extracted by association rule mining; second, by a statistical evaluation in which we checked how effective the system was in predicting new relations; and third, by using the system to discover the candidate gene for a genetic disease.

Medical Meaning of Related Concepts

This evaluation was conducted by a medical doctor (Borut Peterlin, one of the co-authors of this chapter). He used the system to check if the Y concepts, which the system had found to be related to the starting concept X, were medically meaningful. In other words, we wanted to determine whether the association rules are successful in extracting known relations between biomedical concepts from the MEDLINE database. This is very important in our approach, because by combining known relations, our system proposes potentially new relations. If the system failed to extract the known relations, then we could not expect it to discover new relations either.

The doctor selected as a starting concept *multiple sclerosis (MS)*, which can be defined as a demyelinating disease of the central nervous system of putative autoimmune origin. Then he used the system to find the related concepts Y. Below is a list of the first 20 concepts related (associated) with MS ordered by decreasing support, with their type and a short description of the nature of the association:

1. *MRI magnetic resonance imaging (diagnostics).* MRI is nowadays the method of choice to confirm the diagnosis of MS. It has a relatively high sensitivity (80%) and is noninvasive.
2. *Brain (anatomical structure – organ involved).* It simply reflects the anatomical structure of the central nervous system often affected in MS.
3. *Interferon (treatment).* A prophylactic drug nowadays used to reduce the rate of attacks – by 30%.
4. *T-lymphocytes(pathogenesis).* T lymphocytes regulate humoral immune responses and are found in abundance within MS lesions. It is believed that in MS, a T-cell mediated, autoimmune inflammatory reaction, at least as a mechanism for sustaining the inflammation, is involved.
5. *Myelin basic protein (MBP)(pathogenesis).* This structural component of myelin is potentially involved in the pathogenesis of MS. Antibodies to MBP have been found in both the serum and cerebrospinal fluid (CSF) of MS patients, and these antibodies, along with T cells that are reactive to MBP, increase with disease activity.
6. *Optic neuritis (symptoms).* Optic neuritis is one of the most common symptoms of MS (in 40% of patients).
7. *Autoimmune diseases (disease categories).* Due to the involvement of immunocompetent cells, association with certain HLA types, oligoclonal bands in liquor, abnormal subsets of T-cells, and the animal model of MS – EAE (experimental autoimmune encephalomyelitis), MS is an immune mediated disease, and autoimmunity is considered to be an important etiological factor in MS.

8.-20. *Immunosupresives (treatment), IgG (diagnostics), encephalomyelitis (symptoms), cognition disorders (symptoms), VEP (diagnostics), citokines (pathogenesis), TNF (pathogenesis), spinal cord (anatomical structure – organ involved), methylprednison (treatment), receptors-Tcells (pathogenesis), myelin protein (pathogenesis), psychological adjustment (treatment), demyelinating disorders (disease categories).*

Among the 20 concepts analyzed, 6 are related to pathogenesis, 4 to treatment, 3 to diagnostic methods, 3 to symptoms, 2 to target organs-anatomical structures and 2 are related to general disease categories.

We conclude that the concepts found as related by the system are associated with the current main focus of medical endeavors in the field of MS, which is still oriented to treatment and therefore towards better understanding of pathogenesis.

Table 2. The results of the prediction of new relationships between medical concepts in the newer MEDLINE segment (1996-1999) based on the older segment (1990-1995) using the system. The column names ending with 1 are for the AVGS constraint and those ending with 2 for the 2*AVGS constraint. The columns have the following meaning: n - all the relationships that can be predicted; k - new relationships in the newer segment that were not present in the older segment; m - predicted relationships based on the older segment; l - successfully predicted relationships; p - probability of achieving l or more successfully predicted relations by chance; r - the number of successfully predicted relations by chance alone.

Disease	***n***	***k***	***m1***	***l1***	***p1***	***r1***	***m2***	***l2***	***p2***	***r2***
Multiple Sclerosis (MS)	15965	635	6848	521	0	272	3151	366	0	125
Temporal Arteritis (TA)	17190	187	4735	148	0	52	1157	72	0	16
Melanoma (ML)	15336	692	6272	560	0	283	2812	392	0	127
Parkinson Disease (PD)	15966	594	5995	477	0	223	2322	309	0	86
Incontinentia Pigmenti (IP)	17504	44	3435	37	0	9	873	23	0	2
Chondrodysplasia Punctata (CP)	17422	18	2864	15	0	3	1046	9	0.00000016	1
Charcot-Marie-Tooth Disease (CMT)	17355	131	3150	105	0	24	1019	66	0	8
Focal Dermal Hypoplasia (FDH)	17527	23	1511	14	0	2	610	8	0.00000037	1
Noonan Syndrome (NS)	17384	68	3015	59	0	12	536	23	0	2
Ectodermal Dysplasia (ED)	17322	124	3301	96	0	24	967	45	0	7

Statistical Evaluation

The goal of the statistical evaluation was to see how many of the potential discoveries predicted by the system at some point in time become realized at a later time. For us, a potential discovery is a relationship between two concepts proposed by our system, but not present in MEDLINE at some point in time. We consider the potential discovery realized if the two concepts later appear together in a document in the MEDLINE database. In other words, the goal of the evaluation was to see how good our system was in predicting what discoveries would be made in the future.

We approached this goal by first dividing the MEDLINE database and the corresponding association rules into two segments according to the publication date of the documents stored: the older segment is from 1990 to 1995 and the newer segment is from 1996 to 1999. We then analyzed ten diseases, which are listed in Table 2.

Here we will give a discussion of the analysis of *Multiple sclerosis (MS)*. MS appears in 2582 documents in the older segment designated as a major MeSH descriptor. It is related to 1610 distinct concepts. When analyzing the old segment, the system proposed 15617 concepts as potential discoveries. MS is related to 635 new concepts in the new segment that it was not related to in the old segment. Our system successfully predicted 99.5% (632 out of 635) realized discoveries in the new segment. However, only 4% (632 out of 15617) of the proposed potential discoveries got realized. It should be stressed that MS was not related to 15965 out of 17575 distinct concepts appearing in the older segment. The system proposed 97.8% of the concepts MS was not yet related to as potential discoveries. The conclusion is that without using limits on the strength of relationship, the system is very successful at predicting future discoveries, but proposes far too many potential discoveries.

We then repeated the evaluation with two values for thresholds on the support level of the association rules. In one case the threshold was set to the average support of the associations between one concept and the others (AVGS) and in the other case it was set to 2*AVGS. Only associations with support greater than or equal to the threshold were taken into account. The number of proposed potential discoveries dropped from 15617 without thresholds to 6848 for AVGS and to 3151 for the 2*AVGS threshold. The percent of successfully predicted and realized discoveries dropped from 99.5% (632 of 635) without thresholds to 82.0% (521 of 635) for AVGS and to 57.6% (366 of 635) for 2*AVGS. However, the ratio of realized to proposed potential discoveries improved from 4% (632 out of 15617) without thresholds to 7.6% (521 of 6848) for AVGS and to 11.6% (366 of 3151) for 2*AVGS.

The results of the statistical evaluation for the ten selected diseases are in Table 2. The values obtained by our system were tested against the null hypotheses of random hits. Or to put it another way, we wanted to check whether the number of correct predictions obtained by our system could have occurred by chance alone. This is done in the following way.

Let n be the number of all possible relationships that the system could predict based on the older MEDLINE segment. Of these, k actually appear in the new segment (successful predictions), and n-k do not (nonsuccessful predictions). Let m be the number of actual predictions made by the system, of which l are successful, and m-l nonsuccessful. The probability of such an event is

$$\frac{\binom{k}{l}\binom{n-k}{m-l}}{\binom{n}{m}}.$$

This distribution is known as the hypergeometric distribution. The question now is whether l successful predictions represent a statistically significant result. In other words, we need the probability of obtaining l or more successful predictions if we were predicting completely at random. This probability is given by

$$p = \sum_{i=l}^{k} \frac{\binom{k}{i}\binom{n-k}{m-i}}{\binom{n}{m}},$$

or, equivalently

$$p = 1 - \sum_{i=0}^{l-1} \frac{\binom{k}{i}\binom{n-k}{m-i}}{\binom{n}{m}}.$$

Table 2 shows p values for given n, m, k and l for AVGS and 2*AVGS constraints respectively. Zeros in the p value columns actually mean that the probability is less than 10^{-16}. If the predictions were random, we would expect $\frac{m}{n}k$ of them to be successful.

Table 3 shows a summary of the relationship prediction results in terms of precision and recall for the AVGS and 2*AVGS constraints respectively. In our context, we define precision and recall as follows. Precision is the percentage of correctly predicted new relationships among all predicted relationships (***l/m*** from Table 2). Recall is the percentage of correctly predicted new relationships among all new relationships (***l/k*** from Table 2). For the AVGS constraint, the precision ranges from 0.5% to 8.9% with an average of 3.8%, and the recall ranges from 60.9% to 86.8% with an average of 79.5%. For the 2*AVGS constraint, the average precision increases to 6.5% and the average recall falls to 46.2%. However, the increase in precision is equal to the decrease in recall and is around 71%. The column ***Better then random*** in Table 3 shows how much better are the predictions of the system compared with random predictions (***l/r*** from Table 2). The value of this column is 3.8 for the AVGS constraint and 6.9 for 2*AVGS.

Table 3. Summary relationship prediction results for the AVGS and 2*AVGS constraints respectively: **Precision** (correctly predicted among all predicted), **Recall** (correctly predicted among all relationships), **Better then random** (correct predictions of the system divided by random correct predictions)

Disease	*AVGS*			*2*AVGS*		
	Precision	*Recall*	*Better then random*	*Precision*	*Recall*	*Better then random*
MS	7.6%	82.0%	1.9	11.6%	57.6%	2.9
TA	3.1%	79.1%	2.8	6.2%	38.5%	4.5
ML	8.9%	80.9%	2.0	13.9%	56.6%	3.1
PD	8.0%	80.3%	2.1	13.3%	52.0%	3.6
IP	1.1%	84.1%	4.1	2.6%	52.3%	11.5
CP	0.5%	83.3%	5.0	0.9%	50.0%	9.0
CMT	3.3%	80.2%	4.4	6.5%	50.4%	8.3
FDH	0.9%	60.9%	7.0	1.3%	34.8%	8.0
NS	2.0%	86.8%	4.9	4.3%	33.8%	11.5
ED	2.9%	77.4%	4.0	4.7%	36.3%	6.4
Average:	**3.8%**	**79.5%**	**3.8**	**6.5%**	**46.2%**	**6.9**

Disease Candidate Gene (re)Identification

In the preceding section, we showed that our system predicts new relations between medical concepts with statistical significance better than chance. This time we wanted to evaluate the system for the task of candidate gene discovery for a disease. For this purpose, we selected the incontinentia pigmenti disease.

Familial incontinentia pigmenti (IP; MIM 308310) is a genodermatosis that segregates as an X-linked dominant disorder and is usually lethal prenatally in males. In affected females, it causes highly variable abnormalities of the skin, hair, nails, teeth, eyes and central nervous system. The gene for IP was mapped in the terminal part of the long arm of the X chromosome (Xq28) in 1994 [12]. The pathogenesis of the disease is not known yet; however, the immune and haematopoetic systems seem to play an important role [13, 14, 15].

In 2000, the gene NEMO (NF-kappaB essential modulator/IKKγ) mutated in patients with IP was identified via positional cloning approach [16]. NEMO is required for activation of the NF-kappaB transcription factor, which is involved in numerous immune, inflammatory and apoptotic mechanisms.

The NEMO gene was cloned in 1998 and localized in the Xq28 region of the human chromosome X in 1999 [17].

More precisely, our aim regarding IP was to find out whether NEMO could have been identified by our discovery support system using the data that was available before NEMO was officially identified as the gene for IP.

As the starting concept (X) the name of the disease has been entered – Incontinentia pigmenti (Figure 3). Reflecting the fact that the immune system seems

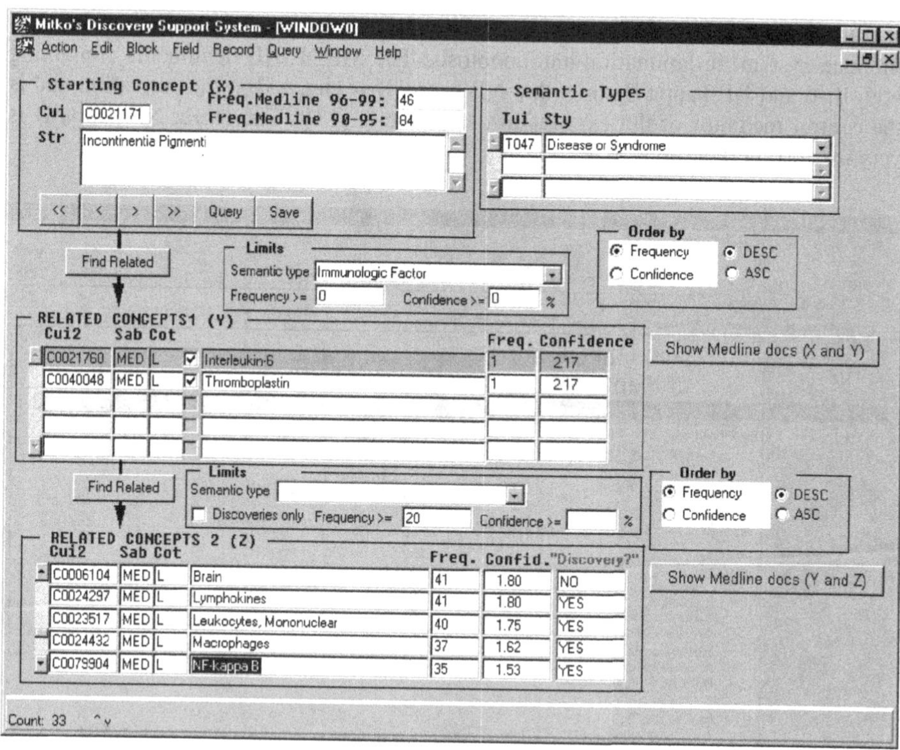

Fig. 3. Searching for the gene candidate for the *Incontinentia Pigmenti* disease. The system has discovered a potentially new relationship between *Incontinentia Pigmenti* and the *NF-kappa B* transcription factor through the intermediate concept *Interleukin-6.*

to play an important role in the pathogenesis of IP, related concepts of the first order (Y) were limited by the semantic type *immune factor*. Only two related concepts were found: *Interleukin-6* (IL-6) and *Thromboplastin*. IL-6 is a cytokine and has an important role in the development of the inflammatory response, the differentiation and activation of cells of the hematopoietic lineage, and the regulation of nerve cell and bone cell functions. By pressing the *Show MEDLINE docs (X and Y)*, the document containing both IP and IL-6 is retrieved and displayed [14]. It reports that increased serum levels of IL-6 have been found in an IP patient, who in addition to IP also demonstrated symptoms of Behcet disease.

In the next step, we searched for concepts related to IL-6 (concepts Z). The column *"Discovery?"* shows if there might be a potentially new relation between the starting concept X and the current concept Z. The value in this field is *YES* if there are no MEDLINE documents containing both the X and Z concepts (as major MeSH descriptors) in the corresponding MEDLINE segment. Among the hits with higher co-occurrence frequency, we identified three hits pathogenetically potentially related to IP: NF-kappa B, apoptosis and vascular endothelium.

NF-kappa B is a transcription factor involved in metabolic pathways related to immune system, inflammation and apoptosis. The MEDLINE documents containing both IL-6 and NF-kappa B show that NF-kappa B is engaged in the activation and is the central mediator of the expression of IL-6 (e.g. [18]). Therefore, NF-kappa B is very interesting regarding IP.

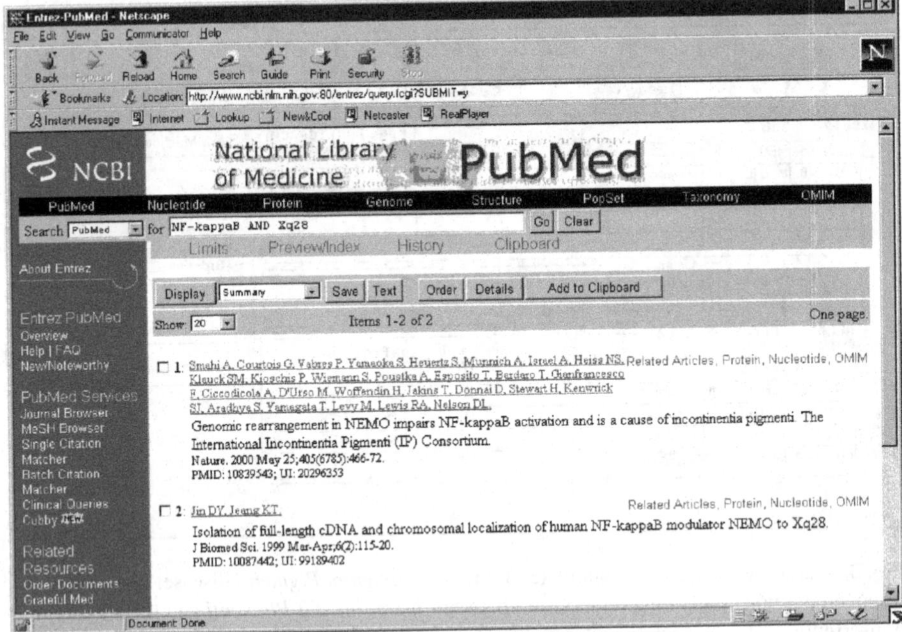

Fig. 4. The MEDLINE articles containing both the transcription factor *NF-kappa B* and the chromosomal region *Xq28*

In the next step, we used the fact that the gene for IP is located at the Xq28 chromosomal region, which has been known since 1994 [12]. When the corresponding button labeled *Show MEDLINE Docs* is pressed, an external Web browser is started and a search request is executed on the PubMed database to display the concepts that are currently under consideration. Figure 4 shows the results of the search request for documents containing the transcription factor *NF-kappa B* and chromosome location *Xq28*. The first article [16] reports the discovery of the NEMO gene as responsible for IP. However, this article was published in 2000. The second article [17] is very important in our procedure because it states that the NEMO gene, which is necessary for the activation of the transcription factor NF-kappa B, is localized in the Xq28 region. This article does not make any reference to IP and was published in 1999. From this we can conclude that by using our system together with some MEDLINE searching, it would have been possible to predict the NEMO gene as responsible for IP in 1999. In this example, the path from the IP disease to the NEMO gene went through two intermediate concepts: IL-6 and NF-kappa B.

6 Discussion and Further Work

We have presented an interactive discovery support system (available at http://www.mf.uni-lj.si/bitola) for the field of medicine. For a given starting medical concept, it discovers new, potentially meaningful relations with other concepts that have not been published in the medical literature before. The proposed relations then need to be evaluated and verified by a qualified medical professional.

As a measure of the relation between concepts, we use association rules calculated from the MEDLINE bibliographic database. We were in a dilemma whether to use the X→Y or Y→X direction of the association rule when finding concepts related to X. Because we have only binary associations, the direction comes into play only when limiting and ordering the related concepts. We selected the X→Y direction based on the intuition that the Y concepts that appear most often in documents regarding X have the strongest relation to X. However, there are cases where some Y concepts might appear infrequently in X documents, but X might appear in almost all Y documents. In this case, there is a strong association in the Y→X direction that should also be considered. We plan to further investigate this issue. One possibility might be to develop a heuristic approach in which sometimes an X→Y association is used and sometimes a Y→X one. Another possibility is to use some kind of a composite measure that takes into account both directions of the associations.

We have not used MEDLINE directly, but rather we use the UMLS. It simplifies the calculations considerably; however, it introduces considerable limitations into our system as well: we can calculate only binary association rules; the association rules are only between major MeSH headings; and we are limited to only two time intervals. Currently in MeSH, only a few dozen genes and other sequences are present. We plan to calculate direct association rules between the MeSH headings and a considerable number of molecular biology sequences and thereby increasing the functionality of the system significantly. To alleviate the above mentioned weaknesses, we are currently developing a new version of the system with these features: it analyzes full MEDLINE, expands the set of concepts with all confirmed human genes, and includes additional domain knowledge, such as the chromosomal location of the diseases and genes.

As part of the evaluation of the system, a medical doctor confirmed that most of the relations between concepts found by our association rules are meaningful. We demonstrated a successful application of the system for predicting a gene candidate for the incontinentia pigmenti disease, by using the information known prior to the discovery of the gene. In the statistical evaluation, the system proved to be successful at predicting future discoveries. However, this came at the expense of generating a large number of potential discoveries that have to be judged and verified by the user of the system. The statistical evaluation also showed that properly set thresholds are crucial for successful use of the system. Thus, we plan to work on setting good default values for the thresholds that can be changed by the user if necessary.

We believe that literature based discovery support systems, such as ours, will help researchers make some important biomedical discoveries in the future.

Acknowledgements: The authors would like to thank Tom Rindflesch and Alan Aronson for providing valuable comments and insights.

References

1. Swanson, D.R.: Fish oil, Raynaud's syndrome, and undiscovered public knowledge. Perspect Biol. Med. 30(1), 7–18 (1986)
2. Swanson, D.R.: Migraine and magnesium: eleven neglected connections. Perspect Biol. Med. 31(4), 526–557 (1988)
3. Swanson, D.R., Smalheiser, N.R.: An interactive system for finding complementary literatures: a stimulus to scientific discovery. Artif. Intell. 91, 183–203 (1997)
4. Gordon, M.D., Lindsay, R.K.: Toward discovery support systems: A replication, re-examination, and extension of Swanson's work on literature-based discovery of a connection between Raynaud's and fish oil. J. Am. Soc. Inf. Sci. 47(2), 116–128 (1996)
5. Weeber, M., Klein, H., Aronson, A.R., Mork, J.G., Jong-Van Den Berg, L., Vos, R.: Text-based discovery in biomedicine: the architecture of the DAD-system. Proc AMIA Symp. J AMIA (Suppl. 20), 903–907 (2000)
6. U.S. National Library of Medicine - MEDLINE. http://www.ncbi.nlm.nih.gov/entrez/query.fcgi?DB=pubmed<28.04.2007>
7. Funk, M.E., Reid, C.A.: Indexing consistency in MEDLINE. Bull. Med. Libr. Assoc. 71(2), 176–183 (1983)
8. Medical Subject Headings - MeSH. http://www.nlm.nih.gov/mesh/MBrowser.html<28.04.2003>
9. Unified Medical Language System – UMLS. http://www.nlm.nih.gov/research/umls/umlsmain.html<28.01.2003>
10. Humphreys, B.L., Lindberg, D.A.B., Schoolman, H.M., Barnett, G.O.: The Unified Medical Language System: an informatics research collaboration. J AMIA 5(1), 1–11 (1998)
11. Agrawal, R., Mannila, H., Srikant, R., Toivonen, H., Inkeri Verkamo, A.: Fast discovery of association rules. In: Fayyad, U., Piatetsky-Shapiro, G., Smyth, P., Uthurusamy, R. (eds.) Advances in Knowledge Discovery and Data Mining, pp. 307–328. MIT Press, Cambridge (1996)
12. Smahi, A., Hyden-Granskog, C., Peterlin, B., Vabres, P., Heuertz, S., Fulchlgnoni-Lataud, M.C., Dahl, N., Labrune, P., Le Marec, B., Piussan, C., Taleb, A., von Koskul, H., Hors-Cayla, M.C.: The gene for the familial form of incontinentia pigmenti (IP2) maps to the distal part of Xq28. Hum. Mol. Genet. 3, 273–278 (1994)
13. Roberts, J.L., Morrow, B., Vega-Rich, C., Salafia, C.M., Nitowsky, H.M.: Incontinentia pigmenti in a newborn male infant with DNA confirmation. Am. J. Med. Genet. 75, 159–163 (1998)
14. Endoh, M., Yokozeki, H., Maruyama, R., Matsunaga, T., Katayama, I., Nishioka, K.: Incontinentia pigmenti and Behcet's disease: a case of impaired neutrophil chemotaxis. Dermatology 192, 285–287 (1996)
15. Dahl, M.V., Matula, G., Leonards, R., Tuffanelli, D.L.: Incontinentia pigmenti and defective neutrophil chemotaxis. Arch. Dermatol. 111, 1603–1605 (1975)
16. The International Incontinentia Pigmenti Consortium: Genomic rearrangement in NEMO impairs NF-κB activation and is a cause of incontinentia pigmenti. Nature 405, 466–472 (2000)
17. Jin, D.Y., Jeang, K.T.: Isolation of full-length cDNA and chromosomal localization of human NF-kappaB modulator NEMO to Xq28. J Biomed. Sci. 6(2), 115–120 (1999)
18. Vanden Berghe, W., De Bosscher, K., Boone, E., Plaisance, S., Haegeman, G.: The nuclear factor-kappaB engages CBP/p300 and histone acetyltransferase activity for transcriptional activation of the interleukin-6 gene promoter. J. Biol. Chem. 274(45), 32091–32098 (1999)

Author Index

Lecture Notes in Artificial Intelligence (LNAI)

Vol. 4369: M. Umeda, A. Wolf, O. Bartenstein, U. Geske, D. Seipel, O. Takata (Eds.), Declarative Programming for Knowledge Management. X, 229 pages. 2006.

Vol. 4342: H. de Swart, E. Orłowska, G. Schmidt, M. Roubens (Eds.), Theory and Applications of Relational Structures as Knowledge Instruments II. X, 373 pages. 2006.

Vol. 4335: S.A. Brueckner, S. Hassas, M. Jelasity, D. Yamins (Eds.), Engineering Self-Organising Systems. XII, 212 pages. 2007.

Vol. 4334: B. Beckert, R. Hähnle, P.H. Schmitt (Eds.), Verification of Object-Oriented Software. XXIX, 658 pages. 2007.

Vol. 4333: U. Reimer, D. Karagiannis (Eds.), Practical Aspects of Knowledge Management. XII, 338 pages. 2006.

Vol. 4327: M. Baldoni, U. Endriss (Eds.), Declarative Agent Languages and Technologies IV. VIII, 257 pages. 2006.

Vol. 4314: C. Freksa, M. Kohlhase, K. Schill (Eds.), KI 2006: Advances in Artificial Intelligence. XII, 458 pages. 2007.

Vol. 4304: A. Sattar, B.-h. Kang (Eds.), AI 2006: Advances in Artificial Intelligence. XXVII, 1303 pages. 2006.

Vol. 4303: A. Hoffmann, B.-h. Kang, D. Richards, S. Tsumoto (Eds.), Advances in Knowledge Acquisition and Management. XI, 259 pages. 2006.

Vol. 4293: A. Gelbukh, C.A. Reyes-Garcia (Eds.), MICAI 2006: Advances in Artificial Intelligence. XXVIII, 1232 pages. 2006.

Vol. 4289: M. Ackermann, B. Berendt, M. Grobelnik, A. Hotho, D. Mladenič, G. Semeraro, M. Spiliopoulou, G. Stumme, V. Svátek, M. van Someren (Eds.), Semantics, Web and Mining. X, 197 pages. 2006.

Vol. 4285: Y. Matsumoto, R.W. Sproat, K.-F. Wong, M. Zhang (Eds.), Computer Processing of Oriental Languages. XVII, 544 pages. 2006.

Vol. 4274: Q. Huo, B. Ma, E.-S. Chng, H. Li (Eds.), Chinese Spoken Language Processing. XXIV, 805 pages. 2006.

Vol. 4265: L. Todorovski, N. Lavrač, K.P. Jantke (Eds.), Discovery Science. XIV, 384 pages. 2006.

Vol. 4264: J.L. Balcázar, P.M. Long, F. Stephan (Eds.), Algorithmic Learning Theory. XIII, 393 pages. 2006.

Vol. 4259: S. Greco, Y. Hata, S. Hirano, M. Inuiguchi, S. Miyamoto, H.S. Nguyen, R. Słowiński (Eds.), Rough Sets and Current Trends in Computing. XXII, 951 pages. 2006.

Vol. 4253: B. Gabrys, R.J. Howlett, L.C. Jain (Eds.), Knowledge-Based Intelligent Information and Engineering Systems, Part III. XXXII, 1301 pages. 2006.

Vol. 4252: B. Gabrys, R.J. Howlett, L.C. Jain (Eds.), Knowledge-Based Intelligent Information and Engineering Systems, Part II. XXXIII, 1335 pages. 2006.

Vol. 4251: B. Gabrys, R.J. Howlett, L.C. Jain (Eds.), Knowledge-Based Intelligent Information and Engineering Systems, Part I. LXVI, 1297 pages. 2006.

Vol. 4248: S. Staab, V. Svátek (Eds.), Managing Knowledge in a World of Networks. XIV, 400 pages. 2006.

Vol. 4246: M. Hermann, A. Voronkov (Eds.), Logic for Programming, Artificial Intelligence, and Reasoning. XIII, 588 pages. 2006.

Vol. 4223: L. Wang, L. Jiao, G. Shi, X. Li, J. Liu (Eds.), Fuzzy Systems and Knowledge Discovery. XXVIII, 1335 pages. 2006.

Vol. 4213: J. Fürnkranz, T. Scheffer, M. Spiliopoulou (Eds.), Knowledge Discovery in Databases: PKDD 2006. XXII, 660 pages. 2006.

Vol. 4212: J. Fürnkranz, T. Scheffer, M. Spiliopoulou (Eds.), Machine Learning: ECML 2006. XXIII, 851 pages. 2006.

Vol. 4211: P. Vogt, Y. Sugita, E. Tuci, C.L. Nehaniv (Eds.), Symbol Grounding and Beyond. VIII, 237 pages. 2006.

Vol. 4203: F. Esposito, Z.W. Raś, D. Malerba, G. Semeraro (Eds.), Foundations of Intelligent Systems. XVIII, 767 pages. 2006.

Vol. 4201: Y. Sakakibara, S. Kobayashi, K. Sato, T. Nishino, E. Tomita (Eds.), Grammatical Inference: Algorithms and Applications. XII, 359 pages. 2006.

Vol. 4200: I.F.C. Smith (Ed.), Intelligent Computing in Engineering and Architecture. XIII, 692 pages. 2006.

Vol. 4198: O. Nasraoui, O. Zaïane, M. Spiliopoulou, B. Mobasher, B. Masand, P.S. Yu (Eds.), Advances in Web Mining and Web Usage Analysis. IX, 177 pages. 2006.

Vol. 4196: K. Fischer, I.J. Timm, E. André, N. Zhong (Eds.), Multiagent System Technologies. X, 185 pages. 2006.

Vol. 4188: P. Sojka, I. Kopeček, K. Pala (Eds.), Text, Speech and Dialogue. XV, 721 pages. 2006.

Vol. 4183: J. Euzenat, J. Domingue (Eds.), Artificial Intelligence: Methodology, Systems, and Applications. XIII, 291 pages. 2006.

Vol. 4180: M. Kohlhase, OMDoc – An Open Markup Format for Mathematical Documents [version 1.2]. XIX, 428 pages. 2006.

Vol. 4177: R. Marín, E. Onaindía, A. Bugarín, J. Santos (Eds.), Current Topics in Artificial Intelligence. XV, 482 pages. 2006.

Vol. 4160: M. Fisher, W. van der Hoek, B. Konev, A. Lisitsa (Eds.), Logics in Artificial Intelligence. XII, 516 pages. 2006.

Vol. 4155: O. Stock, M. Schaerf (Eds.), Reasoning, Action and Interaction in AI Theories and Systems. XVIII, 343 pages. 2006.

Vol. 4149: M. Klusch, M. Rovatsos, T.R. Payne (Eds.), Cooperative Information Agents X. XII, 477 pages. 2006.

Vol. 4140: J.S. Sichman, H. Coelho, S.O. Rezende (Eds.), Advances in Artificial Intelligence - IBERAMIA-SBIA 2006. XXIII, 635 pages. 2006.

Vol. 4139: T. Salakoski, F. Ginter, S. Pyysalo, T. Pahikkala (Eds.), Advances in Natural Language Processing. XVI, 771 pages. 2006.